The Design and Implementation
of the RT-Thread Operating System

The Design and Implementation
of the RT-Thread Operating System

Yi Qiu
Puxiang Xiong
Tianlong Zhu

CRC Press
Taylor & Francis Group
Boca Raton London New York

CRC Press is an imprint of the
Taylor & Francis Group, an **informa** business

AN AUERBACH BOOK

First edition published 2020
by CRC Press

6000 Broken Sound Parkway NW, Suite 300, Boca Raton, FL 33487-2742
and by CRC Press

2 Park Square, Milton Park, Abingdon, Oxon, OX14 4RN

© 2021 Taylor & Francis Group, LLC
CRC Press is an imprint of Taylor & Francis Group, LLC

ISBN: 978-0-367-55486-6 (hbk)
ISBN: 978-1-003-09900-0 (ebk)

Typeset in Times LT Std
by KnowledgeWorks Global Ltd.

Contents

Preface

WHY I WROTE THIS BOOK

Since the release of V0.01 in 2006 to the present V4.0 version, RT-Thread has been developed over a decade. With its great reputation and open source free-of-charge strategy, RT-Thread gained largest embedded open source community in China and gathers hundreds of thousands of software enthusiasts around. RT-Thread has been widely used in energy, automotive, medical, consumer electronics, and many other industries, which has made RT-Thread the most mature and stable open source embedded operating system.

The original purpose of this book is to smooth the learning curve of RT-Thread, so that more people can easily learn and master RT-Thread, participate in the development of RT-Thread, and work together to create an open source, tiny, and beautiful Internet of Things operating system.

Marching on this industry, we deeply feel the rapidly rising trend of the chip industry and the Internet of Things industry in recent years. The industry development needs more talent, especially those who have the ability to master the embedded operating system and other underlying technologies. We're hoping that RT-Thread could make a connection with more people through this book and have more people know the beauty of RT-Thread operating system, so that RT-Thread can help more industries.

College students are a very important group of RT-Thread users; since 2018, RT-Thread has launched a series of college student programs to help students know and learn RT-Thread, which include a book distribution program, training program, curriculum program, a sponsored contest, etc. Writing this book helps us to make things as simple as possible and make RT-Thread easy to use for college students. We're hoping this book will also accelerate the popularizing of RT-Thread in universities.

THIS BOOK'S AUDIENCE

- All developers who program with C/C++
- Students who majored in computer science, electronics engineering, automation, or communication and teachers who teach those majors
- Embedded hardware and software engineers, electronics engineers, and Internet of Things development engineers
- Anyone who has an interest in embedded operating systems

TIPS FOR READING THIS BOOK

To understand this book, you are recommended to learn the C programming language and STM32 programming knowledge first, and it is even better if you already master the knowledge of data structure and object-oriented programming. When reading this book, you are encouraged to practice as you learn. Most chapters come with sample code, which can be actually run, so after reading, you can do some experiments.

All the chapters of this book are divided into two parts: Chapters 1–10 introduce the kernel and Chapters 11–16 introduce components.

Chapters 1–9 introduce the RT-Thread kernel, starting with an overview of RT-Thread, then introduce RT-Thread's thread management, clock management, inter-thread synchronization, inter-thread communication, memory management, and interrupt management. Each chapter has companion sample code. This part of the sample runs in the Keil MDK simulator environment, so there is no hardware to support.

Chapter 10 describes RT-Thread kernel porting; you can try to port RT-Thread to the actual hardware board to run after reading this chapter.

Chapters 11–16 introduce the RT-Thread components, which describe the Env development environment, FinSH console, device management, file system, and network framework, This part of sample can run on the hardware board to achieve peripheral access, file system reading and writing, and network communication functions.

The companion material of this book and its related tools and hardware can be requested from the RT-Thread official Twitter account: https://twitter.com/rt_thread.

EXAMPLE HARDWARE

The hardware of choice is the IoT board, which is jointly developed by RT-Thread and ALIENTEK. It is based on the STM32L475 main chip, and the sample code for the components chapter of this book is based on this IoT board.

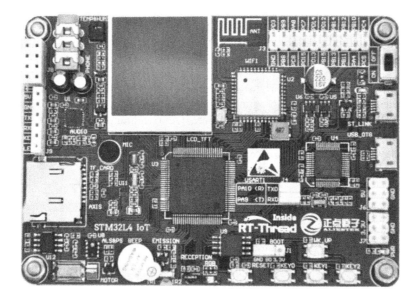

IoT board.

The specifications of the IoT board are as follows:

- Main Chip: STM32L475VET6 (Cortex-M4); LQFP100; SRAM 128KB; FLASH 512KB
- SDIO WIFI Module: AP6181
- 1.3-inch full-color TFT display screen, resolution is 240 × 240
- On-board STM32F103C8T6, integrated ST-LINK V2.1 simulation download function
- On-board high-performance audio decoding chip: ES8388
- On-board SPI Flash: W25Q128
- On-board 6-axis sensor: ICM-20608
- On-board temperature and humidity sensor: AHT10
- On-board light sensor: AP3216C
- On-board motor patch with TC214B drive chip
- On-board infrared transmitter and infrared receiver

Chapter 16 of this book requires the use of the ENC28J60 module to implement the network sample functionality. (This ENC28J60 module is shown in the following image.)

ENC28J60 Network module.

The following are more ways to achieve the learning goals of this book:

- If you do not have a hardware platform, you can view the appendix *Getting Started with QEMU (Ubuntu)* and use QEMU for learning.
- If you have other development boards for stm32, you can view the appendix *Getting Started with RT-Thread Studio (Windows)* to learn RT-Thread based on RT-Thread Studio.

CORRECTIONS AND FEEDBACK

There may be some errors or inaccurate descriptions in this book; you are welcome to send an email to contact@rt-thread.org and tell us. Also during the learning process, if you have any questions or great ideas for RT-Thread, please feel free to contact us. We are looking forward to receiving your feedback. Let's work together to make this technology great.

Acknowledgments

This book was completed with the help of a number of RT-Thread developers. Thanks to Jie Yang, Jiao Luo, Haodi Yu , Yuan Zhang, Cheng Zou, and Jinrun Yao, who helped with writing; Zhanxin Guo, FangLi Han, Guangliang Yang, Panpan Zhao, Wu Yang, Yan Li, Jianjia Ma, and Gang Wang helped with proofreading. And Zhuoran Wang has given us very precious suggestions. Thank you for all the hard work RT-Thread developers contributed.

Yi Qiu

Authors

Yi Qiu is a co-founder of Shanghai Real-Thread Electronic Technology Co., Ltd. He has been working on the development of the RT-Thread open-source operating system since 2006. RT-Thread has now been widely used in many industries, such as energy, vehicle, medical, consumer electronics, and more. It is a mature and stable open-source embedded operating system with a large installed capacity.

Puxiang Xiong is a founder and CEO of Shanghai Real-Thread Electronic Technology Co., Ltd. He created the RT-Thread open-source operating system in 2006 and has led engineers in the design and development of the RT-Thread operating system kernel by integrating object-oriented design concepts and developing many mature and stable software components such as file system, command line, and graphical user interface.

Tianlong Zhu is the CTO of Shanghai Real-Thread Electronic Technology Co., Ltd. He has more than 10 years of RT-Thread open-source development experience and has taken charge of the RT-Thread R&D team, which is committed to researching and developing cutting-edge technology. Also, he is an embedded open-source geek, sharing such open-source software as EasyLogger, EasyFlash, and CmBacktrace.

1 Embedded Real-Time Operating System

The operating system refers to the computer programs used to manage and control the computer hardware and software resources, which are directly run on the computer as the most basic system software—all the other software runs on top of the operating system. According to the application field, the operating system can be divided into several categories, such as desktop operating system, server operating system, mobile operating system, and embedded operating systems.

Desktop operating system refers to the operating system running on the personal computer; the current mainstream desktop operating system is Microsoft's Windows operating system; in addition, Linux and macOS are desktop operating systems.

Server operating system refers to the operating system running on large servers, such as cloud servers, database servers, network servers, etc. The current mainstream server operating system is Linux, but Microsoft's Windows server operating system also occupies a part of the market.

Mobile operating system refers to the operating system running on mobile phones, PADs, and smart TVs. Google's Android and Apple's iOS both belong to the mobile operating system. Traditionally, mobile devices like mobile phones and PADs are also embedded devices, but as the chips used by mobile devices become more powerful, the difference is obviously larger between the mobile operating system and the traditional embedded device, so that the mobile operating system is classified separately.

An embedded operating system is an operating system that is used in an embedded system. Embedded systems are widely used and easier to understand. All computers are embedded devices except servers, personal computers, and mobile devices. Embedded systems are everywhere in our lives, from military equipment to civilian affairs, from industrial control to network applications. The following are some typical embedded device examples; Figure 1.1 also lists some embedded operating system applications.

Consumer electronics: a variety of information appliances, such as digital televisions, set-top boxes, digital cameras, audio equipment, videophones, home network equipment, washing machines, refrigerators, smart toys, etc.

Industrial control: a variety of intelligent measuring instruments, digital control devices, programmable controllers, controlling machines, distributed control systems, field bus instruments and control systems, industrial robots, mechatronic machinery, automotive electronics, etc.

Network applications: including network infrastructure, access equipment, mobile terminal equipment, shared bicycles, water and electrical meters, Internet of Things (IoT) terminal equipment, etc.

Military equipment: for all kinds of weapons control (such as artillery control, missile control, intelligent bomb guidance, and detonating devices); tanks; warships; bombers; and land, sea, and air military electronic equipment; radar; electronic counter-military communications equipment; and a variety of special equipment for field operation command.

Others: all kinds of cash registers, POS systems, electronic scales, bar code readers, commercial terminals, bank money counting machines, IC card input equipment, cash machines, ATMs, automatic service terminals, anti-theft systems, and a variety of professional banking peripherals and medical electronic equipment.

EMBEDDED SYSTEM

An embedded system is a computer system that is fully embedded on a device or inside a device and is designed to meet specific needs, such as the common embedded system in daily life, including TV set-top boxes, routers, refrigerators, microwave ovens, mobile phones, etc. They all have certain

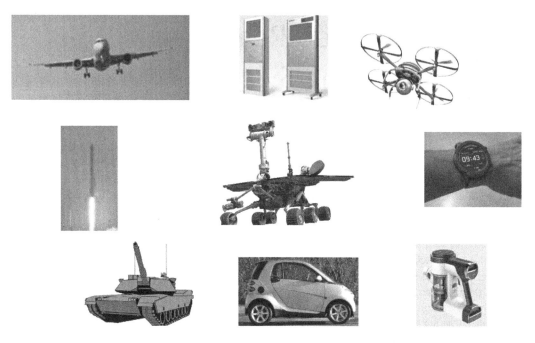

FIGURE 1.1   Applications for common embedded operating systems.

functions: the TV set-top box used to play TV shows on the network, and similarly, the router is used to select the optimal path and forward the network message correctly. Such systems are highly specialized and relatively single in functionality, usually only for specific external input processing, and then give the corresponding results, so that the embedded system only needs to have just enough but a small number of hardware resources to complete the required specific functions, and thus the cost can be effectively controlled.

General computer systems, on the other hand, do not have specific needs, but do everything possible to meet requirements. Even when constructing hardware systems, consider the changes in requirements over the next few years. For example, when people buy computers, they want to get as much high-end performance as possible for multimedia, games, work, etc.

As shown in Figure 1.2, the hardware equipment of the embedded system consists of a number of chips and circuits, including the main control chip, power management, and JTAG interface used

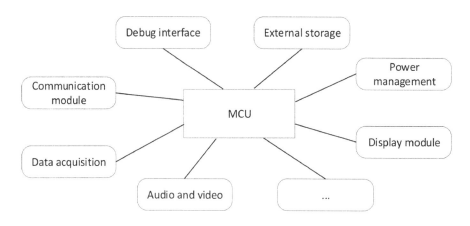

FIGURE 1.2   Hardware block diagram of embedded system.

in the debugging of the development phase, and may also contain some data acquisition modules, communication modules, audio/video modules, and so on.

REAL-TIME SYSTEM

The real-time performance of the system refers to the correct response to external events in a fixed amount of time. During this "time," the system will do some processing, such as input data analysis, calculation, processing, and so on. And outside of this time, the system may be idle and do some spare work. Take a mobile terminal as an example: when a phone dials in, the system should be ringing in a timely manner to inform the host of a call and ask whether to answer, and when the telephone is not dialing in, people can use it to do some other work, such as listening to music, playing games, and so on.

From the earlier example, we can see that the real-time system is a demand-oriented system, for real-time tasks need to respond the first time, and non-real-time tasks can make way for real-time events when they arrive—it is called preempted. So a real-time system can also be thought of as a hierarchical system, and tasks of different importance have different priority levels: important tasks can be prioritized for response, and non-essential tasks can be postponed appropriately.

Real-time calculations can be defined as a type of calculation in which the correctness of the system depends not only on the logical result of the calculation but also on the time at which the result is produced. There are two key points, that is, done correctly and within a given time, and the importance of these two points is equivalent. If the calculation results go wrong, this will not be considered a correct system, and if the calculation results are correct but the calculation has taken time away from the required time, this will also not be considered a correct system. A real-time system is shown in Figure 1.3.

For input signals and events, the real-time system must be able to respond correctly within the specified time, regardless of whether these events are single events or multiple events, synchronous signals, or asynchronous signals.

Example: Suppose a bullet is fired from 20 meters away into a glass, the speed of the bullet is v m/s, then after t1 = 20/v seconds, the bullet will break the glass. And if a protection system takes the

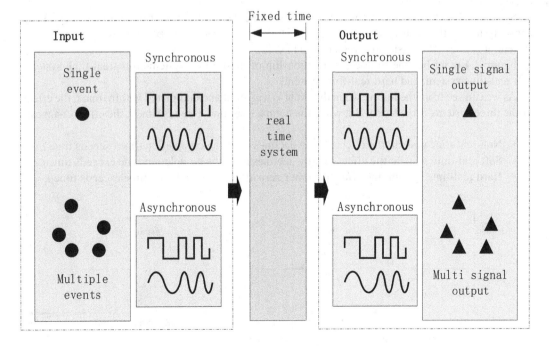

FIGURE 1.3 Real-time system.

glass away after detecting a bullet, assuming the whole process lasts t2 seconds, and if t2 < t1, the glass won't be broken; this is the real-time system.

Similar to embedded systems, there are also certain computing units in real-time systems that can anticipate the environment of the system and its internal applications, which is the determinism that many of the real-time system books have mentioned; that is, the system can respond to a given event within a given time (*t* seconds). The determinism of multiple events and multiple input system responses constitute determinism for the entire real-time system (the real-time system does not mean that it has a real-time response to all input events, but rather completes the response to the event within a specified time). Embedded systems are widely used in applications. We do not require all dedicated functions to have real-time characteristics, only when the system has strict time limits on tasks would we pay attention to its real-time problems. Specific examples include experimental control, process control equipment, robots, air traffic control, telecommunications, military command and control systems, etc. But embedded applications like printers don't have strict time limits; it only has "as fast as possible" requirement, so it's not a real-time system.

SOFT REAL TIME AND HARD REAL TIME

As we described earlier, the real-time system is concerned with two points, namely the correctness of time and the correctness of function. In fact, measuring the correctness of a real-time system is to require the system to perform the corresponding tasks correctly in a given time. But in reality there is also a system that, in most cases, this kind of system is able to perform tasks strictly within a specified time, but occasionally it will complete the task a little beyond the given time, which we often refer to as a soft real-time system. From the system's different requirements for the sensitivity of the specified time, the real-time system can be divided into the hard real-time system and the soft real-time system.

Hard real-time systems are strictly limited to the specified time to complete the task; otherwise, it may lead to disaster. For example, missile intercept systems, automotive engine systems, etc., when these systems cannot meet the required response time, even occasionally, it will lead to car crash or other major disaster.

Soft real-time systems can allow occasional time deviation, but over time, the accuracy of the entire system will decline, for example, a DVD playback system can be seen as a soft real-time system; it is allowed occasional screen or sound delay.

Figure 1.4 describes the time-effect relationship of these three systems (non-real-time system, soft real-time system, and hard real-time system).

As we can see from the figure, when the event is triggered and completed within time *t*, the effect of the three systems is the same. But when the completion time exceeds time *t*, the effect changes:

- Non-real-time system: the effect slowly declines after the completion time exceeds time *t*.
- Soft real-time system: the effect rapidly declines after the completion time exceeds time *t*.
- Hard real-time system: the effect becomes zero after the completion time exceeds time *t*.

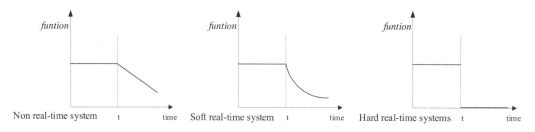

FIGURE 1.4   The relationship between time and utility.

EMBEDDED REAL-TIME OPERATING SYSTEM

In embedded devices, in addition to the embedded operating system, there are bare metal programs. We write a large loop in the main function, the loop is the functional implementation of each task, and all the tasks are performed in a flat order—the next task must wait for the previous task to run completely before starting to run; this running large loop we can call the background program. An interruption can disturb the system's current background task to perform its tasks first; after the interrupted task is complete, the task goes back to the original background where was interrupted to continue to implement the background program. The interrupt process is called the foreground program. Figure 1.5 shows a foreground and background system.

Such a foreground and background system in real-time processing has defects. For example, task 1 is important and needs to be able to respond in a timely manner. In the operation of task 4 when there is an interrupt, even the task 1 implementation conditions are met, and the ideal and quick response is that task 1 should be immediately put into operation, but this cannot be done in the foreground and background program. Because tasks are executed sequentially, even if task 1 is important, it must wait for task 4 to be processed before it can be run.

The embedded real-time operating system is designed as a preemption system, which can solve the earlier real-time problems. It divides the tasks into different priority levels, and high-priority tasks can interrupt the low-priority task when the operating conditions are met, thus greatly improving the real-time system. The real-time operating system performance diagram is shown in Figure 1.6.

Compared with the foreground and background programs, the embedded real-time operating system has progressed in real time; also it provides a complete set of mechanisms in multitasking management, intertask communication, memory management, timer management, device management, etc.,

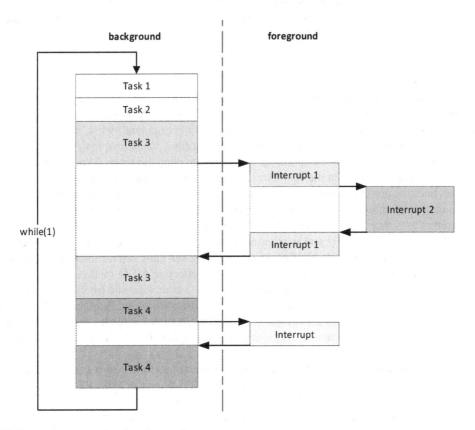

FIGURE 1.5 Foreground and background system.

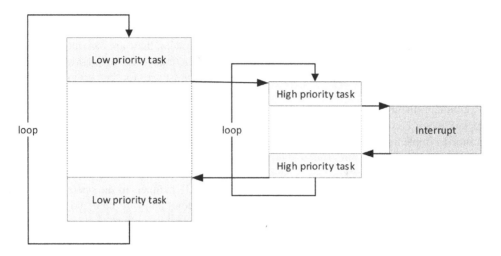

FIGURE 1.6 Real-time operating system.

which greatly facilitates the development, management, and maintenance of embedded applications. If we take the embedded real-time system and desktop operating system for some analogy, then the foreground and background program development is like developing directly with the BIOS, and using the embedded real-time operating system is like developing on Windows.

In general, the embedded operating system is the software used in the embedded system and is used to dock the embedded underlying hardware and upper-level application software. The operating system encapsulates the underlying drivers to provide a functional interface for developers, greatly improving the efficiency of the application development.

Mainstream Embedded Real-Time Operating System

uC/OS is a real-time operating system (RTOS) from the United States, released in 1992. In 2001, Professor Shao of Beihang University translated the book of uC/OS, and this book received a lot of praise. When this book set off an "embedded system development" wind, a large number of college students began to learn embedded systems. At that time, many people took this book as an introduction to learn the embedded operating system and applied the learned knowledge into various projects and products—then the characteristics of uC/OS are gradually revealed. Before 2010, uC/OS has been the preferred choice for most Chinese companies to choose RTOS. But since 2010, open source free-of-charge RTOS has become popular, but uC/OS still charged fees, leading many manufacturers to switch to open source operating systems such as FreeRTOS and RT-Thread.

FreeRTOS was born in 2003 and is released as an open source, free for any commercial and noncommercial projects. In 2004, ARM of the United Kingdom launched its first Cortex-M3 IPcore based on ARMv7-M architecture, focusing on the cost-effective MCU market. And then Texas Instruments launched its first MCU based on the Cortex-M3 core, followed by ST, NXP, Freescale, Atmel, and many other European and American manufacturers who launched Cortex-M core-based MCU. Considering the price and performance, these manufacturers have chosen FreeRTOS as the default embedded operating system for chips, taking advantage of this boom, and FreeRTOS's rapid rise at home and abroad.

RT-Thread is an embedded real-time operating system from China, born in 2006, released as an open source and free-of-charge project. Unlike FreeRTOS and uC/OS, RT-Thread has been positioned not only as an RTOS kernel since day one but also as a middleware platform with components such as network, file system, GUI interface, etc., which embraces the idea of open source, free of charge, and gathers a fast growing and powerful community. After more than a decade of accumulation, RT-Thread has gained a great reputation and is well known as a highly stable and reliable real-time operating system. RT-Thread supports all the mainstream compilation tools available on the market, such as IAR, GCC, Keil, etc., and in terms of hardware support, it completes the porting for more than 50 MCU chips and all major CPU architectures, including ARM, MIPS, C-Sky, Xtensa, Andes, and RISC-V. In industry applications, RT-Thread is widely used in many industries, including security, medical, new energy, in-vehicle, Beidou navigation, and consumer electronics, because of the high reliability and component-rich characteristics.

In the past 2 years, the demand for RT-Thread enterprise projects has increased significantly, and the number of RT-Thread developers also increased a lot. The promotion of RT-Thread has intensified and documentation has grown, thanks to the strong support from ecological partners. In terms of offline activities, RT-Thread regularly organizes a series of technical conferences in first- and second-tier cities with the help of the community and has been welcomed by new learners and developers. Participating in offline technology conferences has become an important channel for developers to learn about RT-Thread and communicate with each other. In addition, the operation of social media, online training lectures, and design competitions have become an important part of ecological construction, promoting the healthy development of the RT-Thread community.

TREND OF DEVELOPMENT

In the traditional embedded era, devices are isolated from each other, systems and applications are simple, and the value of operating systems is relatively low. Each manufacturer uses an open source RTOS core to build and develop its own upper-level software, depending on the different vertical application fields. The production is controllable and can basically meet their own, customer, and industry needs.

As we step into the era of the IoT, the original pattern and mode will be completely broken; the development of networking devices is also increasing in geometric series. Reliability, long standby, low cost, communication mode and transmission protocol, mobile phone compatibility, secondary development, and cloud docking have become a problem to be considered and solved.

For enterprises, the operating system platform with rich middle-tier components and standard API interfaces can undoubtedly greatly reduce the difficulty of networking terminal development, and also simplify docking for multiple cloud platforms, paving the way for the future deployment and update of various IoT service applications.

The rise of the IoT chip industry and the continued strengthening of the industrial chain provide a good opportunity for the success of IoT operating system (IoT OS), as well as great for the development of RT-Thread. The requirements of the IoT terminal for software capabilities greatly increased, RT-Thread as an IoT OS, which contains a wealth of components and highly scalable features are exactly what the market needs, so RT-Thread has been favored by more and more chip manufacturers. They have chosen RT-Thread as their native operating system and bring RT-Thread operating system and chips to the market together.

CHAPTER SUMMARY

This chapter made a brief introduction to the embedded real-time system. The embedded operating system is applied to the embedded system software, and it is used in every aspect of life. The embedded operating system is cataloged into the real-time operating system and the non-real-time operating system. This book will focus on the introduction of the embedded real-time operating system.

2 RT-Thread Introduction

As a beginner to the real-time operating system (RTOS), you might be new to RT-Thread (short for Real Time-Thread). However, with a better understanding of it over time, you will gradually discover the charm of RT-Thread and its advantages over other RTOSs of the same type. RT-Thread is an embedded RTOS. After nearly 14 years of experiences accumulated, along with the rise of the Internet of Things (IoT), it is evolving into a powerful, component-rich Internet of Things Operating System (IoT OS).

RT-THREAD OVERVIEW

RT-Thread, as its name implies, is an embedded real-time multi-threaded operating system. One of its basic properties is to support multitasking. Allowing multiple tasks to run at the same time does not mean that the processor actually performs multiple tasks at the same time. In fact, a processor core can only run one task at a time. Every task is executed shortly and switched by the task scheduler (the scheduler determines the sequence according to priority), and the tasks are switched rapidly, which gives the illusion that multiple tasks are running at the same time. In the RT-Thread system, the task is implemented by threads. The thread scheduler in RT-Thread is the task scheduler mentioned earlier.

RT-Thread is mainly written in the C language, is easy to understand, and is easy to port. It applies the object-oriented programming philosophy to real-time system design, making the code elegant, structured, modular, and very tailorable. For resource-constrained Microcontroller Unit (MCU) systems, RT-Thread NANO version (NANO is a minimum kernel officially released by RT-Thread in July 2017) requires only 3KB of Flash and 1.2KB of RAM resources and can be tailored with easy-to-use tools; for resource-rich IoT devices, RT-Thread can use the online software package management tool, along with system configuration tools, to achieve intuitive and rapid modular cutting, seamlessly importing rich software feature packs, thus achieving complex functions like Android's graphical interface and touch sliding effects, smart voice interaction effects, and so on.

Compared with the Linux operating system, RT-Thread is small in size, low in cost, low in power consumption, and fast in startup. Besides, RT-Thread has high determinism and low resource consumption, which is suitable for various resource constraints (such as cost, power consumption, etc.). Although the 32-bit MCU is its main operating platform, other central processing units (CPUs) like ones with memory management unit (MMU), ones based on ARM9, ARM11, and even the Cortex-A series CPUs are capable of running RT-Thread in specific applications.

LICENSE AGREEMENT

The RT-Thread system is a completely open source system. Versions later than 3.1.0 follow the Apache License 2.0 open source license agreement. The RT-Thread system can be used free of charge in commercial products and does not require opening private code to the public. The 3.1.0 version and earlier versions follow the GPLv2+ open source license agreement.

RT-THREAD ARCHITECTURE

In recent years, the concept of IoT has become widely known, and the IoT market has developed rapidly. The network connection of embedded devices is the trend now. The edge network has greatly increased the complexity of software. The traditional RTOS kernel can hardly meet the needs of

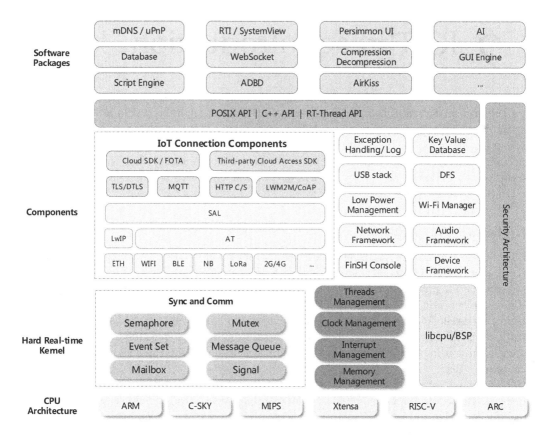

FIGURE 2.1   RT-Thread software framework.

the market. In this case, the concept of the IoT OS has been raised. IoT OS refers to the software platform that is based on an operating system kernel (like RTOS, Linux, etc.) and includes relatively complete middleware components such as file system, graphics library, etc. It is low in consumption and high in security, abides by the Communication Protocol, and has cloud-connect abilities. RT-Thread is an IoT OS.

One of the main differences between RT-Thread and many other RTOSs is that RT-Thread is not only a real-time kernel but also an ecosystem of rich middle-tier components, as shown in Figure 2.1.

It includes:

- Kernel Layer: RT-Thread kernel, the core part of RT-Thread, includes the implementation of objects in the kernel system, such as multi-threading and its scheduling, semaphore, mailbox, message queue, memory management, timer, etc.; libcpu/BSP (Chip Migration Related Files/Board Support Package) is closely related to hardware and consists of peripheral drivers and CPU transport.
- Components and Service Layer: Components are upper-level software on top of the RT-Thread kernel, such as virtual file systems, FinSH command-line interfaces, network frameworks, device frameworks, and more. Its modular design keeps high internal cohesion within the component and low coupling between components.
- RT-Thread Software Package: A general-purpose software component running on the RT-Thread IoT OS platform for different application areas, consisting of description information, source code, or library files. RT-Thread created an open package platform with

official and community packages, which provide developers with a wide choice of reusable packages. Packages are an important part of the RT-Thread ecosystem. The package ecosystem is critical to the choice of an operating system because these packages are highly reusable and modular, making it easy for application developers to build the system they want in the shortest amount of time. RT-Thread supports more than 190 software packages, some of which are listed here:

- IoT-related software packages: Paho MQTT, WebClient, mongoose, WebTerminal, etc.
- Scripting language–related software packages: JerryScript, Lua, and MicroPython are currently supported.
- Multimedia-related software packages: Openmv, mupdf.
- Tools packages: CmBacktrace, EasyFlash, EasyLogger, SystemView.
- System-related software packages: RTGUI, Persimmon UI, lwext4, partition, SQLite, etc.
- Peripheral library and driver software packages: RealTek RTL8710BN SDK.
- Others.

As we can see from Figure 2.1, RT-Thread covers lots of function modules. The book is limited in length, so it won't cover all the modules mentioned earlier. This book is going to mainly introduce the core and basic parts of RT-Thread, such as kernel, FinSH console, I/O device management, DFS virtual file system, network framework, etc. We sincerely hope you can learn more about RT-Thread after reading this book.

ACCESS TO RT-THREAD

LICENSE

RT-Thread is fully open source. RT-Thread version 3.1.0 and its earlier versions follow the GPLv2+ open source license agreement. Versions from 3.1.0 onward follow the Apache License 2.0 open source license agreement. Now, all the RTOS kernel and open source components can be used free of charge for commercial products, and you will not be requested to publish the application source.

GET THE SOURCE CODE

The source code of RT-Thread is hosted on GitHub. You can get the source code from https://github.com/RT-Thread/rt-thread.

This book also has accompanying code material, which is based on RT-Thread 3.1.0.

GETTING STARTED

The main IDE/compilers supported by RT-Thread are:

- KEIL-MDK
- IAR
- GCC
- RT-Thread Studio

You can pick up any of it with RT-Thread.

All the examples mentioned in this book are based on the KEIL-MDK demo. If you also use KEIL-MDK, you will need to install MDK-ARM 5.24 (official or evaluation versions, 5.14 versions and above are all fine), and if you want to remove the 16K compilation code restrictions, you should purchase the MDK-ARM official version.

applications	2018/08/28 15:19
drivers	2018/08/28 15:19
Libraries	2018/08/28 15:19
packages	2018/08/28 15:18
rt-thread	2018/08/28 15:19
project.uvprojx	2018/08/27 18:00
rtconfig.h	2018/08/21 18:40

FIGURE 2.2   Project directory.

FIRST GLANCE AT RT-THREAD

In this section, we begin to learn RT-Thread's routine source code, starting with a streamlined kernel code, which is located in the accompanying materials under the chapter1-9 directory. This code has been tailored based on RT-Thread version 3.1.0 and could help readers get started with RT-Thread quickly. The directory structure of the kernel routine source code used in this book is shown in Figure 2.2.

The file types contained in each directory are described as shown in Table 2.1:

TABLE 2.1
RT-Thread Routine Folder Directory

Directory Name	Description
applications	Application of RT-Thread
rt-thread	Source file of RT-Thread
l- components	Source code of RT-Thread components, such as FinSH, etc.
l- include	Header file of RT-Thread kernel
l- libcpu	Porting code for various types of chips, including the porting file for STM32
l- src	Source file of RT-Thread kernel
drivers	RT-Thread driver, the underlying drivers of different platforms implement specifically
Libraries	STM32 firmware library file
packages	Sample code

Under this chapter1-9 directory, there is an MDK5 project file named project.uvprojx. Double-click on the "project.uvprojx" icon to open this file, as shown in Figure 2.3.

A list of the project's files can be found in the "Project" column on the left side of the project's main window, which is stored in groups, as shown in Table 2.2.

Now click on the button [⚒] in the toolbar above the window, compiling this project.

The results of the compilation are displayed in the Build bar below the window. If everything goes well, the last line will show "0 Error(s), * Warning(s)."

After compilation, we can simulate running RT-Thread through MDK-ARM's simulator. Click the button at the top right of the window [⚲], go to the simulation, and click the [▤] button to start running it.

Then click the [▣] ▾ button on this toolbar or from the menu bar select "View → Serial Windows → UART#1" to open the serial port 1 window. You can see the serial port outputs RT-Thread LOGO and the results of its simulation run, as shown in Figure 2.4.

RT-Thread provides the FinSH function for debugging or viewing system information, and msh in Figure 2.4 indicates that FinSH is running in a traditional command-line mode that can send shell commands just like in dos/bash.

TABLE 2.2
RT-Thread Routine Engineering Catalog

Directory Groups	Descriptions
Applications	The corresponding directory is chapter1-9/applications, which is used to store the user's application code
Drivers	The corresponding directory is chapter1-9/drivers, which is used to store the driver code
STM32_HAL	The corresponding directory is chapter1-9/Libraries/CMSIS/Device/ST32F1xx, which is used to store the firmware library files of STM32
Kernel-samples	Kernel sample code
Kernel	The corresponding directory is chapter1-9/rt-thread/src, which is used to store the RT-Thread core code
CORTEX-M3	The corresponding directory is chapter1-9/rt-thread/libcpu, which is used to store the ARM Cortex-M3 porting code
DeviceDrivers	The corresponding directory is chapter1-9/rt-thread/components/drivers, which is used to store the RT-Thread driver framework source code
finsh	The corresponding directory is chapter1-9/rt-thread/components/finsh, which is used to store the RT-Thread FinSH command line components
libc	The corresponding directory is chapter1-9/rt-thread/components/libc, which is used to store the C library function used by RT-Thread

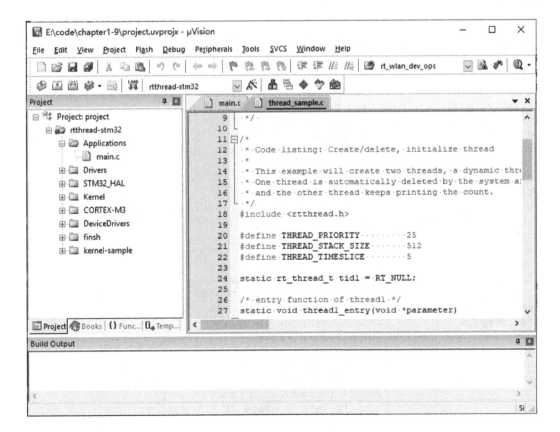

FIGURE 2.3 MDK project.

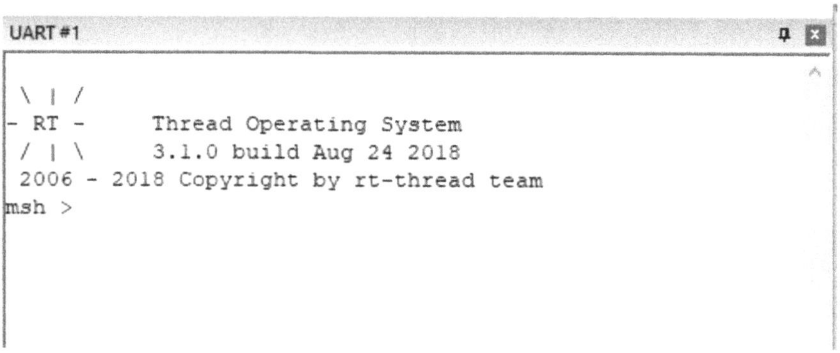

```
UART #1                                                    ⚲ ☒
                                                            ⌃
   \ | /
 - RT -       Thread Operating System
   / | \       3.1.0 build Aug 24 2018
   2006 - 2018 Copyright by rt-thread team
 msh >
```

FIGURE 2.4   MDK simulation.

For example, we can show all the commands supported by the current system by entering "help" and pressing Enter or by pressing Tab directly, as follows:

```
msh >help
RT-Thread shell commands:
thread_sample          - thread sample
timer_sample           - timer sample
semaphore_sample       - semaphore sample
mutex_sample           - mutex sample
event_sample           - event sample
mailbox_sample         - mailbox sample
msgq_sample            - msgq sample
signal_sample          - signal sample
mempool_sample         - mempool sample
dynmem_sample          - dynmem sample
interrupt_sample       - interrupt sample
idle_hook_sample       - idle hook sample
producer_consumer      - producer_consumer sample
timeslice_sample       - timeslice sample
scheduler_hook         - scheduler_hook sample
pri_inversion          - prio_inversion sample
version                - show RT-Thread version information
list_thread            - list thread
list_sem               - list semaphore in system
list_event             - list event in system
list_mutex             - list mutex in system
list_mailbox           - list mail box in system
list_msgqueue          - list message queue in system
list_memheap           - list memory heap in system
list_mempool           - list memory pool in system
list_timer             - list timer in system
list_device            - list device in system
help                   - RT-Thread shell help.
ps                     - List threads in the system.
time                   - Execute command with time.
free                   - Show the memory usage in the system.

msh >
```

At this point, you can enter the commands in the list, such as the input list_thread command, which shows threads currently running on the system. The following example shows that the tshell (FinSH) thread and the tidle (idle) thread are running:

```
msh >list_thread
thread pri  status        sp     stack size max used left tick   error
------ ---  -------    ---------- ---------- ------ ---------- ---
tshell  20  ready      0x00000080 0x00001000   07%  0x0000000a 000
tidle   31  ready      0x00000054 0x00000100   32%  0x00000016 000
msh >
```

FinSH has the function of command auto-complement— Enter some of the characters of the command (the first few letters, note case-sensitivity), press Tab, and the system will find the relevant command registered from the system according to the currently entered character. If the command related to the input is found, the full command will be displayed on the terminal.

For example, to use the version command, you can enter "v" first, and then press Tab. You will find that the system will be completed below the "v" start of the command: version. At this time you only need to press Return to view the execution results of the command.

LED SAMPLE

For a technical engineer working on electronics development, the LED is probably the simplest example, similar to Hello World, the first program that programmers come into contact with in each programming language. So this example starts with the blinking sign, which is updated (on or off) on the LED regularly, as detailed in Code Listing 2.1.

Code listing 2.1 LED Sample

```
#include <rtthread.h>
#include <rtdevice.h>

#define LED_PIN 3

int main(void)
{
    static rt_uint8_t count;
rt_pin_mode(LED_PIN, PIN_MODE_OUTPUT);

    for(count = 0 ; count < 10 ;count++)
      {
            rt_pin_write(LED_PIN, PIN_HIGH);
            rt_kprintf("led on, count : %d\r\n", count);
            rt_thread_mdelay(500);

            rt_pin_write(LED_PIN, PIN_LOW);
            rt_kprintf("led off\r\n");
            rt_thread_mdelay(500);
      }
      return 0;
}
```

This is a simple LED routine, and the "rt_thread_mdelay()" function in the code listing is used to delay milliseconds. The delay in the code causes the LED to be on for 500ms and then off for 500ms. The loop exists after 10 times.

The result of the simulation run is as follows:

```
 \ | /
- RT -      Thread Operating System
/ | \       3.1.0 build Aug 24 2018
2006 - 2018 Copyright by rt-thread team

led on, count : 0
msh >led off
led on, count : 1
led off
led on, count : 2
led off
led on, count : 3
...
led on, count : 9
led off
```

CHAPTER SUMMARY

This chapter gave a brief introduction to RT-Thread, which is not only a real-time kernel but also has a rich set of middle-tier components, characterized by its very small kernel resource consumption, high real-time, tailorable, excellent debugging tool FinSH, and so on. Since RT-Thread is completely open source, you may obtain the source code and documentation from GitHub, and for any problems you encounter during development, don't hesitate to seek assistance and answers from the RT-Thread community.

Please read and reflect on the contents of this chapter. For the coding part, hands-on-experience is most important. From today, we embark on the road to learning about RT-Thread. We hope you have a great journey with it!

3 Kernel Basics

This chapter introduces the basics of the RT-Thread kernel, including an introduction to the kernel, startup process for the system, and some part of the kernel configuration, laying the foundation for the following chapters.

A brief introduction to the RT-Thread kernel, starting with the software architecture to explain the composition and implementation of the real-time kernel, introduces some RT-Thread kernel-related concepts and basic knowledge for beginners, so that beginners have a preliminary understanding of the kernel. After reviewing this chapter, readers will have a basic understanding of the RT-Thread kernel, including what the kernel is composed of, how the system starts up, how memory is distributed, and methods of kernel configuration.

INTRODUCTION TO RT-THREAD KERNEL

The kernel is the most basic and important part of the operating system. Figure 3.1 shows the RT-Thread core architecture diagram. The kernel is on top of the hardware layer. The kernel includes the kernel library and real-time kernel implementation.

The kernel library is a small set of C library–like function implementation subsets to ensure that the kernel can run independently. The built-in C library will be somewhat different depending on the compiler. When using the GNU GCC compiler, it will carry more implementations of the standard C library.

Notes: C library, also called C runtime library, it provides functions like "strcpy" and "memcpy," and some also include the implementation of the "printf" and "scanf" functions. The RT-Thread Kernel Service Library provides only a small portion of the C library function implementation used by the kernel. To avoid duplicate names with the standard C library, the rt_ prefix is added before these functions. CHED

Real-time kernel implementation includes object management, thread management and scheduler, inter-thread communication management, clock management, memory management, etc. The minimum resource occupation of the kernel is 3KB ROM, 1.2KB RAM.

THREAD SCHEDULING

A thread is the smallest scheduling unit in the RT-Thread operating system. The thread scheduling algorithm is a priority-based full preemptive multi-thread scheduling algorithm. Except for the interrupt handler, the code of the scheduler's locked part, and the code that prohibits the interrupt, other parts of the system can be preempted, including the thread scheduler itself. The system supports 256 thread priorities (can also be changed to a maximum of 32 or 8 thread priorities via the configuration file; for STM32 default configuration, it is set as 32 thread priorities), 0 priority represents the highest priority, and the lowest priority is reserved for idle threads; at the same time, it also supports creating multiple threads with the same priority. The same priority threads are scheduled with a time slice round-robin scheduling algorithm, so that each thread runs for the same time; besides, when the scheduler is looking for threads that are at the highest priority thread and ready, the elapsed time is constant. The system does not limit the number of threads; the number of threads is only related to the specific memory of the hardware platform.

Thread management will be covered in detail in the "Thread Management" chapter.

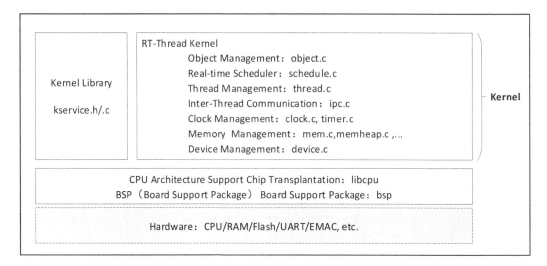

FIGURE 3.1   RT-Thread kernel and its substructure.

CLOCK MANAGEMENT

RT-Thread's clock management is based on a clock tick, which is the smallest clock unit in the RT-Thread operating system. The RT-Thread timer provides two types of timer mechanisms: The first type is a one-shot timer, which triggers only one timer event after startup and then stops automatically. The second type is a periodic trigger timer, which periodically triggers timer events until the user manually stops the timer, or it will continue to operate.

In addition, depending on the context in which the timeout function is executed, the RT-Thread timer can be set to HARD_TIMER mode or SOFT_TIMER mode.

Usually, the timer service is completed using a timer timing callback function (i.e., a timeout function). The user can select the appropriate type of timer according to his own real-time requirements for timing processing.

The timer will be explained in the "Clock Management" chapter.

SYNCHRONIZATION BETWEEN THREADS

RT-Thread uses thread semaphores, mutexes, and event sets to achieve inter-thread synchronization. Threads synchronize through the acquisition and release of semaphores and mutexes; the mutex uses priority inheritance to solve the common priority inversion problem in the real-time system. The thread synchronization mechanism allows threads to wait according to priorities or to acquire semaphores or mutexes following the first-in/first-out method. Threads synchronize through the sending and receiving of events; event sets allows "or trigger" and "and trigger" for multiple events, suitable for situations where threads are waiting for multiple events.

The concepts of semaphores, mutexes, and event sets are detailed in the "Inter-Thread Synchronization" chapter.

INTER-THREAD COMMUNICATION

RT-Thread supports communication mechanisms such as mailbox, message queue, etc. The length of a message in the mailbox is fixed to 4 bytes; the message queue can receive messages of variable length and cache the messages in its own memory space. Compared to the message queue, the mailbox is more efficient. The sending action of the mailbox and message queue can be safely used in the interrupt service routine. The communication mechanism allows threads to wait by priority or to acquire by the first-in/first-out method.

The concept of the mailbox and message queue will be explained in detail in the "Inter-Thread Communication" chapter.

MEMORY MANAGEMENT

RT-Thread allows static memory pool management and dynamic memory heap management. When the static memory pool has available memory space, the time allocated to the memory block will be constant; when the static memory pool is empty, the system will then request for suspending or blocking the thread of the memory block (i.e., the thread will abandon the request and return, if after waiting for a while, the memory block is not obtained, or the thread will abandon and return immediately; the waiting time depends on the waiting time parameter set when the memory block is applied). When other threads release the memory block to the memory pool, if there are threads that are suspending and waiting to be allocated memory blocks, the system will wake up the thread.

Under the circumstances of different system resources, the dynamic memory heap management module provides memory management algorithms for small memory systems and SLAB memory management algorithms for large memory systems.

There is also a dynamic memory heap management called memheap, which is suitable for memory heaps in systems with multiple addresses that can be discontinuous. Using memheap, you can "stick" multiple memory heaps together, letting the user operate as if he was operating a memory heap.

The concept of memory management will be explained in the "Memory Management" chapter.

I/O DEVICE MANAGEMENT

RT-Thread uses PIN, I2C, SPI, USB, UART, etc., as peripheral devices, and is uniformly registered through the device framework. It realized a device management subsystem accessed by the name, and it can access hardware devices according to a unified interface. On the device driver interface, depending on the characteristics of the embedded system, corresponding events can be attached to different devices. When the device event is triggered, the driver notifies the upper application program.

The concept of I/O device management will be explained in the "Device Model" and "General Equipment" chapters.

RT-THREAD STARTUP PROCESS

The understanding of most codes usually starts by learning the startup process. We will first look for the source of the startup. RT-Thread allows multiple platforms and multiple compilers, and the rtthread_startup() function is a uniform entry point specified by RT-Thread. The general execution sequence is as follows: First, the system starts running the startup file, then enters rtthread_startup(), and finally enters the user program entry main(). The startup process of RT-Thread is shown in Figure 3.2.

Taking MDK-ARM as an example, the user program entry for MDK-ARM is the main() function, which is located in the main.c file. The launching of the system starts from the assembly code startup_stm32f103xe.s, then jumps to the C code, initializes the RT-Thread system function, and finally enters the user program entry main().

To complete the RT-Thread system function initialization before entering main(), we used the MDK extensions $Sub$$ and $Super$$. You can add the prefix of $Sub$$ to main to make it a new function. $Sub$$main.$Sub$$main can call some functions to be added before main. (Here, RT-Thread system initialization function is added.) Then, call $Super$$main to the main() function, so that the user does not have to manage the system initialization before main().

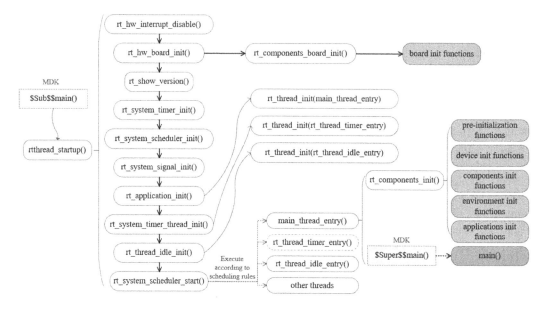

FIGURE 3.2   System startup process.

For more information on the use of the $Sub$$ and $Super$$ extensions, see the ARM®
Compiler v5.06 for μVision®armlink User Guide.

Let's take a look at this code defined in components.c:

```
/* $Sub$$main Function */
int $Sub$$main(void)
{
  rtthread_startup();
  return 0;
}
```

Here, the $Sub$$main function simply calls the rtthread_startup() function. Code for the
rtthread_startup() function is as follows:

```
int rtthread_startup(void)
{
    rt_hw_interrupt_disable();

    /* Board level initialization: system heap initialization is required
inside the function */
    rt_hw_board_init();

    /* Print RT-Thread version information */
    rt_show_version();

    /* Timer initialization */
    rt_system_timer_init();

    /* Scheduler initialization */
    rt_system_scheduler_init();

#ifdef RT_USING_SIGNALS
    /* Signal initialization */
    rt_system_signal_init();
#endif
```

```
/* Create a user main() thread here */
rt_application_init();

/* Timer thread initialization */
rt_system_timer_thread_init();

/* Idle thread initialization */
rt_thread_idle_init();

/* Start scheduler */
rt_system_scheduler_start();

/* Will not execute till here */
return 0;
}
```

This part of the startup code can be roughly divided into four parts:

1. Initialize hardware related to the system;
2. Initialize system kernel objects, such as timers, schedulers, and signals;
3. Create a main thread, and initialize various modules in the main thread one by one;
4. Initialize the timer thread, idle thread, and start the scheduler.

Set the system clock in rt_hw_board_init() to provide heartbeat and serial port initialization for the system, bound the system's input and output terminals to this serial port, and subsequent system operation information will be printed out from the serial port later.

The main() function is the user code entry for RT-Thread, and users can add their own applications to the main() function.

```
int main(void)
{
  /* user app entry */
  return 0;
}
```

RT-THREAD PROGRAM MEMORY DISTRIBUTION

The general MCU contains storage space that includes on-chip Flash and on-chip RAM. RAM is equivalent to memory, and Flash is equivalent to hard disk. The compiler classifies a program into several parts, which are stored in different memory areas of the MCU.

After the Keil project is compiled, there will be a prompt for information for occupied space by the corresponding program, as shown here:

```
linking...
Program Size: Code=48008 RO-data=5660 RW-data=604 ZI-data=2124
After Build - User command \#1: fromelf --bin.\\build\\rtthread-stm32.axf-
-output rtthread.bin
".\\build\\rtthread-stm32.axf" - 0 Error(s), 0 Warning(s).
Build Time Elapsed: 00:00:07
```

The Program Size mentioned earlier contains the following sections:

1. Code: code segment, section of code that stores the program;
2. RO-data: read-only data segment, stores the constants defined in the program;
3. RW-data: read and write data segment, stores global variables initialized to non-zero values;
4. ZI-data: 0-ed data segment, stores uninitialized global variables and initialized-to-0 variables.

After compiling the project, a .map file will be generated that describes the size and address of each function. The last few lines of the file also illustrate the relationship between the above fields:

```
Total RO Size (Code + RO Data)        53668 ( 52.41kB)
Total RW Size (RW Data + ZI Data)      2728 (  2.66kB)
Total ROM Size (Code + RO Data + RW Data) 53780 ( 52.52kB)
```

1. RO Size contains Code and RO-data, indicating the size of the Flash occupied by the program;
2. RW Size contains RW-data and ZI-data, indicating the size of the RAM occupied when operating;
3. ROM Size contains Code, RO Data, and RW Data, indicating the size of the Flash occupied by the programming system.

Before the program runs, the file entity needs to be burned into STM32 Flash—usually it is a bin or hex file. The burned file is called an executable image file. The left figure in Figure 3.3 shows the memory distribution after the executable image file is burned to STM32, which includes two parts: the RO segment and the RW segment. The RO segment stores data of Code and RO-data, and the RW segment holds the data of RW-data. Since ZI-data is 0, it is not included in the image file.

STM32 is launched from Flash by default after powering on. After launching, RW-data (initialized global variable) in the RW segment is transferred to RAM, but the RO segment is not transferred. This means that the execution code of CPU is read from Flash, the ZI segment is allocated according to the ZI address and size that given by the compiler, and the RAM area is cleared.

The dynamic memory heap is unused RAM space, and the memory blocks requested and released by the application come from this space.

This is shown in the following example:

```
rt_uint8_t* msg_ptr;
msg_ptr = (rt_uint8_t*) rt_malloc (128);
rt_memset(msg_ptr, 0, 128);
```

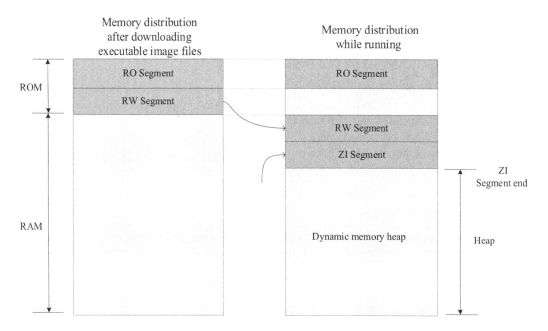

FIGURE 3.3 RT-Thread memory distribution.

The 128-byte memory space pointed to by the msg_ptr pointer in the code is in the dynamic memory heap space.

Some global variables are stored in the RW segment and the ZI segment. The RW segment stores the global variable with the initial value (the global variable in the constant form is placed in the RO segment, which is a read-only property), and the uninitialized global variable is stored in the ZI segment, as in the following example:

```
#include <rtthread.h>

const static rt_uint32_t sensor_enable = 0x000000FE;
rt_uint32_t sensor_value;
rt_bool_t sensor_inited = RT_FALSE;

void sensor_init()
{
    /* ... */
}
```

The sensor_value is stored in the ZI segment and is automatically initialized to zero after system startup (some library functions provided by the user program or compiler are initialized to zero). The sensor_inited variable is stored in the RW segment, and the sensor_enable is stored in the RO segment.

RT-THREAD AUTOMATIC INITIALIZATION MECHANISM

The automatic initialization mechanism means that the initialization function does not need to be invoked by explicit function call. It only needs to be declared by macro definition at the function definition, and it will be executed during system startup.

For example, when calling a macro definition in the serial port driver to inform the function that needs to be called to initialize the system, the code is as follows:

```
int rt_hw_usart_init(void)  /* Serial port initialization function */
{
    ... ...
    /* Register serial port 1 device */
    rt_hw_serial_register(&serial1, "uart1",
                    RT_DEVICE_FLAG_RDWR | RT_DEVICE_FLAG_INT_RX,
                    uart);
    return 0;
}
INIT_BOARD_EXPORT(rt_hw_usart_init);     /* Use component auto-
initialization mechanism */
```

The last part of the sample code INIT_BOARD_EXPORT(rt_hw_usart_init) indicates that the automatic initialization function is used. In this way, rt_hw_usart_init() function is automatically called by the system, so where is it called?

In the system startup flowchart, there are two functions:

rt_components_board_init() and rt_components_init(), subsequent functions inside the box with the background color represent functions that are automatically initialized, where:

1. "board init functions" are all initialization functions declared by INIT_BOARD_EXPORT(fn).
2. "pre-initialization functions" are all initialization functions declared by INIT_PREV_EXPORT(fn).

3. "device init functions" are all initialization functions declared by INIT_DEVICE_ EXPORT(fn).
4. "components init functions" are all initialization functions declared by INIT_ COMPONENT_EXPORT(fn).
5. "environment init functions" are all initialization functions declared by INIT_ENV_ EXPORT(fn).
6. "application init functions" are all initialization functions declared by INIT_APP_ EXPORT(fn).

The rt_components_board_init() function executes earlier, mainly to initialize the relevant hardware environment. When this function is executed, it will traverse the initialization function table declared by INIT_BOARD_EXPORT(fn) and call each function.

The rt_components_init() function is called and executed in the main thread created after the operating system is running. At this time, the hardware environment and the operating system have been initialized, and the application-related code can be executed. The rt_components_init() function will transverse through the remaining few initialization function tables declared by macros.

RT-Thread's automatic initialization mechanism uses a custom RTI symbol segment. It puts the function pointer that needs to be initialized at startup into this segment and forms an initialization function table, which will be traversed during system startup. It calls the functions in the table to achieve the purpose of automatic initialization.

The macro interface definitions used to implement the automatic initialization functions are described in Table 3.1.

TABLE 3.1
Automatic Initialization Functions

Initialization Sequence	Macro Interface	Description
1	INIT_BOARD_EXPORT(fn)	Very early initialization, the scheduler has not started yet
2	INIT_PREV_EXPORT(fn)	Mainly used for pure software initialization, functions without too many dependencies
3	INIT_DEVICE_EXPORT(fn)	Peripheral driver initialization related, such as network card devices
4	INIT_COMPONENT_EXPORT(fn)	Component initialization, such as file system or LWIP
5	INIT_ENV_EXPORT(fn)	System environment initialization, such as mounting file systems
6	INIT_APP_EXPORT(fn)	Application initialization, such as application GUI

The initialization function actively declares through these macro interfaces, such as INIT_BOARD_ EXPORT (rt_hw_usart_init), and the linker will automatically collect all the declared initialization functions placed in the RTI symbol segment, and the symbol segment is located in the RO segment of the memory distribution. All functions in this RTI symbol segment are automatically called when the system is initialized.

RT-THREAD KERNEL OBJECT MODEL

STATIC AND DYNAMIC OBJECTS

The RT-Thread kernel is designed with the object-oriented method. The system-level infrastructures are all kernel objects such as threads, semaphores, mutexes, timers, and so on. Kernel objects fall into two categories: static kernel objects and dynamic kernel objects. Static kernel objects are usually placed in RW and ZI segments and are initialized in the program after system startup; dynamic kernel objects are created from the memory heap and then manually initialized.

Code Listing 3.1 is an example of static threads and dynamic threads.

Code listing 3.1 An Example of Static Threads and Dynamic Threads

```
/* Thread 1 object and stack used while running */
static struct rt_thread thread1;
static rt_uint8_t thread1_stack[512];

/* Thread 1 entry */
void thread1_entry(void* parameter)
{
    int i;

    while (1)
    {
        for (i = 0; i < 10; i ++)
        {
            rt_kprintf("%d\n", i);

            /* Delay 100ms */
            rt_thread_mdelay(100);
        }
    }
}

/* Thread 2 entry */
void thread2_entry(void* parameter)
{
    int count = 0;
    while (1)
    {
        rt_kprintf("Thread2 count:%d\n", ++count);

        /* Delay 50ms */
        rt_thread_mdelay(50);
    }
}

/* Thread routine initialization */
int thread_sample_init()
{
    rt_thread_t thread2_ptr;
    rt_err_t result;

    /* Initialize thread 1 */
    /* The thread entry is thread1_entry and the parameter is RT_NULL
     * Thread stack is thread1_stack
     * Priority is 200 and time slice is 10 OS Tick
     */
    result = rt_thread_init(&thread1,
                            "thread1",
                            thread1_entry, RT_NULL,
                            &thread1_stack[0], sizeof(thread1_stack),
                            200, 10);
```

```
    /* Start thread */
    if (result == RT_EOK) rt_thread_startup(&thread1);

    /* Create thread 2 */
    /* The thread entry is thread2_entry and the parameter is RT_NULL
     *  Stack space is 512, priority is 250, and time slice is 25 OS Tick
     */
    thread2_ptr = rt_thread_create("thread2",
                                   thread2_entry, RT_NULL,
                                   512, 250, 25);

    /* Start thread */
    if (thread2_ptr != RT_NULL) rt_thread_startup(thread2_ptr);

    return 0;
}
```

In this example, thread1 is a static thread object and thread2 is a dynamic thread object. The memory space of the thread1 object, including the thread control block thread1 and the stack space thread1_stack, is determined while compiling, because there is no initial value in the code, and they are uniformly placed in the uninitialized data segment. The space used by thread2 is dynamically allocated and includes the thread control block (the content pointed to by thread2_ptr) and the stack space.

Static objects take up RAM space and are not dependent on the memory heap manager. When allocating static objects, the time required to allocate static objects is determined. Dynamic objects depend on the memory heap manager. It requests RAM space while running. When the object is deleted, the occupied RAM space is released. These two methods have their own advantages and disadvantages, and can be selected according to actual needs.

KERNEL OBJECT MANAGEMENT STRUCTURE

RT-Thread uses the kernel object management system to access/manage all kernel objects. Kernel objects contain most of the facilities in the kernel. These kernel objects can be statically allocated static objects and dynamic objects allocated from the system memory heap.

Because of this design for kernel object, RT-Thread cannot depend on the specific memory allocation method, and the flexibility of the system is greatly improved.

RT-Thread kernel objects include threads, semaphores, mutexes, events, mailboxes, message queues and timers, memory pools, device drivers, and so on. The object container contains information about each type of kernel object, including object type, size, and so on. The object container assigns a linked list to each type of kernel object. All kernel objects are linked to the linked list. The kernel object container and linked list of RT-Thread are shown in Figure 3.4.

Figure 3.5 shows the derivation and inheritance relationships of various kernel objects in RT-Thread. For each specific kernel object and object control block, in addition to the basic structure, they have their own extended attributes (private attributes). Take the thread control block as example: the base object is extended, and attributes like thread state, precedence, and so on are added. These attributes are not used in the operation of the base class object and are only used in operations related to a specific thread. Therefore, from the object-oriented point of view, each concrete object can be considered a derivative of an abstract object, inheriting the attributes of the base object and extending the attributes related to itself.

In the object management module, a common data structure is defined to store the common attributes of various objects. Each specific object only needs to add some special attributes of its own, and its feature will be clearly expressed.

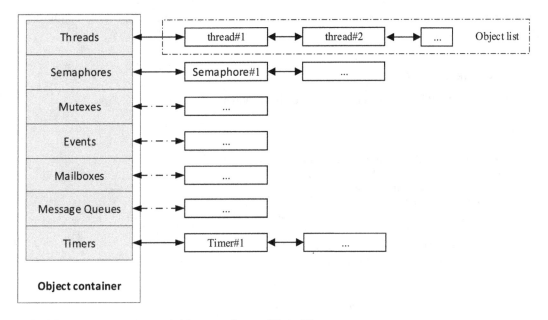

FIGURE 3.4 RT-Thread kernel object container and linked list.

The advantages of this design approach are:

1. Improve the reusability and scalability of the system. It is easy to add new object catego-ries, just inherit the attributes of the general object and add a small amount of extension.
2. Provide a unified object operation mode, simplify the operation of various specific objects, and improve the reliability of the system.

Derivations from the object control block rt_object in the above figure include thread object, mem-ory pool object, timer object, device object, and IPC object (IPC: Inter-Process Communication. In RT-Thread real-time operating system, IPC objects are used for synchronization and communication

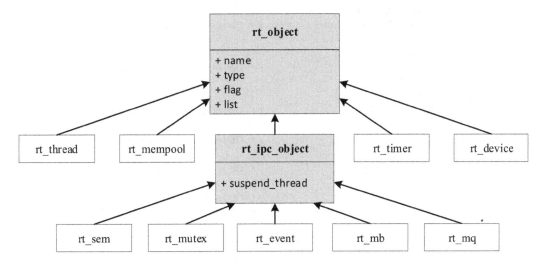

FIGURE 3.5 RT-Thread kernel object inheritance relationship.

between threads); derivations from IPC objects include semaphores, mutexes, events, mailboxes, message queues, signals, etc.

OBJECT CONTROL BLOCK

The data structure of kernel object control block is as shown:

```
struct rt_object
{
    /* Kernel object name     */
    char      name[RT_NAME_MAX];
    /* Kernel object type     */
    rt_uint8_t  type;
    /* Parameters to the kernel object   */
    rt_uint8_t  flag;
    /* Kernel object management linked list */
    rt_list_t    list;
};
```

Types currently supported by kernel objects are as follows:

```
enum rt_object_class_type
{
    RT_Object_Class_Thread = 0,        /* Object is thread type */
#ifdef RT_USING_SEMAPHORE
    RT_Object_Class_Semaphore,         /* Object is semaphore type */
#endif
#ifdef RT_USING_MUTEX
    RT_Object_Class_Mutex,             /* Object is mutex type */
#endif
#ifdef RT_USING_EVENT
    RT_Object_Class_Event,             /* Object is event type */
#endif
#ifdef RT_USING_MAILBOX
    RT_Object_Class_MailBox,           /* Object is mailbox type */
#endif
#ifdef RT_USING_MESSAGEQUEUE
    RT_Object_Class_MessageQueue,      /* Object is message queue type */
#endif
#ifdef RT_USING_MEMPOOL
    RT_Object_Class_MemPool,           /* Object is memory pool type */
#endif
#ifdef RT_USING_DEVICE
    RT_Object_Class_Device,            /* Object is device type */
#endif
    RT_Object_Class_Timer,             /* Object is timer type */
#ifdef RT_USING_MODULE
    RT_Object_Class_Module,            /* Object is module */
#endif
    RT_Object_Class_Unknown,           /* Object is unknown */
    RT_Object_Class_Static = 0x80      /* Object is a static object */
};
```

From the above type specification, we can see that if it is a static object, the highest bit of the object type will be 1 (which is the OR operation of RT_Object_Class_Static and other object types and operations). Otherwise, it will be a dynamic object, and the maximum number of object classes that the system can accommodate is 127.

Kernel Object Management

The data structure of the kernel object container is as follows:

```
struct rt_object_information
{
    /* Object type */
    enum rt_object_class_type type;
    /* Object linked list */
    rt_list_t object_list;
    /* Object size */
    rt_size_t object_size;
};
```

A class of objects is managed by an rt_object_information structure, and each practical instance of each type of object is mounted to the object_list in the form of a linked list. The memory block size of this type of object is identified by object_size (the memory block of each practical instance of each type of object is the same size).

Initialization Object

An uninitialized static object must be initialized before it can be used. The initialization object uses the following interfaces:

```
void rt_object_init(struct   rt_object*   object ,
                    enum rt_object_class_type  type ,
                    const char* name)
```

When this function is called to initialize the object, the system will place the object into the object container for management, then initialize some parameters of the object, and then insert the object node into the object linked list of the object container. Input parameters of the function are described in Table 3.2.

TABLE 3.2
rt_object_init

Parameters	Description
object	The object pointer that needs to be initialized must point to a specific object memory block, not a null pointer or a wild pointer.
type	The type of the object must be an enumeration type listed in rt_object_class_type, RT_Object_Class_Static excluded. (For static objects, or objects initialized with the rt_object_init interface, the system identifies it as an RT_Object_Class_Static type.)
name	Name of the object. Each object can be set to a name, and the maximum length for the name is specified by RT_NAME_MAX. The system does not care if it uses '\θ' as a terminal symbol.

Detach Object

The following interfaces are used to detach objects from the object kernel manager:

```
void rt_object_detach(rt_object_t object);
```

Calling this interface makes a static kernel object be detached from the kernel object container, meaning the corresponding object node is deleted from the kernel object container linked list. After the object is detached, the memory occupied by the object will not be released.

Allocate Object

The earlier descriptions are interfaces of object initialization and detachment, both of which are when object-oriented memory blocks already exist. But dynamic objects can be requested when needed. The memory space is freed for other applications when not needed. To request assigning new objects, you can use the following interfaces:

```
rt_object_t rt_object_allocate(enum  rt_object_class_type type ,
                               const  char*  name)
```

When calling the above interface, the system first needs to obtain object information according to the object type (especially the size information of the object type for the system to allocate the correct size of the memory data block) and then allocate memory space corresponding to the size of the object from the memory heap. Next, the system starts the necessary initialization for the object, and finally inserts it into the object container linked list in which it is located. The input parameters for this function are described in Table 3.3.

TABLE 3.3
rt_object_allocate

Parameters	Description
type	The type of the allocated object can only be of type rt_object_class_type other than RT_Object_Class_Static. In addition, the type of object allocated through this interface is dynamic, not static.
name	Name of the object. Each object can be set to a name, and the maximum length for the name is specified by RT_NAME_MAX. The system does not care if it uses '\θ'as a terminal symbol.
Return	——
object handle allocated successfully	Allocate successfully.
RT_NULL	Fail to allocate.

Delete Object

For a dynamic object, when it is no longer used, you can call the following interface to delete the object and release the corresponding system resources:

```
void rt_object_delete(rt_object_t object);
```

When the above interface is called, the object is first detached from the object container linked list, and then the memory occupied by the object is released. Table 3.4 describes the input parameters of the function:

TABLE 3.4
rt_object_delete

Parameter	Description
object	Object handle

Identify Objects

The following interface is used to identify whether the object is a system object (static kernel object):

```
rt_err_t rt_object_is_systemobject(rt_object_t object);
```

Calling the rt_object_is_systemobject interface can help identify whether an object is a system object. In the RT-Thread operating system, a system object is also a static object: RT_Object_Class_Static bit is set to 1 on the object type identifier. Usually, objects that are initialized using the rt_object_init() method are system objects. The input parameters for this function are described in Table 3.5.

TABLE 3.5
rt_object_is_systemobject

Parameter	Description
object	Object handle

RT-THREAD KERNEL CONFIGURATION EXAMPLE

An important feature of RT-Thread is its high degree of tailorability, which allows for fine-tuning of the kernel and flexible removal of components.

Configuration is mainly done by modifying the file under project directory—rtconfig.h. A user can conditionally compile the code by opening/closing the macro definition in the file, and finally achieve the purpose of system configuration and cropping, as follows:

1. RT-Thread Kernel Part

```
/* Indicates the maximum length of the name of the kernel object. If the
maximum length of the name of the object in the code is greater than the
length of the macro definition,
* the extra part will be cut off. */
#define RT_NAME_MAX 8

/* Set the number of aligned bytes when bytes are aligned. Usually use
ALIGN(RT_ALIGN_SIZE) for byte alignment.*/
#define RT_ALIGN_SIZE 4

/* Define the number of system thread priorities; usually define the
priority of idle threads with RT_THREAD_PRIORITY_MAX-1 */
#define RT_THREAD_PRIORITY_MAX 32

/* Define the clock beat. When it is 100, it means 100 tick per second,
and a tick is 10ms. */
#define RT_TICK_PER_SECOND 100

/* Check if the stack overflows, if not defined, close. */
#define RT_USING_OVERFLOW_CHECK

/* Define this macro to enable debug mode, if not defined, close. */
#define RT_DEBUG
/* When debug mode is enabled: When the macro is defined as 0, the print
component initialization information is turned off. When it is defined
as 1, it is enabled. */
```

```
#define RT_DEBUG_INIT 0
/* When debug mode is enabled: When the macro is defined as 0, the print
thread switching information is turned off. When it is defined as 1,
it is enabled. */
#define RT_DEBUG_THREAD 0

/* Defining this macro means the use of the hook function is started,
if not defined, close. */
#define RT_USING_HOOK

/* Defines the stack size of idle threads. */
#define IDLE_THREAD_STACK_SIZE 256
```

2. Inter-thread synchronization and communication part, the objects that will be used in this part, are semaphores, mutexes, events, mailboxes, message queues, signals, and so on.

```
/* Define this macro to enable the use of semaphores, if not defined,
close. */
#define RT_USING_SEMAPHORE

/* Define this macro to enable the use of mutexes, if not defined,
close. */
#define RT_USING_MUTEX

/* Define this macro to enable the use of events, if not defined,
close. */
#define RT_USING_EVENT

/* Define this macro to enable the use of mailboxes, if not defined,
close. */
#define RT_USING_MAILBOX

/* Define this macro to enable the use of message queues, if not
defined, close. */
#define RT_USING_MESSAGEQUEUE

/* Define this macro to enable the use of signals, if not defined,
close. */
#define RT_USING_SIGNALS
```

3. Memory Management Part

```
/* Start the use of static memory pool */
#define RT_USING_MEMPOOL

/* Define this macro to start the concatenation of two or more memory
heap , if not defined, close. */
#define RT_USING_MEMHEAP

/* Start algorithm for small memory management */
#define RT_USING_SMALL_MEM

/* Turn off SLAB memory management algorithm */
/* #define RT_USING_SLAB */

/* Start the use of heap */
#define RT_USING_HEAP
```

4. Kernel Device Object

```
/* Indicates the start of useing system devices */
#define RT_USING_DEVICE

/* Define this macro to start the use of system console devices, if
not defined, close. */
#define RT_USING_CONSOLE
/* Define the buffer size of the console device. */
#define RT_CONSOLEBUF_SIZE 128
/* Name of the console device. */
#define RT_CONSOLE_DEVICE_NAME "uart1"
```

5. Automatic Initialization Method

```
/* Define this macro to enable automatic initialization mechanism,
if not defined, close. */
#define RT_USING_COMPONENTS_INIT

/* Define this macro to set application entry as main function */
#define RT_USING_USER_MAIN
/* Define the stack size of the main thread */
#define RT_MAIN_THREAD_STACK_SIZE 2048
```

6. FinSH

```
/* Define this macro to start the use of the system FinSH debugging
tool, if not defined, close. */
#define RT_USING_FINSH

/* While starting the system FinSH: the thread name is defined as tshell
*/
#define FINSH_THREAD_NAME "tshell"

/* While turning the system FinSH: use history commands. */
#define FINSH_USING_HISTORY
/* While turning the system FinSH: define the number of historical
command lines. */
#define FINSH_HISTORY_LINES 5

/* While turning the system FinSH: define this macro to open the Tab
key, if not defined, close. */
#define FINSH_USING_SYMTAB

/* While turning the system FinSH: define the priority of the thread. */
#define FINSH_THREAD_PRIORITY 20
/* While turning the system FinSH: define the stack size of the thread.
*/
#define FINSH_THREAD_STACK_SIZE 4096
/* While turning the system FinSH: define the length of command
character. */
#define FINSH_CMD_SIZE 80

/* While turning the system FinSH: define this macro to enable the MSH
function. */
#define FINSH_USING_MSH
```

```
/* While turning the system FinSH: when MSH function is enabled, macro
is defined to use the MSH function by default. */
#define FINSH_USING_MSH_DEFAULT
/* While turning the system FinSH: define this macro to use only the MSH
function. */
#define FINSH_USING_MSH_ONLY
```

7. About MCU

```
/* Define the MCU used in this project is STM32F103ZE; the system
defines the chip pins by defining the chip type. */
#define STM32F103ZE

/* Define the clock source frequency. */
#define RT_HSE_VALUE 8000000

/* Define this macro to enable the use of UART1. */
#define RT_USING_UART1
```

In practice, the system configuration file rtconfig.h is automatically generated by configuration tools and does not need to be changed manually.

COMMON MACRO DEFINITION DESCRIPTION

Macro definitions are often used in RT-Thread. For example, some common macro definitions in the Keil compilation environment include:

1. rt_inline, the definition of which follows. The static keyword is used to make the function only available for use in the current file. After creating the modification using static, the compiler is recommended to perform an inline expansion when calling the function.

```
#define rt_inline                      static __inline
```

2. RT_USED, the definition of which follows. The purpose of this macro is to explain to the compiler that this code is useful; the compilation needs to be saved even if it is not called in the function. For example, RT-Thread auto-initialization uses custom segments; using RT_USED will retain custom code snippets.

```
#define RT_USED                        __attribute__((used))
```

3. RT_UNUSED, the definition of which follows, indicates that a function or variable may not be used. This attribute prevents the compiler from generating warnings.

```
#define RT_UNUSED                      __attribute__((unused))
```

4. RT_WEAK, the definition of which follows, is often used to define functions. When linking the function, the compiler will link the function without this keyword prefix first and then link the function modified by weak if it can't find those functions.

```
#define RT_WEAK                        __weak
```

5. ALIGN(n), the definition of which follows, is used to align its stored address with n bytes when allocating address space to an object. Here, n can be the power of 2.

Byte alignment not only facilitates quick CPU access but also saves memory space if properly used.

```
#define ALIGN(n)                          __attribute__((aligned(n)))
```

6. RT_ALIGN(size, align), the definition of which follows, is used to increase the size to a multiple of an integer defined by aligning. For example, RT_ALIGN(13,4) will return 16.

```
#define RT_ALIGN(size, align)  (((size) + (align) - 1) & ~((align) - 1))
```

CHAPTER SUMMARY

This chapter described RT-Thread's kernel, kernel object model, system configuration, and system startup process. This chapter can be summarized as follows:

- The implementation of the RT-Thread kernel includes thread management and scheduler, clock management, inter-thread communication management, memory management, etc., which are the core of the operating system.
- The system configuration file of rtconfig.h is automatically generated by the configuration tool and does not require manual changes in general.
- When using the auto-initialization function, you need to arrange the execution sequence of each function according to the actual requirements.

4 Thread Management

When we are facing a big task in our daily life, we usually break it down into a number of simple, easy-to-manage smaller tasks. Then, we would deal with these smaller tasks one by one, gradually; the big task is worked out. In a multi-threaded operating system, developers also need to break down a complex application into multiple small, schedulable, and serialized program units. When tasks are reasonably divided and properly executed, this design allows the system to meet the capacity and time requirements of the real-time system. For example, to have the embedded system perform such tasks, the system would collect data through sensors and display the data on the screen. In a multi-threaded real-time system, the task can be decomposed into two subtasks. The subtask, as shown in Figure 4.1, reads the sensor data continuously and writes the data into the shared memory. The other subtask periodically reads the data from the shared memory and outputs the sensor data onto the screen.

In RT-Thread, the application entity corresponding to the above subtask is the thread. The thread is the carrier of the task. It is the most basic scheduling unit in RT-Thread. It describes the running environment of task execution. It also describes the priority level of the task—the important task can be set a relatively high priority, the non-important task can be set a lower priority, and different tasks can also be set the same priority and take turns to run.

When a thread is running, it thinks it is hogging the CPU. The runtime environment where the thread executes in is called the context; the contexts are all variables and data, including all the register variables, stacks, memory information, and so on.

This chapter will be divided into five sections to introduce thread management of RT-Thread. After reading this chapter, readers will have a deeper understanding of the thread management mechanism of RT-Thread. They will have clear answers to questions like how many states does a thread have, how to create a thread, why does an idle thread exist, etc.

THREAD MANAGEMENT FEATURES

The main function of RT-Thread thread management is to manage and schedule threads. There are two types of threads in the system, namely system thread and user thread. System thread is the thread created by RT-Thread kernel. User thread is the thread created by application. Both types of threads will allocate thread objects from the kernel object container. When the thread is deleted, it will also be deleted from the object container. As shown in Figure 4.2, each thread has important attributes, such as thread control block, thread stack, entry function, and so on.

The thread scheduler of RT-Thread is preemptive, and its main job is to find the highest priority thread from the list of ready threads to ensure that the highest-priority thread can be run. Once the highest-priority task is ready, the thread can always get the CPU usage right.

When a running thread makes the running condition ready for another thread with a higher priority, then the current running thread's CPU usage right is deprived, or in other words, released, and the high-priority thread immediately gets the CPU usage right.

If it is the interrupt service routine that makes the running condition ready for the thread with a higher priority, when the interrupt is completed, the interrupted thread is suspended, and the thread with the higher priority starts running.

When the scheduler schedules threads and switches them, the current thread context is first saved. When it is switched back to this thread, the scheduler restores the context information of the thread.

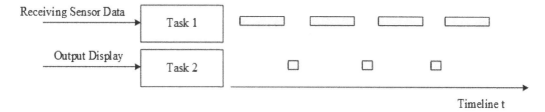

FIGURE 4.1   Switching execution of sensor's data receiving task and display task.

WORKING MECHANISM OF THREAD

THREAD CONTROL BLOCK

In RT-Thread, the thread control block is represented by the structure struct rt_thread, which is a data structure used by the operating system to manage threads. It stores information about the thread, such as priority, thread name, thread status, etc. It also includes a linked list structure for connecting threads, event collection of thread waiting, etc., which are defined as follows:

```
/* Thread Control Block */
struct rt_thread
{
    /* rt Object */
    char        name[RT_NAME_MAX];      /* Thread Name */
    rt_uint8_t  type;                   /* Object Type */
    rt_uint8_t  flags;                  /* Flag Position */

    rt_list_t   list;                   /* Object List */
    rt_list_t   tlist;                  /* Thread List */

    /* Stack Pointer and Entry pointer */
    void        *sp;                    /* Stack Pointer */
    void        *entry;                 /* Entry Function Pointer */
    void        *parameter;             /* Parameter */
    void        *stack_addr;            /* Stack Address Pointer, */
    rt_uint32_t stack_size;             /* Stack Size */
```

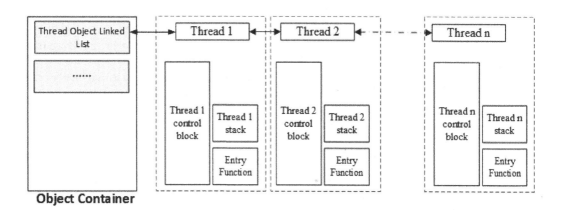

FIGURE 4.2   Object container and thread object.

```
    /* Error Code */
    rt_err_t     error;                         /* Thread Error Code */
    rt_uint8_t   stat;                          /* Thread State */

    /* Priority */
    rt_uint8_t   current_priority;              /* Current Priority */
    rt_uint8_t   init_priority;                 /* Initial Priority */
    rt_uint32_t number_mask;


    ......

    rt_ubase_t   init_tick;                     /* Thread Initialization Count
Value */
    rt_ubase_t   remaining_tick;                /* Thread Remaining Count Value
*/

    struct rt_timer thread_timer;                  /* Built-in Thread Timer
*/

    void (*cleanup)(struct rt_thread *tid);     /* Thread Exit Clear
Function */
    rt_uint32_t user_data;                         /* User Data */
};
```

init _ priority is the thread priority specified when the thread was created, and will not be changed while the thread is running (unless the user executes the thread control function to manually adjust the thread priority). cleanup will be called back by the idle thread when the thread exits to perform the user-setup cleanup site and so on. The last member, user _ data, can be mounted by the user with some data information into the thread control block to provide an implementation similar to thread private data.

IMPORTANT ATTRIBUTES OF THREAD

Thread Stack

RT-Thread's thread has an independent stack. When the thread is switched out, the context of the current thread is stored in the stack. When the thread resumes running, the context information is read from the stack and recovered.

The thread stack is also used to store local variables in functions: local variables in functions are applied from the thread stack space; local variables in functions are initially allocated from registers (take the ARM architecture); when this function calls another function, these local variables will be placed on the stack.

For the first run of a thread, context can be constructed manually to set the initial environment like entry function (PC register), entry parameter (R0 register), return position (LR register), and current machine operating status (CPSR register).

The growth direction of the thread stack is closely related to the chip architecture. Versions before RT-Thread 3.1.0 only allow the stack to grow from a high address to low address. For ARM Cortex-M architecture, the thread stack can be constructed as shown in Figure 4.3.

When setting the size for thread stack, a larger thread stack can be designed for an MCU with a relatively large resource, or a larger stack can be set initially, for example, a size of 1K or 2K bytes, then in FinSH, use the list _ thread command to check the size of the stack used by the thread during the running of the thread. With this command, you can see the maximum stack depth used by the thread from the start of the thread to the current point in time, and then add the appropriate margin to form the final thread stack size, and finally modify the size of the stack space.

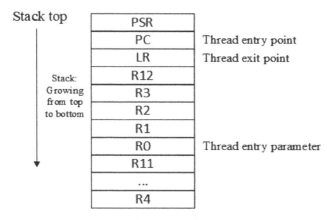

FIGURE 4.3   Thread stack.

Thread State

When the thread is running, only one thread is allowed to run in the processor at the same time. Through its lifecycle, a thread has various operating states, such as initial state, suspended state, ready state, etc. In RT-Thread, a thread has five states, and the operating system will automatically adjust its state based on its running condition.

The five states of a thread in RT-Thread are shown in Table 4.1.

TABLE 4.1
Thread States

States	Description
Initial State	Thread is in the initial state when it has just been created but has not started running; in initial state, the thread does not participate in scheduling. In RT-Thread, the macro definition of this state is RT_THREAD_INIT.
Ready State	In the ready state, the thread is queued according to priority, waiting to be executed; the processor is available again once the current thread is finished, and the operating system will then immediately find the ready thread with the highest priority to run. In RT-Thread, the macro definition of this state is RT_THREAD_READY.
Running State	Thread is currently running. In a single-core system, only the thread returned from the rt_thread_self() function is running; in a multi-core system, more than one thread may be running. In RT-Thread, the macro definition of this state is RT_THREAD_RUNNING.
Suspended State	Also known as the blocking state. It may be suspended and paused because the resource is unavailable, or the thread is suspended because it is voluntarily delayed. In the suspended state, threads do not participate in scheduling. In RT-Thread, the macro definition of this state is RT_THREAD_SUSPEND.
Closed State	It will be turned to a closed state when the thread finishes running. The thread in the closed state does not participate in the thread's scheduling. In RT-Thread, the macro definition of this state is RT_THREAD_CLOSE.

Thread Priority

The priority of the RT-Thread thread indicates the thread's priority of being scheduled. Each thread has its priority. The more important the thread, the higher priority it should be given and the bigger the chance of being scheduled.

RT-Thread supports a maximum of 256 thread priorities (0 ~ 255). The lower the number, the higher the priority, and 0 is the highest priority. In some systems with tight resources, you can choose system configurations that only support 8 or 32 priorities according to the actual situation; for the ARM Cortex-M series, 32 priorities are commonly used. The lowest priority is assigned to idle threads by default and is not used by users. In the system, when a thread with a higher priority is ready, the current thread with the lower priority will be swapped out immediately, and the high-priority thread will preempt the processor.

Time Slice

Each thread has a time slice parameter, but time slice is only effective for ready-state threads of the same priority. The system schedules the ready-state threads with the same priority using a time slice rotation scheduling method. In this case, time slice plays the role of the constraining thread's single running time, and the unit is a system tick (OS Tick). Suppose there are two ready-state threads, A and B, with the same priority. The time slice of A thread is set to 10, and the time slice of B thread is set to 5. When there is no ready-state thread with a higher priority than A in the system, the system will switch back and forth between thread A and B. Each time the system performs 10 OS Ticks on thread A and 5 OS Ticks on thread B, as shown in Figure 4.4.

Thread Entry Function

"Entry" in Thread Control Block is the thread's entry function, which is a function for the thread to achieve the intended functionality. The thread's entry function is designed by the user. There are generally two forms of code.

Infinite Loop Mode

In real-time systems, threads are usually passive. This is determined by the characteristics of the real-time system, which usually waits for external events to occur and then performs the appropriate services:

```
void thread_entry(void* parameter)
{
    while (1)
    {
    /* waiting for an event to occur */

    /* Serve and process events */
    }
}
```

It seems that threads have no restrictions on program execution and that all operations can be performed. But as a real-time system with clear priorities, if a program in a thread is stuck in an infinite

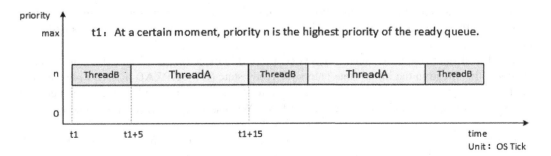

FIGURE 4.4 Same priority time slice round robin.

loop, then threads with lower priorities will never be executed. Therefore, one thing that must be noted in the real-time operating system is that the thread cannot be stuck in an infinite loop operation, and there must be an action to relinquish the use of the CPU, such as calling a delay function in the loop or actively suspending. The purpose of the user designing an infinite loop thread is to let this thread be continuously scheduled and run by the system and never be deleted.

Sequential Execution or Finite-Cycle Mode

Exemplified by simple sequential statements, such as do while() or for() loop, etc., these threads will not cycle or not loop forever. They can be described as a "one-off" thread and will surely be executed. After the execution is complete, the thread will be automatically deleted by the system.

```
static void thread_entry(void* parameter)
{
    /* Processing Transaction #1 */
    ...
    /* Processing Transaction #2 */
    ...
    /* Processing Transaction #3 */
}
```

Thread Error Code

One thread is one execution scenario. The error code is closely related to the execution environment, so each thread is equipped with a variable to store the error code. The error code of the thread includes:

```
#define RT_EOK          0 /* No error        */
#define RT_ERROR        1 /* Regular error      */
#define RT_ETIMEOUT     2 /* Timeout error      */
#define RT_EFULL        3 /* Resource is full     */
#define RT_EEMPTY       4 /* No resource       */
#define RT_ENOMEM       5 /* No memory       */
#define RT_ENOSYS       6 /* System does not support      */
#define RT_EBUSY        7 /* System busy       */
#define RT_EIO          8 /* IO error        */
#define RT_EINTR        9 /* Interrupt system call    */
#define RT_EINVAL      10 /* Invalid Parameter       */
```

SWITCHING THREAD STATE

RT-Thread provides a set of operating system call interfaces that make the state of a thread switch back and forth between these five states. The conversion relationship between these states is shown in Figure 4.5.

The thread enters the initial state (RT_THREAD_INIT) by calling the function rt_thread_create/init(); the thread in the initial state enters the ready state (RT_THREAD_READY) by calling the function rt_thread_startup(); the thread in the ready state is scheduled by the scheduler and enters the running state (RT_THREAD_RUNNING). When a running thread calls a function such as rt_thread_delay(), rt_sem_take(), rt_mutex_take(), or rt_mb_recv(), or fails to get resources, it will enter the suspended state (RT_THREAD_SUSPEND); if threads in the suspended state waited until timeout and still didn't acquire the resources, or other threads released the resources, it will return to the ready state.

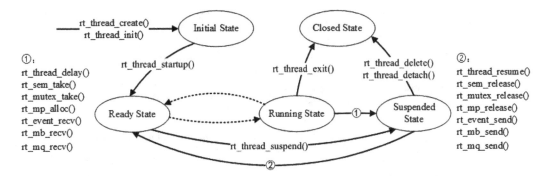

FIGURE 4.5 Thread state switching diagram.

If a thread in a suspended state calls function rt_thread_delete/detach(), it will switch to the closed state (RT_THREAD_CLOSE); as for a thread in the running state, if operation is completed, the function rt_thread_exit() will be executed at the last part of the thread to change the state into a closed state.

Notes: In RT-Thread, the thread does not actually have a running state; the ready state and the running state are equivalent.

System Thread

As mentioned previously, a system thread refers to a thread created by the system and a user thread is a thread created by the user program calling the thread management interface. The system thread in RT-Thread kernel includes idle thread and main thread.

Idle Thread

An idle thread is the lowest-priority thread created by the system, and its thread state is always ready. When no other ready thread exists in the system, the scheduler will schedule the idle thread, which is usually an infinite loop and can never be suspended. In addition, idle threads have special functions in RT-Thread.

If a thread finishes running, the system will automatically delete the thread: automatically execute function rt_thread_exit(), first remove the thread from the system ready queue, then change the state of the thread to closed state which means it no longer participates in system scheduling, and then suspend it into the rt_thread_defunct queue (a thread queue that is not reclaimed and in a closed state). Lastly, idle thread reclaims the resources of the deleted thread.

Idle thread also provides an interface to run the hook function set by the user. The hook function is called when the idle thread is running, which is suitable for operations like hooking into power management, watchdog feeding, etc.

Main Thread

When the system starts, the system will create the main thread. Its entry function is main_thread_entry(). User's application entry function main() starts from here. After the system scheduler starts, the main thread starts running. The process is as shown in Figure 4.6. Users can add their own application initialization code to the main(). function.

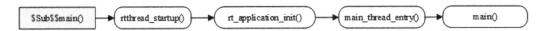

FIGURE 4.6 Main thread calling process.

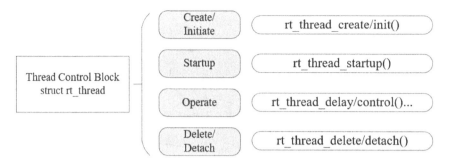

FIGURE 4.7 Thread-related operations.

THREAD MANAGEMENT

The first two sections of this chapter conceptually explain the function and working mechanism of threads. I believe that we all are familiar with threads now. This section will dive into the various interfaces of the thread and give some source code to help the reader better understand threads.

Figure 4.7 depicts related operations to threads, including create/initialize threads, start threads, run threads, and delete/detach threads. You can use rt_thread_create() to create a dynamic thread and rt_thread_init() to initialize a static thread. The difference between a dynamic thread and a static thread is that for a dynamic thread, the system automatically allocates stack space and thread handles from the dynamic memory heap (only after initializing the heap can a dynamic thread be created). As for a static thread, it is the user that allocates the stack space and the thread handle.

CREATE AND DELETE THREAD

To become an executable object, a thread must be created by the kernel of the operating system. You can create a dynamic thread through the following interface:

```
rt_thread_t rt_thread_create(const char* name,
                             void (*entry)(void* parameter),
                             void* parameter,
                             rt_uint32_t stack_size,
                             rt_uint8_t priority,
                             rt_uint32_t tick);
```

When this function is called, the system will allocate a thread handle from the dynamic heap memory and allocate the corresponding space from the dynamic heap memory according to the stack size specified in the parameter. The allocated stack space is aligned in RT_ALIGN_SIZE mode configured in rtconfig.h. The parameters and return values of the thread creation rt_thread_create() are as shown in Table 4.2.

For some threads created with rt_thread_create(), when not needed or when an error occurs, we can use the following function interface to completely remove the thread from the system:

```
rt_err_t rt_thread_delete(rt_thread_t thread);
```

After calling this function, the thread object will be moved out of the thread list and removed from the kernel object manager. Consequently, the stack space occupied by the thread will also be freed, and the reclaimed space will be reused for other memory allocations. In fact, using the rt_thread_delete() function to delete the thread interface is just changing the corresponding thread state to RT_THREAD_CLOSE state and then putting it into rt_thread_defunct queue; the actual delete

TABLE 4.2
rt_thread_create

Parameters	Description
name	The name of the thread; the maximum length of the thread name is specified by the macro RT_NAME_MAX in rtconfig.h, and the extra part is automatically truncated.
entry	Thread entry function.
parameter	Thread entry function's parameter.
stack_size	Thread stack size in bytes.
priority	Priority of the thread. The priority range is based on the system configuration (macro definition RT_THREAD_PRIORITY_MAX in rtconfig.h). If 256-level priority is supported, then the range is from 0 to 255. The smaller the value, the higher the priority, and 0 is the highest priority.
tick	The time slice size of the thread. The unit of the time slice (tick) is the tick of the operating system. When there are threads with the same priority in the system, this parameter specifies the maximum length of time of a thread for one schedule. At the end of this time slice run, the scheduler automatically selects the next ready state of the same priority thread to run.
Return	— —
thread	Thread creation succeeds, return thread handle.
RT_NULL	Failed to create thread.

action (release the thread control block and release the thread stack) needs to be completed later by an idle thread when it is being executed. The parameters and return values of the thread-deleting rt_thread_delete() interface are shown in Table 4.3.

TABLE 4.3
rt_thread_delete

Parameter	Description
thread	Thread handles to delete
Return	— —
RT_EOK	Delete thread successfully
-RT_ERROR	Failed to delete thread

This function is only valid when the system dynamic heap is enabled (meaning RT_USING_ HEAP macro definition is already defined).

INITIALIZE AND DETACH THREAD

The initialization of a thread can be done using the following function interface to initialize a static thread object:

```
rt_err_t rt_thread_init(struct rt_thread* thread,
                        const char* name,
                        void (*entry)(void* parameter), void* parameter,
                        void* stack_start, rt_uint32_t stack_size,
                        rt_uint8_t priority, rt_uint32_t tick);
```

The thread handle of the static thread (in other words, the thread control block pointer) and the thread stack are provided by the user. A static thread means that the thread control block and the thread running stack are generally set to global variables, which are determined and allocated when compiling. The kernel is not responsible for dynamically allocating memory space. It should be noted that the user-provided stack starting address needs to be system aligned (e.g., 4-byte alignment is required on ARM). The parameters and return values of the thread initialization interface rt_thread_init() are shown in Table 4.4.

TABLE 4.4
rt_thread_init

Parameter	Description
thread	Thread handle. The thread handle is provided by the user and points to the corresponding thread control block memory address.
name	Name of the thread; the maximum length of the thread name is specified by the RT_NAME_MAX macro defined in rtconfig.h, and the extra part is automatically truncated.
entry	Thread entry function.
parameter	Thread entry function parameter.
stack_start	Thread stack start address.
stack_size	Thread stack size in bytes. Stack space address alignment is required in most systems (e.g., alignment to 4-byte addresses in the ARM architecture).
priority	The priority of the thread. The priority range is based on the system configuration (macro definition RT_THREAD_PRIORITY_MAX in rtconfig.h). If 256 levels of priority are supported, the range is from 0 to 255. The smaller the value, the higher the priority, and 0 is the highest priority.
tick	The time slice size of the thread. The unit of the time slice (tick) is the tick of the operating system. The unit of the time slice (tick) is the tick of the operating system. When there are threads with the same priority in the system, this parameter specifies the maximum length of time of a thread for one schedule. At the end of this time slice run, the scheduler automatically selects the next ready state of the same priority thread to run.
Return	— —
RT_EOK	Thread creation succeeds.
-RT_ERROR	Failed to create thread.

For threads initialized with rt_thread_init(), using rt_thread_detach() will cause the thread object to be detached from the thread queue and kernel object manager. The detach thread function is as follows:

```
rt_err_t rt_thread_detach (rt_thread_t thread);
```

Parameters and return values of the thread detached from the interface rt_thread_detach() are shown in Table 4.5.

TABLE 4.5
rt_thread_detach

Parameters	Description
thread	Thread handle, which should be the thread handle initialized by rt_thread_init
Return	— —
RT_EOK	Thread detached successfully
-RT_ERROR	Thread detachment failed

This function interface corresponds to the rt_thread_delete() function. The object operated by the rt_thread_delete() function is the handle created by rt_thread_create(), and the object operated by the rt_thread_detach() function is the thread control block initialized with the rt_thread_init() function. Again, the thread itself should not call this interface to detach thread itself.

START THREAD

The thread created (initialized) is in the initial state and does not enter the scheduling queue of the ready thread. We can call the following function interface to make the thread enter the ready state after the thread is initialized/created successfully:

```
rt_err_t rt_thread_startup(rt_thread_t thread);
```

When this function is called, the state of the thread is changed to the ready state and placed in the corresponding priority queue for scheduling. If the newly started thread has a higher priority than the current thread, it will immediately switch to the new thread. The parameters and return values of the thread start interface rt_thread_startup() are shown in Table 4.6.

TABLE 4.6
rt_thread_startup

Parameter	Description
thread	Thread handle
Return	— —
RT_EOK	Thread started successfully
-RT_ERROR	Thread start failed

OBTAINING CURRENT THREAD

During the running of the program, the same piece of code may be executed by multiple threads. At the time of execution, the currently executed thread handle can be obtained through the following function interface:

```
rt_thread_t rt_thread_self(void);
```

The return value of this interface is shown in Table 4.7.

TABLE 4.7
rt_thread_self

Return	Description
thread	The currently running thread handle
RT_NULL	Failed, the scheduler has not started yet

MAKING THREAD RELEASE PROCESSOR RESOURCES

When the current thread's time slice runs out or the thread actively requests to release the processor resource, it will no longer occupy the processor, and the scheduler will select the next thread of the same priority to execute. After the thread calls this interface, the thread is still in the ready queue. The thread gives up the processor using the following function interface:

```
rt_err_t rt_thread_yield(void);
```

After calling this function, the current thread first removes itself from its ready priority thread queue, then suspends itself to the end of the priority queue list, and then activates the scheduler for thread context switching (if there is no other thread with the same priority, then this thread continues to execute without context switching action).

The rt_thread_yield() function is similar to the rt_schedule() function, but the behavior of the system is completely different when other ready-state threads of the same priority exist. After executing the rt_thread_yield() function, the current thread is swapped out and the next ready thread of the same priority will be executed. After the rt_schedule() function is executed, the current thread is not necessarily swapped out. Even if it is swapped out, it will not be placed at the end of the ready thread list. Instead, the thread with the highest priority is selected in the system and executed. (If there is no thread in the system with a higher priority than the current thread, the system will continue to execute the current thread after the rt_schedule() function is executed.)

THREAD SLEEP

In practical applications, we sometimes need to delay the current thread running for a period of time and re-run it at a specified time. This is called "thread sleep." Thread sleep can use the following three function interfaces:

```
rt_err_t rt_thread_sleep(rt_tick_t tick);
rt_err_t rt_thread_delay(rt_tick_t tick);
rt_err_t rt_thread_mdelay(rt_int32_t ms);
```

These three function interfaces have the same effect. Calling them can cause the current thread to suspend for a specified period. After that period, the thread will wake up and enter the ready state again. This function accepts a parameter that specifies the sleep time of the thread. The parameters and return values of the thread sleep interface rt_thread_sleep/delay/mdelay() are shown in Table 4.8.

TABLE 4.8
rt_thread_sleep/delay/mdelay

Parameters	Description
tick/ms	Thread sleep time
	The input parameter tick of rt_thread_sleep/ rt_thread_delay is in units of 1 OS Tick
	The input parameter ms of rt_thread_mdelay is in units of 1 ms
Return	— —
RT_EOK	Successful operation

SUSPEND AND RESUME THREAD

When a thread calls rt_thread_delay(), the thread will voluntarily suspend; when a function such as rt_sem_take() or rt_mb_recv() is called, the resource is not available for use and will cause the thread to suspend. A thread in a suspended state, if it waits for resource overtime (over the set time), will no longer wait for these resources and will return to the ready state; or when other threads release the resource the thread is waiting for, the thread will also return to the ready state.

The thread suspends using the following function interface:

```
rt_err_t rt_thread_suspend (rt_thread_t thread);
```

The parameters and return values of the thread suspend interface rt_thread_suspend() are shown in Table 4.9.

TABLE 4.9
rt_thread_suspend

Parameters	Description
thread	Thread handle
Return	— —
RT_EOK	Thread suspends successfully
-RT_ERROR	Thread suspension failed because the thread is not in the ready state

Notes: Generally, you should not use this function to suspend the thread itself; if you really need to use rt_thread_suspend() to suspend the current task, rt_schedule() needs to be called immediately after calling function rt_thread_suspend().

The function's context switch is achieved manually. The user only needs to understand the role of the interface, which is not recommended.

Resuming a thread is to let the suspended thread re-enter the ready state and put the thread into the system's ready queue; if the thread recovered is in the first place of the priority list, then the system will start context switching. Thread resumption uses the following function interface:

```
rt_err_t rt_thread_resume (rt_thread_t thread);
```

The parameters and return values of the thread recovery interface rt_thread_resume() are shown in Table 4.10.

TABLE 4.10
rt_thread_resume

Parameter	Description
thread	Thread handle
Return	— —
RT_EOK	Thread resumed successfully
-RT_ERROR	Thread recovery failed because the state of the thread is not RT_THREAD_SUSPEND state

CONTROL THREAD

When you need other control over a thread, such as dynamically changing the priority of a thread, you can call the following function interface:

```
rt_err_t rt_thread_control(rt_thread_t thread, rt_uint8_t cmd, void*
arg);
```

The parameters and return values of the thread control interface rt_thread_control() are shown in Table 4.11.

TABLE 4.11
rt_thread_control

Function Parameters	Description
thread	Thread handle
Cmd	Control command demand
Arg	Control parameter
Return	— —
RT_EOK	Control execution is correct
-RT_ERROR	Failure

Demands supported by control command demand Cmd include:

- RT_THREAD_CTRL_CHANGE_PRIORITY: Dynamically change the priority of a thread;
- RT_THREAD_CTRL_STARTUP: Start running a thread, equivalent to the rt_thread_startup() function call;
- RT_THREAD_CTRL_CLOSE: Close a thread, equivalent to the rt_thread_delete() function call.

SET AND DELETE IDLE HOOKS

The idle hook function is a hook function of the idle thread. If the idle hook function is set, the idle hook function can be automatically executed to perform other things, such as the LED of the system indicator, when the system executes the idle thread. The interface for setting/deleting idle hooks is as follows:

```
rt_err_t rt_thread_idle_sethook(void (*hook)(void));
rt_err_t rt_thread_idle_delhook(void (*hook)(void));
```

Input parameters and return values of setting the idle hook function rt_thread_idle_sethook() are shown in Table 4.12.

TABLE 4.12
rt_thread_idle_sethook

Function Parameters	Description
hook	Set hook function
Return	— —
RT_EOK	Set successfully
-RT_EFULL	Set fail

Input parameters and return values of deleting the idle hook function rt_thread_idle_delhook() are shown in Table 4.13.

TABLE 4.13
rt_thread_idle_delhook

Function Parameters	Description
hook	Deleted hook function
Return	— —
RT_EOK	Successfully deleted
-RT_ENOSYS	Failed to delete

Notes: An idle thread is a thread whose state is always ready. Therefore, the hook function must ensure that idle threads will not be suspended at any time. Functions like rt_thread_delay(), rt_sem_take(), etc., can't be used because they may cause the thread to suspend.

SET THE SCHEDULER HOOK

When the system is running, it is in the process of thread running, interrupt triggering, responding to interrupts, switching to other threads, and switching between threads. In other words, context switching is the most common event in the system. Sometimes the user may want to know what kind of thread switch has occurred at times. You can set a corresponding hook function by calling the following function interface. This hook function will be called when the system thread switches:

```
void rt_scheduler_sethook(void (*hook)(struct rt_thread* from, struct
rt_thread* to));
```

Input parameters for setting the scheduler hook function are shown in Table 4.14.

TABLE 4.14
rt_scheduler_sethook

Function Parameters	Description
hook	Represents a user-defined hook function pointer

The hook function hook() is declared as follows:

```
void hook(struct rt_thread* from, struct rt_thread* to);
```

Input parameters for the scheduler hook function hook() are shown in Table 4.15.

TABLE 4.15
Hook

Function Parameters	Description
from	Indicates the thread control block pointer that the system wants to switch out
to	Indicates the thread control block pointer that the system wants to switch to

Notes: Please carefully compile your hook function. Any carelessness is likely to cause the entire system to run abnormally (in this hook function, it is basically not allowed to call the system interface and should not cause the current running context to suspend).

THREAD APPLICATION SAMPLE

An application example in the Keil simulator environment is given later.

CREATE THREAD SAMPLE

This sample is creating a dynamic thread and initializing a static thread. A thread is automatically deleted by the system after it has finished running. The other thread is always printing the counts, as shown in the Code Listing 4.1.

Code listing 4.1 Thread Sample

```
#include <rtthread.h>

#define THREAD_PRIORITY          25
#define THREAD_STACK_SIZE        512
#define THREAD_TIMESLICE         5

static rt_thread_t tid1 = RT_NULL;

/* Entry Function for Thread 1 */
static void thread1_entry(void *parameter)
{
    rt_uint32_t count = 0;

    while (1)
    {
        /* Thread 1 runs with low priority and prints the count value all
the time */
        rt_kprintf("thread1 count: %d\n", count ++);
        rt_thread_mdelay(500);
    }
}


ALIGN(RT_ALIGN_SIZE)
static char thread2_stack[1024];
static struct rt_thread thread2;
/* Entry for Thread 2 */
static void thread2_entry(void *param)
{
    rt_uint32_t count = 0;

    /* Thread 2 has a higher priority to preempt thread 1 and get
executed */
    for (count = 0; count < 10 ; count++)
    {
        /* Thread 2 prints count value */
        rt_kprintf("thread2 count: %d\n", count);
    }
    rt_kprintf("thread2 exit\n");
    /* Thread 2 will also be automatically detached from the system after
it finishes running. */
}

/* Thread Sample */
int thread_sample(void)
{
```

```
    /* Create thread 1, Name is thread1,Entry is thread1_entry */
    tid1 = rt_thread_create("thread1",
                            thread1_entry, RT_NULL,
                            THREAD_STACK_SIZE,
                            THREAD_PRIORITY, THREAD_TIMESLICE);

    /* Start this thread if you get the thread control block */
    if (tid1 != RT_NULL)
        rt_thread_startup(tid1);

    /* Init thread 2, Name is thread2,Entry is thread2_entry */
    rt_thread_init(&thread2,
                   "thread2",
                   thread2_entry,
                   RT_NULL,
                   &thread2_stack[0],
                   sizeof(thread2_stack),
                   THREAD_PRIORITY - 1, THREAD_TIMESLICE);
    rt_thread_startup(&thread2);

    return 0;
}

/* Export to msh command list */
MSH_CMD_EXPORT(thread_sample, thread sample);
```

The results are as follows:

```
\ | /
- RT -     Thread Operating System
 / | \     3.1.0 build Aug 24 2018
 2006 - 2018 Copyright by rt-thread team

msh >thread_sample
msh >thread2 count: 0
thread2 count: 1
thread2 count: 2
thread2 count: 3
thread2 count: 4
thread2 count: 5
thread2 count: 6
thread2 count: 7
thread2 count: 8
thread2 count: 9
thread2 exit
thread1 count: 0
thread1 count: 1
thread1 count: 2
thread1 count: 3
...
```

When thread 2 counts to a certain value, it will stop running. Then thread 2 is automatically deleted by the system, and therefore the counting stops. Thread 1 prints the count all the time.

Notes: About deleting threads: Most threads are executed cyclically without needing to be deleted. For threads that can finish running, RT-Thread automatically deletes the thread after the thread finishes running, and deletes it in rt_thread_exit(). The user only needs to

understand the role of the interface. It is not recommended to use this interface (this interface can be called by other threads or you can call this interface in the timer timeout function to delete a thread, which is not used very often).

THREAD TIME SLICE ROUND-ROBIN SCHEDULING SAMPLE

This sample is creating two threads that will always print counts when executing, as shown in Code Listing 4.2.

Code listing 4.2 Time Slice Sample

```
#include <rtthread.h>

#define THREAD_STACK_SIZE    1024
#define THREAD_PRIORITY      20
#define THREAD_TIMESLICE     10

/* Thread Entry */
static void thread_entry(void* parameter)
{
    rt_uint32_t value;
    rt_uint32_t count = 0;

    value = (rt_uint32_t)parameter;
    while (1)
    {
        if(0 == (count % 5))
        {
            rt_kprintf("thread %d is running ,thread %d count = %d\n",
value , value , count);

            if(count> 200)
                return;
        }
         count++;
    }
}

int timeslice_sample(void)
{
    rt_thread_t tid = RT_NULL;
    /* Create Thread 1 */
    tid = rt_thread_create("thread1",
                        thread_entry, (void*)1,
                        THREAD_STACK_SIZE,
                        THREAD_PRIORITY, THREAD_TIMESLICE);
    if (tid != RT_NULL)
        rt_thread_startup(tid);


    /* Create Thread 2 */
    tid = rt_thread_create("thread2",
                        thread_entry, (void*)2,
                        THREAD_STACK_SIZE,
                        THREAD_PRIORITY, THREAD_TIMESLICE-5);
```

```
        if (tid != RT_NULL)
            rt_thread_startup(tid);
        return 0;
}

/* Export to msh command list */
MSH_CMD_EXPORT(timeslice_sample, timeslice sample);
```

The results are as follows:

```
 \ | /
- RT -        Thread Operating System
 / | \        3.1.0 build Aug 27 2018
 2006 - 2018 Copyright by rt-thread team

msh >timeslice_sample
msh >thread 1 is running ,thread 1 count = 0
thread 1 is running ,thread 1 count = 5
thread 1 is running ,thread 1 count = 10
thread 1 is running ,thread 1 count = 15
...
thread 1 is running ,thread 1 count = 125
thread 1 is rthread 2 is running ,thread 2 count = 0
thread 2 is running ,thread 2 count = 5
thread 2 is running ,thread 2 count = 10
thread 2 is running ,thread 2 count = 15
thread 2 is running ,thread 2 count = 20
thread 2 is running ,thread 2 count = 25
thread 2 is running ,thread 2 count = 30
thread 2 is running ,thread 2 count = 35
thread 2 is running ,thread 2 count = 40
thread 2 is running ,thread 2 count = 45
thread 2 is running ,thread 2 count = 50
thread 2 is running ,thread 2 count = 55
thread 2 is running ,thread 2 count = 60
thread 2 is running ,thread 2 cunning ,thread 2 count = 65
thread 1 is running ,thread 1 count = 135
...
thread 2 is running ,thread 2 count = 205
```

As can be seen from the running count results, thread 2 runs half the time of thread 1.

THREAD SCHEDULER HOOK SAMPLE

When a thread is scheduling a switch, it executes the scheduling. We can set a scheduler hook so that we can do other things when the thread is being switched. This sample is printing switch information between the threads in the scheduler hook function, as shown in Code Listing 4.3.

Code listing 4.3 The Scheduler Hook Sample

```
#include <rtthread.h>

#define THREAD_STACK_SIZE    1024
#define THREAD_PRIORITY      20
#define THREAD_TIMESLICE     10

/* Counter for each thread */
volatile rt_uint32_t count[2];
```

```c
/* Threads 1, 2 share an entry, but the entry parameters are different */
static void thread_entry(void* parameter)
{
    rt_uint32_t value;

    value = (rt_uint32_t)parameter;
    while (1)
    {
        rt_kprintf("thread %d is running\n", value);
        rt_thread_mdelay(1000); // Delay for a while
    }
}

static rt_thread_t tid1 = RT_NULL;
static rt_thread_t tid2 = RT_NULL;

static void hook_of_scheduler(struct rt_thread* from, struct rt_thread*
to)
{
    rt_kprintf("from: %s -->  to: %s \n", from->name , to->name);
}

int scheduler_hook(void)
{
    /* Set the scheduler hook */
    rt_scheduler_sethook(hook_of_scheduler);

    /* Create Thread 1 */
    tid1 = rt_thread_create("thread1",
                            thread_entry, (void*)1,
                            THREAD_STACK_SIZE,
                            THREAD_PRIORITY, THREAD_TIMESLICE);
    if (tid1 != RT_NULL)
        rt_thread_startup(tid1);

    /* Create Thread 2 */
    tid2 = rt_thread_create("thread2",
                            thread_entry, (void*)2,
                            THREAD_STACK_SIZE,
                            THREAD_PRIORITY,THREAD_TIMESLICE - 5);
    if (tid2 != RT_NULL)
        rt_thread_startup(tid2);
    return 0;
}

/* Export to msh command list */
MSH_CMD_EXPORT(scheduler_hook, scheduler_hook sample);
```

The results are as follows:

```
 \ | /
- RT -     Thread Operating System
 / | \     3.1.0 build Aug 27 2018
 2006 - 2018 Copyright by rt-thread team

msh > scheduler_hook
msh >from: tshell -->  to: thread1
```

```
thread 1 is running
from: thread1 --> to: thread2
thread 2 is running
from: thread2 --> to: tidle
from: tidle --> to: thread1
thread 1 is running
from: thread1 --> to: tidle
from: tidle --> to: thread2
thread 2 is running
from: thread2 --> to: tidle
...
```

The simulation result shows that when the threads are being switched, the scheduler hook function installed is working normally and printing the information of the thread switching, including switching to the idle thread.

CHAPTER SUMMARY

This chapter described the concept of the thread, the functional characteristics of thread, the working mechanism of thread, thread interfaces, etc., and here is a review of this chapter that summarizes the following key points:

1. The thread must have an action to relinquish the use of the CPU.
2. Thread scheduling is done in a priority preemptive manner; threads with the same priority will be executed through time slice round-robin.
3. Dynamic thread creation and deletion calls interface rt_thread_create() and rt_thread_delete(); the static thread initialization and disengagement calls interface rt_thread_init() and rt_thread_detach(). Note that the thread system can automatically delete the executed threads, so users are not recommended to use the deletion/disengagement thread interface.

Apart from what was mentioned earlier, you are encouraged to create your own thread to handle some tasks, using simulation tools to check the correctness of the results.

5 Clock Management

The concept of time is very important. You need to set a time to go out with friends, and it takes time to complete tasks. Life is inseparable from time. The same is true for operating systems, which require time to regulate the execution of their tasks. The smallest time unit in an operating system is the clock tick (OS Tick). This chapter focuses on the introduction of clock ticks and clock-based timers. After reading this chapter, we will learn how clock ticks are generated and how to use RT-Thread timers.

CLOCK TICK (OS TICK)

Any operating system needs to provide a clock tick for the system to handle all time-related events, such as thread latency, thread time slice rotation scheduling, timer timeout, etc. A clock tick is a specific periodic interrupt. This interrupt can be regarded as the system's heartbeat. The time interval between interrupts depends on different applications—generally it is 1–100ms. The faster the clock tick rate, the greater the overhead of the system. The number of clock ticks counted from the start of the system is called the system time.

In RT-Thread, the duration of a clock tick can be adjusted according to the value of RT_TICK_PER_SECOND, which is equal to 1/RT_TICK_PER_SECOND second.

CLOCK TICK IMPLEMENTATION

The clock tick is generated by a hardware timer configured in interrupt trigger mode. `void rt_tick_increase(void)` will be called when an interrupt occurs, notifying the operating system that a system clock has passed; different hardware drivers have different timer interrupt implementations. Here is an example of using STM32 `SysTick_Handler` to achieve a clock tick.

```
void SysTick_Handler(void)
{
    /* Entry Interrupt*/
    rt_interrupt_enter();
    ......
    rt_tick_increase();
    /* Leave Interrupt */
    rt_interrupt_leave();
}
```

Call `rt_tick_increase()` in interrupt function to self-add the global variable rt_tick. The code is as follows:

```
void rt_tick_increase(void)
{
    struct rt_thread *thread;

    /* Global variable rt_tick self-add */
    ++ rt_tick;

    /* Check time slice */
    thread = rt_thread_self();
```

```
   -- thread->remaining_tick;
   if (thread->remaining_tick == 0)
   {
       /* Re-assign initial value */
       thread->remaining_tick = thread->init_tick;

       /* Thread suspends */
       rt_thread_yield();
   }

   /* Check time slice */
   rt_timer_check();
}
```

You can see that global variable rt_tick is incremented by one on every clock tick. The value of rt_tick indicates the total number of clock ticks that the system has elapsed since it started, that is, system time. In addition, on every clock tick, whether the current thread's time slice is exhausted and whether there is a timer, timeout will be checked.

Notes: In interrupt, rt_timer_check() is used to check the system hardware timer linked list. If there is a timer timeout, the corresponding timeout function will be called. All timers are removed from the timer linked list if it timed out, and for periodic timer, it will be added back to the timer linked list when it is started again.

Obtain a Clock Tick

Since the global variable rt_tick is incremented by one on every clock tick, the value of the current rt_tick will be returned by calling `rt _ tick _ get`, which is the current clock tick value. This interface can be used to record the duration of time a system is running or to measure the time it takes for a task to run. The interface function is as follows:

```
rt_tick_t rt_tick_get(void);
```

Table 5.1 describes the return values of `rt _ tick _ get()` function.

TABLE 5.1
rt_tick_get

Return	Description
rt_tick	Current clock tick value

TIMER MANAGEMENT

A timer triggers an event after a certain specified time from a specified moment, for example, setting a timer to wake up you the next morning. Timers include hardware timers and software timers:

1. **Hardware timer** is the timing function provided by the chip itself. The hardware timer can be used by configuring the timer module into a timer mode and setting the time. A hardware timer is accurate to nanosecond precision, and is in interrupt trigger mode.
2. **Software timer** is a type of system interface provided by the operating system. It is built on the basis of the hardware timer to enable the system to provide a timer service with no constraint on numbers.

The RT-Thread operating system provides software-implemented timers in units of clock tick (OS Tick); that is, the timing value must be an integer multiple of OS Tick. For example, an OS Tick is 10ms; then the software timer can only be timed 10ms, 20ms, 100ms, etc., but not 15ms. RT-Thread timer is also based on the clock tick, providing timing capabilities based on integral multiples of the clock tick.

RT-THREAD TIMER INTRODUCTION

RT-Thread timer provides two types of timer mechanisms: the first type is a one-shot timer, which only triggers a timer event for onetime after startup, and then the timer stops automatically. The second type is a periodic triggering timer, which periodically triggers a timer event until the user manually stops it; otherwise, it will continue to execute forever.

In addition, according to the context in which the timeout function is executed, RT-Thread timer can be divided into HARD_TIMER mode and SOFT_TIMER mode, as shown in Figure 5.1.

HARD_TIMER Mode

The timer timeout function of HARD_TIMER mode is executed in the interrupt context and can be specified with the parameter RT_TIMER_FLAG_HARD_TIMER when initializing or creating the timer.

When executed in interrupt context, the requirements for timeout function are the same as those for the interrupt service routine: execution time should be as short as possible, and the execution should not cause the current context to suspend and wait. For example, a timeout function executed in an interrupt context should not attempt to apply for dynamic memory, free dynamic memory, etc.

The default mode of RT-Thread timer is HARD_TIMER mode, which means after the timer timeout, the timeout function runs in the context of the system clock interrupt. The execution mode in the interrupt context determines that the timer's timeout function should not call any system function that will cause the current context to suspend; nor can it be executing for a very long time; otherwise, the response time of other interrupts will be lengthened or the running time of other threads will be preempted.

SOFT_TIMER Mode

The SOFT_TIMER mode is configurable, and the macro RT_USING_TIMER_SOFT is used to determine whether the mode should be enabled. When this mode is enabled, the system will create

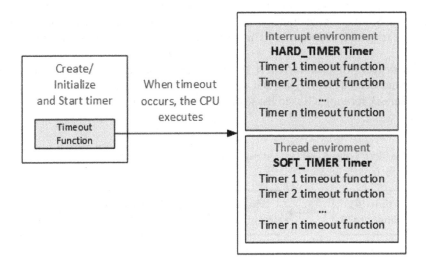

FIGURE 5.1 Timer context.

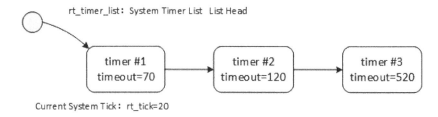

FIGURE 5.2   Timer linked list diagram.

a timer thread at initialization, and then the timeout function of the timer in SOFT_TIMER mode will be executed in the context of the timer thread. SOFT_TIMER mode can be specified using the parameter RT_TIMER_FLAG_SOFT_TIMER when initializing or creating the timer.

Timer Working Mechanism

The following is an example to illustrate the working mechanism of the RT-Thread timer. Two important global variables are maintained in the RT-Thread timer module:

1. Elapsed tick time of the current system rt_tick (when hardware timer interrupt comes, it will add 1).
2. Timer linked list rt _ timer _ list. Newly created and activated timers of the system are inputted into the rt _ timer _ list linked list in a timeout-ordered manner.

As shown in Figure 5.2, the current tick value of the system is 20. In the current system, three timers have been created and started, which are Timer1 with 50 ticks set time, Timer2 with 100 ticks, and Timer3 with 500 ticks. The current time of the system rt_tick=20 is added on these three timers, and they are linked from small to large in the rt_timer_list linked list, forming a timer linked list structure shown in Figure 5.2.

Along with the trigger of the hardware timer, rt _ tick has been increasing (rt_tick variable is incremented by 1 every time the hardware timer is interrupted). After 50 ticks, rt_tick is increased from 20 to 70, which is equal to the timeout value of Timer1. Then, timeout functions associated with the Timer1 timer will be triggered, and Timer1 from the rt_timer_list linked list will be removed. Similarly, after 100 ticks and 500 ticks, timeout functions associated with Timer2 and Timer3 are triggered, and timers of Timer2 and Timer3 will be removed from the rt_timer_list linked list.

If after the system's 10 ticks (current rt_tick=30), a new task has created Timer4 with 300 ticks, the Timer4's timeout is equal to rt_tick and adds 300, that is 330. Timer4 will be inserted in between Timer2 and Timer3, forming a linked list structure shown in Figure 5.3.

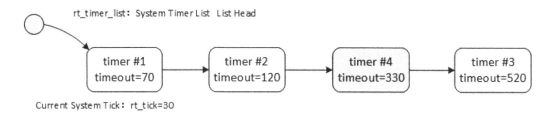

FIGURE 5.3   Timer linked list insertion diagram.

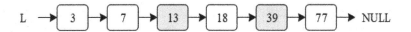

FIGURE 5.4 Ordered linked list.

Timer Control Block

In the RT-Thread operating system, timer control block is defined by the structure `struct rt_timer` and forms a timer kernel object, which is then linked to the kernel object container for management. It is a data structure used by the operating system to manage timers. It stores information about timers, such as the initial number of ticks, the number of timeout ticks, the linked list structure used to connect timers, timeout callback functions, etc.

```
struct rt_timer
{
    struct rt_object parent;
    rt_list_t row[RT_TIMER_SKIP_LIST_LEVEL];    /* Timer Linked List Node */

    void (*timeout_func)(void *parameter);    /* Timeout Function */
    void        *parameter;                   /* Parameters of Timeout
Function    */
    rt_tick_t init_tick;                      /* Timer Initial Timeout
Ticks */
    rt_tick_t timeout_tick;                   /* Number of ticks when the
timer actually times out */
};
typedef struct rt_timer *rt_timer_t;
```

The timer control block is defined by the structure `struct rt_timer` and forms a timer kernel object, which is then linked to the kernel object container for management. The `list` member is used to link an active (already started) timer to the `rt_timer_list` linked list.

Timer Skip List Algorithm

In the introduction of the working mechanics of the timer above, we talked about the newly created and activated timers inserted into the rt_timer_list linked list in the order of the timeout; that is, the rt_timer_list linked list is an ordered list. RT-Thread uses a skip list algorithm to speed up the search for linked list elements.

A skip list is a data structure based on a parallel linked list, which is simple to implement, and the time complexity of insertion, deletion, and search is O(log n). A skip list is a kind of linked list, but it adds a "skip" function to the linked list. It is this function that enables the skip list to have the time complexity of O(log n) when looking for elements, for example.

An ordered list, as shown in Figure 5.4, searches for elements {13, 39} from the ordered list. The number of comparisons is {3, 5} and the total number of comparisons is 3 + 5 = 8 times.

After using the skip list algorithm, a method similar to the binary search tree can be used to extract some nodes as indexes, and then the structure shown in Figure 5.5 is obtained.

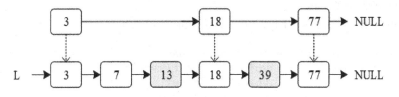

FIGURE 5.5 Ordered linked list index.

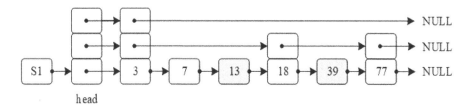

FIGURE 5.6   Three-layer skip list.

In this structure, {3, 18, 77} is extracted as the first-level index, so that the number of comparisons can be reduced when searching. For example, there are only three times of comparisons when searching for 39 (by comparing 3, 18, and 39). Of course, we can also extract some elements from the first-level index as a secondary index, which can speed up the element search. Figure 5.6 is a three-layer skip list.

Therefore, the timer skip list can pass the index of the upper layer, reducing the number of comparisons during the search and improving the efficiency of the search. This is an algorithm of "space in exchange of time"; the macro definition RT_TIMER_SKIP_LIST_LEVEL is used to configure the number of layers in the skip list. The default value is 1, which means that the ordered linked list algorithm for the first-order ordered list graph is used. Each additional one means that another level of index is added to the original list.

TIMER MANAGEMENT

The RT-Thread timer was introduced in the previous sections, and the working mechanism of the timer was conceptually explained. This section will go deeper into the various interfaces of the timer to help the reader understand the RT-Thread timer at the code level.

The timer management system needs to be initialized at system startup. This can be done through the following function interface:

```
void rt_system_timer_init(void);
```

If you need to use SOFT_TIMER, the following function interface should be called when the system is initialized:

```
void rt_system_timer_thread_init(void);
```

The timer control block contains important parameters related to the timer and acts as a link between various states of the timer. Relevant operations of the timer are shown in Figure 5.7. Relevant operations of the timer include creating/initializing the timer, starting the timer, running the timer, and deleting/detaching the timer. All the timers will be moved from the timer linked list after their timings expire. However, the periodic timer is added back to the timer linked list when it is started again, which is related to timer parameter settings. Each time an operating system clock interrupt occurs, a change is made to the timer status parameter that has timed out.

Create and Delete a Timer

When dynamically creating a timer, the following function interface can be used:

```
rt_timer_t rt_timer_create(const char* name,
                           void (*timeout)(void* parameter),
                           void* parameter,
                           rt_tick_t time,
                           rt_uint8_t flag);
```

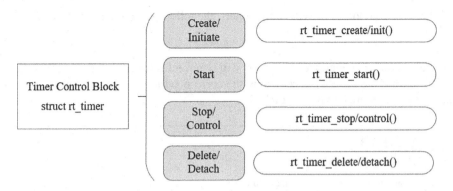

FIGURE 5.7 Timer-related operations.

After calling the function interface, the kernel first allocates a timer control block from the dynamic memory heap and then performs basic initialization on the control block. The description of each parameter and return value is detailed in Table 5.2.

TABLE 5.2
rt_timer_create

Parameters	Description
name	Name of the timer.
void (timeout) (void parameter)	Timer timeout function pointer (this function is called when the timer expires).
parameter	Entry parameter of the timer timeout function (when the timer expires, calling the timeout callback function will pass this parameter as the entry parameter to the timeout function).
time	Timeout of the timer, the unit is the clock tick.
flag	Parameters when the timer is created. The supported values include one-shot timing, periodic timing, hardware timer, software timer, etc. (You can use multiple values with "OR.")
Return	— —
RT_NULL	Creation failed (usually returning RT_NULL due to insufficient system memory).
Timer Handle	Timer was created successfully.

In include/rtdef.h, some timer-related macros are defined, as follows:

```
#define RT_TIMER_FLAG_ONE_SHOT     0x0    /* One shot timing     */
#define RT_TIMER_FLAG_PERIODIC     0x2    /* Periodic timing     */

#define RT_TIMER_FLAG_HARD_TIMER   0x0    /* Hardware timer  */
#define RT_TIMER_FLAG_SOFT_TIMER   0x4    /* Software timer  */
```

The above two sets of values can be assigned to the flag in an "or" logical manner. When the specified flag is RT_TIMER_FLAG_HARD_TIMER, if the timer expires, the timer's callback function will be called in the context of the service routine of the clock interrupt; when the specified flag is RT_TIMER_FLAG_SOFT_TIMER, if the timer expires, the timer's callback function will be called in the context of the system clock timer thread.

When the system no longer uses dynamic timers, the following function interface can be used:

```
rt_err_t rt_timer_delete(rt_timer_t timer);
```

After calling this function interface, the system will remove this timer from the rt_timer_list linked list and then release the memory occupied by the corresponding timer control block. The parameters and return values are detailed in Table 5.3.

TABLE 5.3
rt_timer_delete

Parameters	Description
timer	Timer handle, pointing at the timer that needs to be deleted
Return	— —
RT_EOK	Deletion is successful (if the parameter timer handle is RT_NULL, it will result in an ASSERT assertion)

Initialize and Detach a Timer

When creating a timer statically, the timer can be initialized by using the rt _ timer _ init interface. The function interface is as follows:

```
void rt_timer_init(rt_timer_t timer,
                   const char* name,
                   void (*timeout)(void* parameter),
                   void* parameter,
                   rt_tick_t time, rt_uint8_t flag);
```

When using this function interface, the corresponding timer control block, the corresponding timer name, timer timeout function, etc., will be initialized. The description of each parameter and return value is shown in Table 5.4.

TABLE 5.4
rt_timer_init

Parameters	Description
timer	Timer handle, pointing to the to-be-initialized timer control block.
name	Name of the timer.
void (timeout) (void parameter)	Timer timeout function pointer (this function is called when the timer expires).
parameter	Entry parameter of the timer timeout function (when the timer expires, the call timeout callback function will pass this parameter as the entry parameter to the timeout function).
time	Timeout of the timer; the unit is a clock tick.
flag	Parameters of when the timer is created. The supported values include one-shot timing, periodic timing, hardware timer, and software timer (multiple values can be taken with OR). For details, see Create and Delete a Timer.

When a static timer does not need to be used again, you can use the following function interface:

```
rt_err_t rt_timer_detach(rt_timer_t timer);
```

When detaching a timer, the system will detach the timer object from the kernel object container, but the memory occupied by the timer object will not be released. The parameters and return values are detailed in Table 5.5.

TABLE 5.5
rt_timer_detach

Parameters	Description
timer	Timer handle, pointing to the to-be-detached timer control block
Return	— —
RT_EOK	Successfully detached

Start and Stop a Timer

When the timer is created or initialized, it will not be started immediately. It will start after the timer function interface is called. The timer function interface is started as follows:

```
rt_err_t rt_timer_start(rt_timer_t timer);
```

After the timer start function interface is called, the state of the timer is changed to the activated state (RT_TIMER_FLAG_ACTIVATED) and inserted into the rt_timer_list queue linked list. The parameters and return values are detailed in Table 5.6.

TABLE 5.6
rt_timer_start

Parameters	Description
timer	Timer handle, pointing to the to-be-initialized timer control block
Return	— —
RT_EOK	Successful startup

For an example of starting the timer, please refer to the sample code.

After starting the timer, if you want to stop it, you can use the following function interface:

```
rt_err_t rt_timer_stop(rt_timer_t timer);
```

After the timer stop function interface is called, the timer state will change to the stop state and will be detached from the rt_timer_list linked list without participating in the timer timeout check. When a (periodic) timer expires, this function interface can also be called to stop the (periodic) timer itself. The parameters and return values are detailed in Table 5.7.

TABLE 5.7
rt_timer_stop

Parameters	Description
timer	Timer handle, pointing to the to-be-stopped timer control block
Return	— —
RT_EOK	Timer successfully stopped
- RT_ERROR	Timer is in stopped state

Control Timer

In addition to some of the programming interfaces provided earlier, RT-Thread provides a timer control function interface to obtain or set more timer information. The control timer function interface is as follows:

```
rt_err_t rt_timer_control(rt_timer_t timer, rt_uint8_t cmd, void* arg);
```

The control timer function interface can view or change the setting of the timer according to the parameters of the command type. The description of each parameter and return value is shown in Table 5.8.

TABLE 5.8
rt_timer_control

Parameters	Description
timer	Timer handle, pointing to the to-be-stopped timer control block.
cmd	The command for controlling the timer currently supports four command interfaces, which are setting timing, viewing the timing time, setting a one shot trigger, and setting the periodic trigger.
arg	Control command parameters corresponding to cmd. For example, when cmd is the set timeout time, the timeout time parameter can be set by arg.
Return	— —
RT_EOK	Successful.

Commands supported by function parameters cmd are as follows:

```
#define RT_TIMER_CTRL_SET_TIME     0x0    /* Set Timeout value     */
#define RT_TIMER_CTRL_GET_TIME     0x1    /* Obtain Timer Timeout
Time     */
#define RT_TIMER_CTRL_SET_ONESHOT  0x2    /* Set the timer as a one-shot
timer.   */
#define RT_TIMER_CTRL_SET_PERIODIC 0x3    /* Set the timer as a periodic
timer */
```

See "Code Listing 5.1—a dynamic timer sample" for code that uses the timer control interface.

TIMER APPLICATION SAMPLE

This is an example of creating a timer that creates two dynamic timers—one for one-shot timing, another one for periodic timing—and for the periodic timer to run for a while and then stop running, as shown in Code Listing 5.1.

Code listing 5.1 Dynamic Timer

```
#include <rtthread.h>

/* Timer Control Block */
static rt_timer_t timer1;
static rt_timer_t timer2;
static int cnt = 0;

/* Timer 1 Timeout Function */
static void timeout1(void *parameter)
{
    rt_kprintf("periodic timer is timeout %d\n", cnt);
```

```
      /* On the 10th time, stops periodic timer */
      if (cnt++>= 9)
      {
          rt_timer_stop(timer1);
          rt_kprintf("periodic timer was stopped! \n");
      }
}

/* Timer 2 Timeout Function */
static void timeout2(void *parameter)
{
    rt_kprintf("one shot timer is timeout\n");
}

int timer_sample(void)
{
    /* Create Timer 1 Periodic Timer */
    timer1 = rt_timer_create("timer1", timeout1,
                        RT_NULL, 10,
                        RT_TIMER_FLAG_PERIODIC);

    /* Start Timer 1*/
    if (timer1 != RT_NULL) rt_timer_start(timer1);

    /* Create Timer 2 One Shot Timer */
    timer2 = rt_timer_create("timer2", timeout2,
                        RT_NULL,  30,
                        RT_TIMER_FLAG_ONE_SHOT);

    /* Start Timer 2 */
    if (timer2 != RT_NULL) rt_timer_start(timer2);
    return 0;
}

/* Export to msh command list */
MSH_CMD_EXPORT(timer_sample, timer sample);
```

The results are as follows:

```
 \ | /
- RT -      Thread Operating System
 / | \      3.1.0 build Aug 24 2018
 2006 - 2018 Copyright by rt-thread team

msh >timer_sample
msh >periodic timer is timeout 0
periodic timer is timeout 1
one shot timer is timeout
periodic timer is timeout 2
periodic timer is timeout 3
periodic timer is timeout 4
periodic timer is timeout 5
periodic timer is timeout 6
periodic timer is timeout 7
periodic timer is timeout 8
periodic timer is timeout 9
periodic timer was stopped!
```

The timeout function of periodic timer1 gets to run every 10 OS Ticks for 10 times (after 10 times, rt_timer_stop is called to stop timer1); the timeout function of the one-shot timer2 runs once on the 30th OS Tick.

The example of initializing a timer is similar to the example of creating a timer. This program initializes two static timers: one is one-shot timing and one is periodic timing, as shown in Code Listing 5.2.

Code listing 5.2 One-shot Timer and Periodic Timer

```c
#include <rtthread.h>

/* Timer Control Block */
static struct rt_timer timer1;
static struct rt_timer timer2;
static int cnt = 0;

/* Timer 1 Timeout Function */
static void timeout1(void* parameter)
{
    rt_kprintf("periodic timer is timeout\n");
    /* Run for 10 times */
    if (cnt++>= 9)
    {
        rt_timer_stop(&timer1);
    }
}

/* Timer 2 Timeout Function */
static void timeout2(void* parameter)
{
    rt_kprintf("one-shot timer is timeout\n");
}

int timer_static_sample(void)
{
    /* Initialize Timer */
    rt_timer_init(&timer1, "timer1",  /* Timer name is timer1 */
                timeout1, /* Callback handler for timeout */
                RT_NULL, /* Entry parameter of the timeout function */
                10, /* Timing length in OS Tick, 10 OS Tick */
                RT_TIMER_FLAG_PERIODIC); /* Periodic timer */
    rt_timer_init(&timer2, "timer2",   /* Timer name is timer2 */
                timeout2, /* Callback handler for timeout */
                  RT_NULL, /* Entry parameter of the timeout function
*/
                  30, /* Timing length is 30 OS Tick */
                RT_TIMER_FLAG_ONE_SHOT); /* One-shot timer */

    /* Start Timer */
    rt_timer_start(&timer1);
    rt_timer_start(&timer2);
    return 0;
}
/* Export to msh command list */
MSH_CMD_EXPORT(timer_static_sample, timer_static sample);
```

The results are as follows:

```
 \ | /
- RT -      Thread Operating System
 / | \      3.1.0 build Aug 24 2018
 2006 - 2018 Copyright by rt-thread team

msh >timer_static_sample
msh >periodic timer is timeout
periodic timer is timeout
one shot timer is timeout
periodic timer is timeout
periodic timer is timeout
periodic timer is timeout
periodic timer is timeout
periodic timer is timeout
periodic timer is timeout
periodic timer is timeout
periodic timer is timeout
```

The timeout function of periodic timer1 runs per 10 OS Ticks for 10 times (after 10 times rt_timer_stop is called to stop timer1); the timeout function of the one-shot timer2 runs once on the 30th OS Tick.

HIGH-PRECISION DELAY

The minimum precision of the RT-Thread timer is determined by the system clock tick (1 OS Tick = 1/RT_TICK_PER_SECOND second, RT_TICK_PER_SECOND value is defined in the rtconfig.h file), and the timer must be set to an integer multiple of the OS Tick. At times it is necessary to implement system timing for a shorter time length—for example, the OS Tick is 10ms but the program needs to implement a timing or delay of 1ms. In this case, the operating system timer can't meet the requirements. This problem can be solved by reading the counter of a hardware timer of the system or using the hardware timer directly.

In Cortex-M series, SysTick has been used by RT-Thread as an OS Tick. It is configured to trigger an interrupt after 1/RT_TICK_PER_SECOND seconds. The interrupt handler uses the Cortex-M3 default name SysTick _ Handler. In the Cortex-M3 CMSIS (Cortex Microcontroller Software Interface Standard) specification, SystemCoreClock represents the dominant frequency of the chip. So, based on SysTick and SystemCoreClock, we can use SysTick to obtain an accurate delay function, as shown in the following example, with the Cortex-M3 SysTick-based precision delay (requires the system to enable SysTick).

The high-precision delay routine is as follows:

```
#include <board.h>
void rt_hw_us_delay(rt_uint32_t us)
{
    rt_uint32_t delta;
    /* Obtain the number of ticks of the delay */
    us = us * (SysTick->LOAD/(1000000/RT_TICK_PER_SECOND));
    /* Obtain current time */
    delta = SysTick->VAL;
    /* Loop to obtain the current time until the specified time elapses
and exits the loop */
    while (delta - SysTick->VAL< us);
}
```

The entry parameter "us" indicates the number of microseconds that need to be delayed. This function can only support delays shorter than 1 OS Tick; otherwise, the SysTick will overflow and not be able to obtain the specified delay time.

CHAPTER SUMMARY

This chapter introduced the timer and its working mechanism, timer management interface, and some considerations. Here's a look back at some key points for using the timer:

1. The timeout function of the HARD_TIMER timer is executed in the interrupt context and can be specified by using the parameter RT_TIMER_FLAG_HARD_TIMER in the timer control block. The timeout function of the HARD_TIMER timer has the same requirements as the interrupt service routine: the execution time should be as short as possible, and there should be no action that causes the current context to suspend and wait.
2. The timeout function of the SOFT_TIMER timer is executed in the thread's context and can be specified by using parameter RT_TIMER_FLAG_SOFT_TIMER in the timer control block.
3. Dynamic timer creation and deletion call the interface rt_timer_create and rt_timer_delete; static timer initialization and disengagement call the interface rt_timer_init and rt_timer_detach.
4. When there is a need to achieve a shorter period of system timing, such as if the OS Tick is 10ms and the program needs to implement 1ms of timing or delay, at this point, the operating system timer will not be able to meet the requirements. A shorter period of system timing can only be achieved by reading the counter of a hardware timer in the system or by using the hardware timer directly.

You can write some timers and simulate them to experience the meaning of the parameters in the timer interface.

6 Inter-Thread Synchronization

In a multi-threaded real-time system, the completion of a task can often be accomplished through the coordination of multiple threads. So how do these threads collaborate well with each other to perform without errors? Here is an example.

There are two threads in one task: one thread receives data from the sensor and writes the data to shared memory, while another thread periodically reads data from the shared memory and sends it to the display. Figure 6.1 depicts the data transfer between the two threads.

If access to shared memory is not exclusive, then it may be accessed simultaneously by both threads, which causes data consistency issues. For example, before thread 2 (the thread that can display data) attempts to display data, thread 1 (the thread that can receive data) has not yet completed the writing in of data; then the display will contain data sampled at different times, causing the display data to be disordered.

Thread 1 that writes the sensor data to the shared memory block and thread 2 that reads the sensor data from the shared memory block access the same memory block. To prevent data errors, the actions of the two threads must be mutually exclusive. One of threads should only be allowed after another thread completes its operation on the shared memory block. This way, thread 1 and thread 2 can work properly to execute this task correctly.

Synchronization refers to running in a predetermined order. Thread synchronization refers to multiple threads controlling the execution order between threads through specific mechanisms (such as mutex, event object, and critical section). In other words, they establish a relationship of execution order by synchronization between threads, and if there is no synchronization, the threads will be out of order.

Multiple threads operate and access the same area (code). This block of code is called the critical section, and the shared memory block in the earlier example is the critical section. Thread mutual exclusion refers to the exclusiveness of access to critical section resources. When multiple threads use critical section resources, only one thread is allowed at a specific moment. Other threads that want to use the resource must wait until the resource occupant releases the resource. Mutual exclusion between threads can be regarded as a special thread synchronization.

There are many ways to synchronize threads. The core idea is that **only one (or one kind of) thread is allowed to run when accessing the critical section.** There are many ways to enter and exit the critical section:

1. Call rt_hw_interrupt_disable() to enter the critical section, and call rt_hw_interrupt_enable() to exit the critical section; see "Enable and disable global interrupt" in the Interrupt Management chapter for details.
2. Call rt_enter_critical() to enter the critical section, and call rt_exit_critical() to exit the critical section.

This chapter introduces several synchronization methods: **semaphore, mutex**, and **event**. After reading this chapter, you will know how to use semaphore, mutex, and event to synchronize threads.

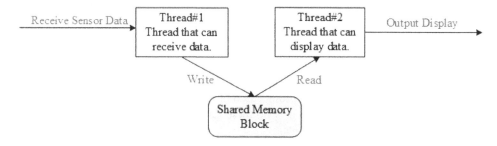

FIGURE 6.1   Diagram of data transfer between threads.

SEMAPHORE

Take a parking lot as an example to understand the concept of the semaphore:

1. When the parking lot is empty, the administrator of the parking lot finds that there are a lot of empty parking spaces. And then cars outside will enter the parking lot and get parking spaces.
2. When the parking space of the parking lot is full, the administrator finds that there is no empty parking space. As a result, cars outside will be prohibited from entering the parking lot, and they will be waiting in line.
3. When cars are leaving the parking lot, the administrator finds that there are empty parking spaces for cars outside to enter the parking lot; after the empty parking spaces are taken, cars outside are prohibited from entering.

In this example, the administrator is equivalent to the semaphore. The number of empty parking spaces that the administrator is in charge of is the value of the semaphore (non-negative, dynamic change); the parking space is equivalent to the common resource (critical section), and the cars are equivalent to the threads. Cars access the parking spaces by obtaining permission from the administrator, which is similar to a thread accessing public resources by obtaining the semaphore.

SEMAPHORE WORKING MECHANISM

A semaphore is a light-duty kernel object that can solve the problems of synchronization between threads. By obtaining or releasing semaphores, a thread can achieve synchronization or mutual exclusion.

A schematic diagram of a semaphore is shown in Figure 6.2. Each semaphore object has a semaphore value and a thread waiting queue. The semaphore value corresponds to the actual number of instances of the semaphore object and the number of resources. If the semaphore value is 5, it means that there are 5 semaphore instances (resources) that can be used. If the number of semaphore

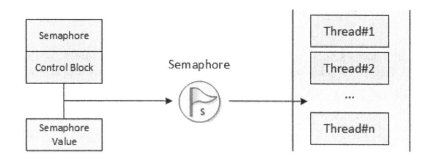

FIGURE 6.2   Schematic diagram of semaphore.

instances is zero, the thread that is applying for the semaphore will be suspended on the waiting queue of the semaphore, waiting for available semaphore instances (resources).

SEMAPHORE CONTROL BLOCK

In RT-Thread, the semaphore control block is a data structure used by the operating system to manage semaphores, represented by struct rt rt_semaphore. Another C expression is rt_sem_t, which represents the handle of the semaphore, and the implementation in C language is a pointer to the semaphore control block. The detailed definition of semaphore control block structure is as follows:

```
struct rt_semaphore
{
  struct rt_ipc_object parent;   /* Inherited from the ipc_object class */
  rt_uint16_t value;             /* Semaphore Value */
};
/* rt_sem_t is the type of pointer pointing to semaphore structure */
typedef struct rt_semaphore* rt_sem_t;
```

The rt_semaphore object is derived from rt_ipc_object and is managed by the IPC container. The maximum semaphore is 65535.

SEMAPHORE MANAGEMENT

The semaphore control block contains important parameters related to the semaphore and acts as a link between various states of the semaphore. Interfaces related to semaphore are shown in Figure 6.3. Operations on a semaphore include creating/initializing the semaphore, obtaining the semaphore, releasing the semaphore, and deleting/detaching the semaphore.

Create and Delete a Semaphore

When creating a semaphore, the kernel first creates a semaphore control block, then performs basic initialization on the control block. The following function interface is used to create a semaphore:

```
rt_sem_t rt_sem_create(const char *name,
                       rt_uint32_t value,
                       rt_uint8_t flag);
```

When this function is called, the system will first allocate a semaphore object from the object manager and initialize the object, and then initialize the parent class IPC object and semaphore-related

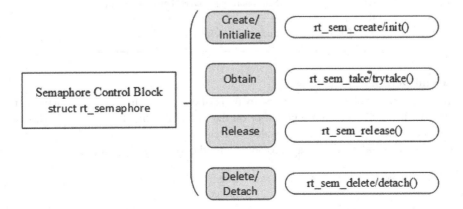

FIGURE 6.3   Interfaces related to semaphore.

parts. Among parameters specified in the creation of semaphore, the semaphore flag parameter determines the queuing way of how multiple threads wait when the semaphore is not available. When the RT_IPC_FLAG_FIFO (first-in, first-out) mode is selected, the waiting thread queue will be queued in a first-in, first-out manner. The first thread that goes in will first obtain the waiting semaphore. When the RT_IPC_FLAG_PRIO (priority waiting) mode is selected, the waiting threads will be queued in order of priority. Threads waiting with the highest priority will get the wait semaphore first. Table 6.1 describes the input parameters and return values for this function.

TABLE 6.1
rt_sem_create

Parameters	Description
name	Semaphore name
value	Semaphore initial value
flag	Semaphore flag, which can be the following values: RT_IPC_FLAG_FIFO or RT_IPC_FLAG_PRIO
Return	——
RT_NULL	Creation failed
semaphore control block pointer	Creation successful

For dynamically created semaphores, when the system no longer uses the semaphore, they can be removed to release system resources. To delete a semaphore, use the following function interface:

```
rt_err_t rt_sem_delete(rt_sem_t sem);
```

When this function is called, the system will delete this semaphore. If there is a thread waiting for this semaphore when it is being deleted, the delete operation will first wake up the thread waiting on the semaphore (return value of the waiting thread is -RT_ERROR), then release the semaphore's memory resources. Table 6.2 describes the input parameters and return values for this function.

TABLE 6.2
rt_sem_delete

Parameters	Description
sem	Semaphore object created by rt_sem_create()
Return	——
RT_EOK	Successfully deleted

Initialize and Detach a Semaphore

For a static semaphore object, its memory space is allocated by the compiler during compiling and placed on the read-write data segment or on the uninitialized data segment. In this case, the rt_sem_create() interface is no longer needed to create the semaphore to use it—just initialize it before using it. To initialize the semaphore object, use the following function interface:

```
rt_err_t rt_sem_init(rt_sem_t      sem,
                const char     *name,
                rt_uint32_t     value,
                rt_uint8_t      flag)
```

When this function is called, the system will initialize the semaphore object, then initialize the IPC object and parts related to the semaphore. The flag mentioned earlier in semaphore function creation can be used as the semaphore flag here. Table 6.3 describes the input parameters and return values for this function.

TABLE 6.3
rt_sem_init

Parameters	Description
sem	Semaphore object handle
name	Semaphore name
value	Semaphore initial value
flag	Semaphore flag, which can be the following values: RT_IPC_FLAG_FIFO or RT_IPC_FLAG_PRIO
Return	——
RT_EOK	Initialization successful

For statically initialized semaphore, detaching the semaphore is letting the semaphore object detach from the kernel object manager. To detach the semaphore, use the following function interface:

```
rt_err_t rt_sem_detach(rt_sem_t sem);
```

After using this function, the kernel wakes up all threads suspended in the semaphore wait queue and then detaches the semaphore from the kernel object manager. The waiting thread that was originally suspended on the semaphore will get the return value of -RT_ERROR. Table 6.4 describes the input parameters and return values for this function.

TABLE 6.4
rt_sem_detach

Parameters	Description
sem	Semaphore object handle
Return	——
RT_EOK	Successfully detached

Obtain Semaphore

The thread obtains semaphore resource instances by obtaining semaphores. When the semaphore value is greater than zero, the thread will get the semaphore, and the corresponding semaphore value will be reduced by 1. The semaphore is obtained using the following function interface:

```
rt_err_t rt_sem_take (rt_sem_t sem, rt_int32_t time);
```

When calling this function, if the value of the semaphore is zero, it means the current semaphore resource instance is not available, and the thread applying for the semaphore will choose according to the time parameters to either return directly, or suspend for a period of time, or wait forever.

While waiting, if other threads or ISR(Interrupt Service Routines) released the semaphore, then the thread will stop the waiting. If the semaphore is still not available after the parameter specified time, the thread will time out and return: the return value is -RT_ETIMEOUT. Table 6.5 describes the input parameters and return values for this function.

TABLE 6.5
rt_sem_take

Parameters	Description
sem	Semaphore object handle
Time	Specified wait time; unit is operating system clock tick (OS Tick)
Return	——
RT_EOK	Semaphore obtained successfully
-RT_ETIMEOUT	Did not received semaphore after timeout
-RT_ERROR	Other errors

Obtain Semaphore without Waiting

When a developer does not want to suspend the thread on the applied semaphore and wait, the semaphore can be obtained using the wait-free mode, and the following function interface is used for obtaining the semaphore without waiting:

```
rt_err_t rt_sem_trytake(rt_sem_t sem);
```

This function has the same effect as rt_sem_take(sem, 0), which means when the semaphore resource instance requested by the thread is not available, it will not wait on the semaphore; instead, it returns to -RT_ETIMEOUT directly. Table 6.6 describes the input parameters and return values for this function.

TABLE 6.6
rt_sem_trytake

Parameter	Description
sem	Semaphore object handle
Return	——
RT_EOK	Semaphore successfully obtained
-RT_ETIMEOUT	Semaphore obtainment failed

Release Semaphore

The semaphore is released to wake up the thread that suspends on the semaphore. To release the semaphore, use the following function interface:

```
rt_err_t rt_sem_release(rt_sem_t sem);
```

For example, when the semaphore value is zero and a thread is waiting for this semaphore, releasing the semaphore will wake up the first thread waiting in the thread queue of the semaphore, and this thread will obtain the semaphore; otherwise, the value of the semaphore will add 1. Table 6.7 describes the input parameters and return values of the function.

TABLE 6.7
rt_sem_release

Parameters	Description
sem	Semaphore object handle
Return	——
RT_EOK	Semaphore successfully released

SEMAPHORE APPLICATION SAMPLE

This is a sample of a semaphore usage routine. This routine creates a dynamic semaphore and initializes two threads: one thread sends the semaphore, and one thread receives the semaphore and performs the corresponding operations. The entire operation is shown in Code Listing 6.1.

Code listing 6.1 Semaphore Application Sample

```c
#include <rtthread.h>

#define THREAD_PRIORITY         25
#define THREAD_TIMESLICE        5

/* pointer to semaphore */
static rt_sem_t dynamic_sem = RT_NULL;

ALIGN(RT_ALIGN_SIZE)
static char thread1_stack[1024];
static struct rt_thread thread1;
static void rt_thread1_entry(void *parameter)
{
    static rt_uint8_t count = 0;

    while(1)
    {
        if(count <= 100)
        {
            count++;
        }
        else
            return;

        /* release semaphore every 10 counts */
         if(0 == (count % 10))
        {
            rt_kprintf("t1 release a dynamic semaphore.\n");
            rt_sem_release(dynamic_sem);
        }
    }
}

ALIGN(RT_ALIGN_SIZE)
static char thread2_stack[1024];
static struct rt_thread thread2;
static void rt_thread2_entry(void *parameter)
{
    static rt_err_t result;
```

```
    static rt_uint8_t number = 0;
    while(1)
    {
        /* permanently wait for the semaphore; once obtain the semaphore,
perform the number self-add operation */
        result = rt_sem_take(dynamic_sem, RT_WAITING_FOREVER);
        if (result != RT_EOK)
        {
            rt_kprintf("t2 take a dynamic semaphore, failed.\n");
            rt_sem_delete(dynamic_sem);
            return;
        }
        else
        {
            number++;
            rt_kprintf("t2 take a dynamic semaphore. number = %d\n"
            ,number);
        }
    }
}

/* initialization of the semaphore sample */
int semaphore_sample(void)
{
    /* create a dynamic semaphore with an initial value of 0 */
    dynamic_sem = rt_sem_create("dsem", 0, RT_IPC_FLAG_FIFO);
    if (dynamic_sem == RT_NULL)
    {
        rt_kprintf("create dynamic semaphore failed.\n");
        return -1;
    }
    else
    {
        rt_kprintf("create done. dynamic semaphore value = 0.\n");
    }

    rt_thread_init(&thread1,
                    "thread1",
                    rt_thread1_entry,
                    RT_NULL,
                    &thread1_stack[0],
                    sizeof(thread1_stack),
                    THREAD_PRIORITY, THREAD_TIMESLICE);
    rt_thread_startup(&thread1);

    rt_thread_init(&thread2,
                    "thread2",
                    rt_thread2_entry,
                    RT_NULL,
                    &thread2_stack[0],
                    sizeof(thread2_stack),
                    THREAD_PRIORITY-1, THREAD_TIMESLICE);
    rt_thread_startup(&thread2);

    return 0;
}
```

```
/* export to msh command list */
MSH_CMD_EXPORT(semaphore_sample, semaphore sample);
```

The results are as follows:

```
 \ | /
- RT -        Thread Operating System
 / | \        3.1.0 build Aug 27 2018
 2006 - 2018 Copyright by rt-thread team

msh >semaphore_sample
create done. dynamic semaphore value = 0.
msh >t1 release a dynamic semaphore.
t2 take a dynamic semaphore. number = 1
t1 release a dynamic semaphore.
t2 take a dynamic semaphore. number = 2
t1 release a dynamic semaphore.
t2 take a dynamic semaphore. number = 3
t1 release a dynamic semaphore.
t2 take a dynamic semaphore. number = 4
t1 release a dynamic semaphore.
t2 take a dynamic semaphore. number = 5
t1 release a dynamic semaphore.
t2 take a dynamic semaphore. number = 6
t1 release a dynamic semaphore.
t2 take a dynamic semaphore. number = 7
t1 release a dynamic semaphore.
t2 take a dynamic semaphore. number = 8
t1 release a dynamic semaphore.
t2 take a dynamic semaphore. number = 9
t1 release a dynamic semaphore.
t2 take a dynamic semaphore. number = 10
```

As the result of the operation above, thread 1 sends a semaphore when the count is a multiple of 10 (the thread exits after the count reaches 100), and thread 2 adds one on top of the number after receiving the semaphore.

Another semaphore application routine is shown in Code Listing 6.2. This sample will use two threads and three semaphores to implement an example of producers and consumers.

1. sem_lock: This semaphore acts as a lock because both threads operate on the same array, which means this array is a shared resource and sem_lock is used to protect this shared resource.
2. sem_empty: Its value is used to indicate the number of "warehouse" available seats, and the value of sem_empty is initialized to 5, indicating that there are 5 "empty seats."
3. sem_full: Its value is used to indicate the number of "full seats" in the "warehouse," and the value of sem_full is initialized to 0, indicating that there are 0 "full seats."

The two threads in the example are:

1. Producer thread: After obtaining the available seat (sem_empty value minus 1), generate a number, loop into the array, and then release a "full seat" (sem_full value plus 1).
2. Consumer thread: After getting the "full seat" (the value of sem_full is decremented by 1), the contents of the array are read and added, and then an "empty seat" is released (the value of sem_empty is increased by 1).

Code listing 6.2 Producer Consumer Routine

```
#include <rtthread.h>

#define THREAD_PRIORITY        6
#define THREAD_STACK_SIZE      512
#define THREAD_TIMESLICE       5

/* Define a maximum of 5 elements to be generated */
#define MAXSEM 5

/* An array of integers used to place production */
rt_uint32_t array[MAXSEM];

/* Point to the producer and consumer's read-write position in the array */
static rt_uint32_t set, get;

/* Pointer to the thread control block */
static rt_thread_t producer_tid = RT_NULL;
static rt_thread_t consumer_tid = RT_NULL;

struct rt_semaphore sem_lock;
struct rt_semaphore sem_empty, sem_full;

/* Pointer to the thread control block */
void producer_thread_entry(void *parameter)
{
    int cnt = 0;

    /* Run for 10 times*/
    while (cnt < 10)
    {
        /* Obtain one vacancy */
        rt_sem_take(&sem_empty, RT_WAITING_FOREVER);

        /* Modify array content, lock */
        rt_sem_take(&sem_lock, RT_WAITING_FOREVER);
        array[set % MAXSEM] = cnt + 1;
        rt_kprintf("the producer generates a number: %d\n", array[set %
MAXSEM]);
        set++;
        rt_sem_release(&sem_lock);

        /* publish one item filled */
        rt_sem_release(&sem_full);
        cnt++;

        /* Pause for a while */
        rt_thread_mdelay(20);
    }

    rt_kprintf("the producer exit!\n");
}

/* Consumer thread entry */
void consumer_thread_entry(void *parameter)
{
    rt_uint32_t sum = 0;
```

```
    while (1)
    {
        /* obtain a "full seat" */
        rt_sem_take(&sem_full, RT_WAITING_FOREVER);

        /* Critical region, locked for operation */
        rt_sem_take(&sem_lock, RT_WAITING_FOREVER);
        sum += array[get % MAXSEM];
        rt_kprintf("the consumer[%d] get a number: %d\n", (get % MAXSEM),
array[get % MAXSEM]);
        get++;
        rt_sem_release(&sem_lock);

        /* Release one vacancy */
        rt_sem_release(&sem_empty);

        /* The producer produces up to 10 numbers, stops, and the consumer
thread stops accordingly */
        if (get == 10) break;

        /* Pause for a while */
        rt_thread_mdelay(50);
    }

    rt_kprintf("the consumer sum is: %d\n", sum);
    rt_kprintf("the consumer exit!\n");
}

int producer_consumer(void)
{
    set = 0;
    get = 0;

    /* Initialize 3 semaphores */
    rt_sem_init(&sem_lock, "lock",    1,       RT_IPC_FLAG_FIFO);
    rt_sem_init(&sem_empty, "empty",   MAXSEM, RT_IPC_FLAG_FIFO);
    rt_sem_init(&sem_full, "full",     0,       RT_IPC_FLAG_FIFO);

    /* Create producer thread */
    producer_tid = rt_thread_create("producer",
                                    producer_thread_entry, RT_NULL,
                                    THREAD_STACK_SIZE,
                                    THREAD_PRIORITY - 1,
                                    THREAD_TIMESLICE);
    if (producer_tid != RT_NULL)
    {
        rt_thread_startup(producer_tid);
    }
    else
    {
        rt_kprintf("create thread producer failed");
        return -1;
    }

    /* Create consumer thread */
    consumer_tid = rt_thread_create("consumer",
```

```
                                       consumer_thread_entry, RT_NULL,
                                       THREAD_STACK_SIZE,
                                       THREAD_PRIORITY + 1,
                                       THREAD_TIMESLICE);
    if (consumer_tid != RT_NULL)
    {
        rt_thread_startup(consumer_tid);
    }
    else
    {
        rt_kprintf("create thread consumer failed");
        return -1;
    }

    return 0;
}

/* Export to msh command list */
MSH_CMD_EXPORT(producer_consumer, producer_consumer sample);
```

The results of this routine are as follows:

```
\ | /
- RT -     Thread Operating System
 / | \     3.1.0 build Aug 27 2018
 2006 - 2018 Copyright by rt-thread team

msh >producer_consumer
the producer generates a number: 1
the consumer[0] get a number: 1
msh >the producer generates a number: 2
the producer generates a number: 3
the consumer[1] get a number: 2
the producer generates a number: 4
the producer generates a number: 5
the producer generates a number: 6
the consumer[2] get a number: 3
the producer generates a number: 7
the producer generates a number: 8
the consumer[3] get a number: 4
the producer generates a number: 9
the consumer[4] get a number: 5
the producer generates a number: 10
the producer exit!
the consumer[0] get a number: 6
the consumer[1] get a number: 7
the consumer[2] get a number: 8
the consumer[3] get a number: 9
the consumer[4] get a number: 10
the consumer sum is: 55
the consumer exit!
```

This routine can be understood as the process of producers producing products and putting them into the warehouse and the consumer taking the products from the warehouse.

1. Producer thread:
2. Obtain 1 "empty seat" (put product number), now the number of "empty seats" is decremented by 1.

3. Lock protection; the generated number value is cnt+1, and the value is looped into the array and then unlocked.

4. Release 1 "full seat" (put one product into the warehouse, there will be one more "full seat" in the warehouse), add 1 to the number of "full seats."

5. Consumer thread:
   i. Obtain 1 "full seat" (take product number), then the number of "full seats" is decremented by 1.
   ii. Lock protection: read the number produced by the producer from the array and add it to the last number, then unlock it.
   iii. Release 1 "empty seat" (take one product from the warehouse; then there is one more "empty seat" in the warehouse), add 1 to the number of "empty seats."

The producer generates 10 numbers in turn, and the consumers take them away in turn and sum the values of the 10 numbers. A semaphore lock protects an array of critical region resources, ensuring the exclusivity of number taking for the consumers each time, and achieving inter-thread synchronization.

SEMAPHORE USAGE SCENARIO

Semaphores are a very flexible way to synchronize and can be used in a variety of situations, like forming locks, synchronization, resource counts, etc. It can also be conveniently used for synchronization between threads and threads and interrupts and threads.

Thread Synchronization

Thread synchronization is one of the simplest types of semaphore applications. For example, using semaphores to synchronize between two threads, the value of the semaphore is initialized to 0, indicating that there are 0 semaphore resource instances; the thread attempting to acquire the semaphore will wait directly on this semaphore.

When the thread holding the semaphore completes the work it is processing, it will release this semaphore. The thread waiting on this semaphore can be awakened, and it can then perform the next part of the work. This occasion can also be seen as using the semaphore for the work completion flag: the thread holding the semaphore completes its own work and then notifies the thread waiting for the semaphore to continue the next part of the work.

Lock

A single lock is often applied to multiple threads accessing the same shared resource (in other words, a critical region). When a semaphore is used as a lock, the semaphore resource instance should normally be initialized to 1, indicating that the system has one resource available by default. Because the semaphore value always varies between 1 and 0, this type of lock is also called a binary semaphore. As shown in Figure 6.4, when a thread needs to access a shared resource, it needs to obtain the resource lock first. When this thread successfully obtains the resource lock, other threads that intend to access the shared resource will suspend because they cannot obtain the resource. This is because it is already locked (semaphore value is 0) when other threads are trying to obtain the lock. When the thread holding the semaphore is processed and exiting the critical region, it will release the semaphore and unlock the lock, and the first waiting thread that is suspending on the lock will be awakened to gain access to the critical region.

Synchronization between Thread and Interrupt

Semaphore can also be easily applied to synchronize between the thread and interrupt, such as an interrupt trigger. When interrupting a service routine, the thread needs to be notified to perform corresponding data processing. At this time, the initial value of the semaphore can be set to 0. When

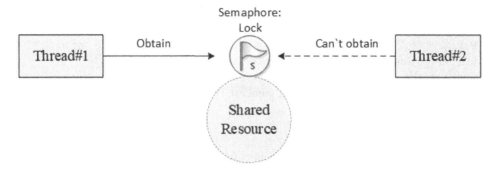

FIGURE 6.4   Lock.

the thread tries to hold this semaphore, since the initial value of the semaphore is 0, the thread will then suspend on this semaphore until the semaphore is released. When the interrupt is triggered, hardware-related actions are performed first, such as reading corresponding data from the hardware I/O port, and confirming the interrupt to clear the interrupt source, and then releasing a semaphore to wake up the corresponding thread for subsequent data processing. For example, the processing of FinSH threads is shown in Figure 6.5.

The value of the semaphore is initially 0. When the FinSH thread attempts to obtain the semaphore, it will be suspended because the semaphore value is 0. When the console device has data input, an interrupt is generated to enter the interrupt service routine. In the interrupt service routine, FinSH threadreads the data of the console device, puts the read data into the UART buffer for buffering, and then releases the semaphore. The semaphore release will wake up the shell thread. After the interrupt service routine has finished, if there are no ready threads with a higher priority than the shell thread in the system, the shell thread will hold the semaphore and run, obtaining the input data from the UART buffer.

Notes: The mutual exclusion between interrupts and threads cannot be done by means of semaphores (locks); use disabling-enabling interrupts instead.

Resource Count

A semaphore can also be considered as an incrementing or decrementing counter. It should be noted that the semaphore value is non-negative. For example, if the value of a semaphore is initialized to 5, then the semaphore can be reduced by a maximum of five consecutive times until the counter is reduced to zero. Resource count is suitable for the case where the processing speeds between threads do not match. At this time, the semaphore can be counted as the number of completed tasks of the previous thread, and when dispatched to the next thread, it can also be used in a continuous manner, handling multiple events each time. For example, in the producer and consumer problem,

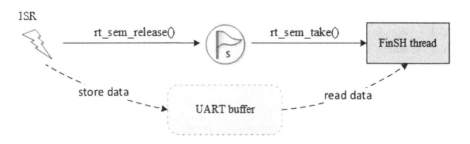

FIGURE 6.5   Sync between ISR and FinSH thread.

the producer can release the semaphore multiple times, and then the consumer can process multiple semaphore resources each time when dispatched.

Notes: Generally, resource count is mostly inter-thread synchronization in a hybrid mode because there are still multiple accesses from threads for a single resource processing, which requires accessing and processing for a single resource and lock mutex operation.

MUTEX

Mutex, also known as mutually exclusive semaphores, are a special binary semaphore. The mutex is similar to a parking lot with only one parking space: when one car enters, the parking lot gate is locked and other vehicles are waiting outside. When the car inside comes out, the parking lot gate will open and the next car can enter.

MUTEX WORKING MECHANISM

The difference between a mutex and a semaphore is that the thread with a mutex has ownership of the mutex; the mutex supports recursive access and prevents the thread priority from reversing; also, the mutex can only be released by the thread holding it, whereas a semaphore can be released by any thread.

There are only two states for mutex: unlocked and locked (two state values). When a thread holds it, then the mutex is locked and its ownership is obtained by this thread. Conversely, when this thread releases it, it unlocks the mutex and loses its ownership. When a thread is holding a mutex, other threads will not be able to unlock this mutex or hold it. The thread holding the mutex can also acquire the lock again without being suspended, as shown in Figure 6.6. This feature is quite different from the general binary semaphore: in semaphore, because there is no instance, the thread will suspend if the thread recursively holds the semaphore (which eventually leads to a deadlock).

Another potential problem with using semaphores is thread priority inversion. The so-called priority inversion is when a high-priority thread attempts to access the shared resource through the semaphore mechanism. If the semaphore is already held by a low-priority thread that may happen to be preempted by other medium-priority threads while running, this leads to high-priority threads being blocked by many lower-priority threads, which means determinism is difficult to guarantee. As shown in Figure 6.7, there are three threads with the priority levels A, B, and C: priority A > B > C. Threads A and B are in the suspended state, waiting for an event to trigger; thread C is running, and thread C starts using a shared resource M. While using the resource, the event thread A is waiting for occurs, and thread A switches to the ready state because it has higher priority than thread C, so it executes immediately. But when thread A wants to use shared resource M, because it is being used by thread C, thread A is suspended and thread C is running. If the event thread B is waiting for occurs, thread B switches to the ready state. Since thread B has a higher priority than thread C, thread B starts running; thread C won't run until thread B finishes. Thread A is only executed when

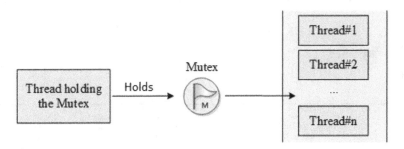

FIGURE 6.6 Mutex working mechanism diagram.

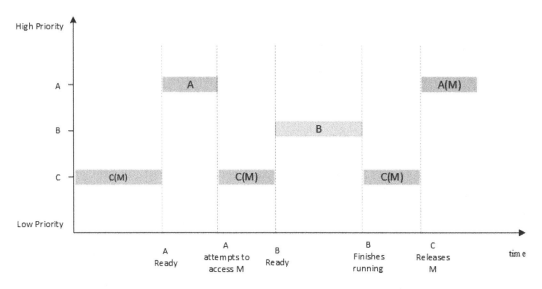

FIGURE 6.7 Priority inversion.

thread C releases the shared resource M. In this case, the priority has been reversed: thread B runs before thread A. This does not guarantee the response time for high-priority threads.

In the RT-Thread operating system, mutex can solve the priority inversion problem and implement the priority inheritance algorithm. Priority inheritance solves the problems caused by priority inversion by raising the priority of thread C to the priority of thread A during the period when thread A is suspended trying to access the shared resource. This prevents C (indirectly preventing A) from being preempted by B, as shown in Figure 6.8. Priority inheritance refers to raising the priority of a low-priority thread that occupies a certain resource, making the priority level of the low-priority thread equal to the priority of the thread with the highest priority level among all threads waiting for the resource, and then executes. When this low-priority thread releases the resource, the priority level returns to the initial setting. Therefore, threads that inherit priority help prevent the system resources from being preempted by any intermediate-priority thread.

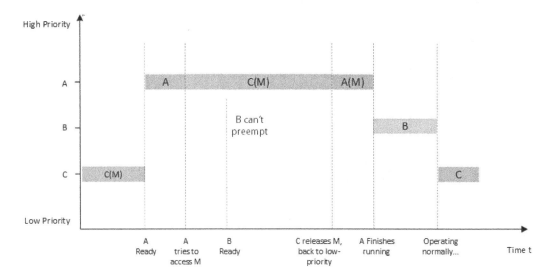

FIGURE 6.8 Priority inheritance.

Notes: After the mutex is obtained, release the mutex as soon as possible. During the time when holding the mutex, you must not change the priority of the thread holding the mutex.

Mutex Control Block

In RT-Thread, the mutex control block is a data structure used by the operating system to manage mutex, represented by the struct rt rt_mutex. Another C expression, rt_mutex_t, represents the handle of the mutex, and the implementation in C language refers to the pointer of the mutex control block. See the following code for a detailed definition of the mutex control block structure:

```
struct rt_mutex
    {
        struct rt_ipc_object parent;            /* inherited from the
ipc_object class */

        rt_uint16_t      value;                 /* mutex value */
        rt_uint8_t       original_priority;     /* hold the original
priority of the thread */
        rt_uint8_t       hold;                  /* number of times
holding the threads */
        struct rt_thread *owner;                /* thread that
currently owns the mutex */
    };
    /* rt_mutext_t pointer type of the one pointer pointing to the mutex
structure */
    typedef struct rt_mutex* rt_mutex_t;
```

The rt_mutex object is derived from the rt_ipc_object and is managed by the IPC container.

Mutex Management

The mutex control block contains important parameters related to the mutex, and it plays an important role in the implementation of the mutex function. The mutex-related interface is shown in Figure 6.9. The operation of a mutex includes creating/initiating a mutex, obtaining a mutex, releasing a mutex, and deleting/detaching a mutex.

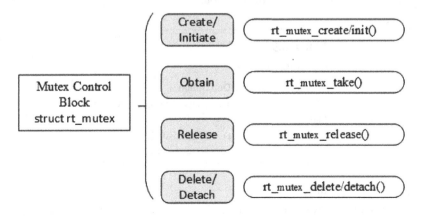

FIGURE 6.9 Mutex related interface.

Create and Delete a Mutex

When creating a mutex, the kernel first creates a mutex control block and then completes the initialization of the control block. Create a mutex using the following function interface:

```
rt_mutex_t rt_mutex_create (const char* name, rt_uint8_t flag);
```

You can call the rt_mutex_create function to create a mutex designated by name. When this function is called, the system will first allocate a mutex object from the object manager, initialize the object, and then initialize the parent class IPC object and the mutex-related part. The flag of the mutex is set to RT_IPC_FLAG_PRIO, which means that when multiple threads are waiting for resources, the resources will be accessed by the thread with a higher priority. The flag is set to RT_IPC_FLAG_FIFO, which means that when multiple threads are waiting for resources, resources are being accessed in first-come/first-served order. Table 6.8 describes the input parameters and return values for this function.

TABLE 6.8
rt_mutex_create

Parameters	Description
name	Mutex name
flag	Mutex flag, which can take the following values: RT_IPC_FLAG_FIFO or RT_IPC_FLAG_PRIO
Return	——
Mutex handle	Created successfully
RT_NULL	Creation failed

For dynamically created mutex, when the mutex is no longer used, the system resource is released by removing the mutex. To remove a mutex, use the following function interface:

```
rt_err_t rt_mutex_delete (rt_mutex_t mutex);
```

When a mutex is deleted, all threads waiting for this mutex will be woken up; the return value for the waiting threads is -RT_ERROR. The system then removes the mutex from the kernel object manager linked list and releases the memory space occupied by the mutex. Table 6.9 describes the input parameters and return values for this function.

TABLE 6.9
rt_mutex_delete

Parameters	Description
mutex	The handle of the mutex object
Return	——
RT_EOK	Deleted successfully

Initialize and Detach a Mutex

The memory of a static mutex object is allocated by the compiler during system compilation, and is usually placed in a read-write data segment or an uninitialized data segment. Before using such static mutex objects, you need to initialize them first. To initialize the mutex, use the following function interface:

```
rt_err_t rt_mutex_init (rt_mutex_t mutex, const char* name, rt_uint8_t flag);
```

When using this function interface, you need to specify the handle of the mutex object (i.e., the pointer to the mutex control block), the mutex name, and the mutex flag. The mutex flag can be the flags mentioned in the creation of mutex function above. Table 6.10 describes the input parameters and return values for this function.

TABLE 6.10
rt_mutex_init

Parameters	Description
mutex	The handle of the mutex object, which is provided by the user and points to the memory block of the mutex object
name	Mutex name
flag	Mutex flag, which can take the following values: RT_IPC_FLAG_FIFO or RT_IPC_FLAG_PRIO
Return	——
RT_EOK	Initialization successful

For a statically initialized mutex, detaching the mutex means to remove the mutex object from the kernel object manager. To detach the mutex, use the following function interface:

```
rt_err_t rt_mutex_detach (rt_mutex_t mutex);
```

After using this function interface, the kernel wakes up all threads suspended on the mutex (the return value of the thread is -RT_ERROR), and then the system detaches the mutex from the kernel object manager. Table 6.11 describes the input parameters and return values for this function.

TABLE 6.11
rt_mutex_detach

Parameters	Description
mutex	The handle of the mutex object
Return	——
RT_EOK	Successful

Obtain Mutex

Once the thread obtains the mutex, the thread has ownership of the mutex; that is, a mutex can only be held by one thread at a time. To obtain the mutex, use the following function interface:

```
rt_err_t rt_mutex_take (rt_mutex_t mutex, rt_int32_t time);
```

If the mutex is not controlled by another thread, the thread requesting the mutex will successfully obtain the mutex. If the mutex is already controlled by the current thread, then add 1 to the number of holds for the mutex, and the current thread will not be suspended to wait. If the mutex is already occupied by another thread, the current thread suspends and waits on the mutex until another thread

TABLE 6.12
rt_mutex_take

Parameters	Description
mutex	The handle of the mutex object
time	Specified waiting time
Return	——
RT_EOK	Successfully obtained mutex
-RT_ETIMEOUT	Timeout
-RT_ERROR	Failed to obtain

releases it or until the specified timeout elapses. Table 6.12 describes the input parameters and return values for this function.

Release Mutex

When a thread completes the access to a mutually exclusive resource, it should release the mutex it occupies as soon as possible, so that other threads can obtain the mutex in time. To release the mutex, use the following function interface:

```
rt_err_t rt_mutex_release(rt_mutex_t mutex);
```

When using this function interface, only threads that already have control of the mutex can release it. Each time the mutex is released, its holding count is reduced by 1. When the mutex's holding count is zero (i.e., the holding thread has released all holding operations), it becomes available and threads waiting on the semaphore are awakened. If the thread's priority is increased by the mutex, then when the mutex is released, the thread reverts to the priority level before holding the mutex. Table 6.13 describes the input parameters and return values for this function.

TABLE 6.13
rt_mutex_release

Parameters	Description
mutex	The handle of the mutex object
Return	——
RT_EOK	Success

MUTEX APPLICATION SAMPLE

This is a mutex application routine, and a mutex lock is a way to protect shared resources. When a thread has the mutex lock, it can protect shared resources from being destroyed by other threads. The following example can be used to illustrate. There are two threads: thread 1 and thread 2. Thread 1 adds 1 to each of the two numbers; thread 2 also adds 1 to each of the two numbers. A mutex is used to ensure that the operation of the thread changing the values of the two numbers is not interrupted, as shown in Code Listing 6.3:

Code listing 6.3 Mutex Sample

```
#include <rtthread.h>

#define THREAD_PRIORITY        8
#define THREAD_TIMESLICE       5
```

```
/* Pointer to the mutex */
static rt_mutex_t dynamic_mutex = RT_NULL;
static rt_uint8_t number1,number2 = 0;


ALIGN(RT_ALIGN_SIZE)
static char thread1_stack[1024];
static struct rt_thread thread1;
static void rt_thread_entry1(void *parameter)
{
      while(1)
      {
          /* After thread 1 obtains the mutex, it adds 1 to number1 and
number2, and then releases the mutex. */
          rt_mutex_take(dynamic_mutex, RT_WAITING_FOREVER);
          number1++;
          rt_thread_mdelay(10);
          number2++;
          rt_mutex_release(dynamic_mutex);
      }
}


ALIGN(RT_ALIGN_SIZE)
static char thread2_stack[1024];
static struct rt_thread thread2;
static void rt_thread_entry2(void *parameter)
{
      while(1)
      {
          /* After thread 2 obtains the mutex, check whether the values of
number1 and number2 are the same. If they are the same, it means the mutex
successfully played the role of a lock. */
          rt_mutex_take(dynamic_mutex, RT_WAITING_FOREVER);
          if(number1 != number2)
          {
            rt_kprintf("not protect.number1 = %d, mumber2 = %d \n",number1
,number2);
          }
          else
          {
            rt_kprintf("mutex protect ,number1 = mumber2 is %d\n",number1);
          }

           number1++;
           number2++;
           rt_mutex_release(dynamic_mutex);

          if(number1>=50)
              return;
      }
}

/* Initialization of the mutex sample */
int mutex_sample(void)
{
    /* Create a dynamic mutex */
```

```
    dynamic_mutex = rt_mutex_create("dmutex", RT_IPC_FLAG_FIFO);
    if (dynamic_mutex == RT_NULL)
    {
        rt_kprintf("create dynamic mutex failed.\n");
        return -1;
    }

    rt_thread_init(&thread1,
                   "thread1",
                   rt_thread_entry1,
                   RT_NULL,
                   &thread1_stack[0],
                   sizeof(thread1_stack),
                   THREAD_PRIORITY, THREAD_TIMESLICE);
    rt_thread_startup(&thread1);

    rt_thread_init(&thread2,
                   "thread2",
                   rt_thread_entry2,
                   RT_NULL,
                   &thread2_stack[0],
                   sizeof(thread2_stack),
                   THREAD_PRIORITY-1, THREAD_TIMESLICE);
    rt_thread_startup(&thread2);
    return 0;
}

/* Export to the MSH command list */
MSH_CMD_EXPORT(mutex_sample, mutex sample);
```

Both thread 1 and thread 2 use mutex to protect the operation on the two numbers (if the obtain and release mutex statements in thread 1 are commented out, thread 1 will no longer protect the number), and the simulation results are as follows:

```
\ | /
- RT -     Thread Operating System
 / | \     3.1.0 build Aug 24 2018
 2006 - 2018 Copyright by rt-thread team

msh >mutex_sample
msh >mutex protect ,number1 = mumber2 is 1
mutex protect ,number1 = mumber2 is 2
mutex protect ,number1 = mumber2 is 3
mutex protect ,number1 = mumber2 is 4
...
mutex protect ,number1 = mumber2 is 48
mutex protect ,number1 = mumber2 is 49
```

Threads use mutex to protect the operation on the two numbers, keeping the number values consistent.

Another example of a mutex is shown in Code Listing 6.4. This example creates three dynamic threads to check if the priority level of the thread holding the mutex is adjusted to the highest-priority level among the waiting threads.

Code listing 6.4 Prevent Priority Inversion Routine

```
#include <rtthread.h>
```

```c
/* Pointer to the thread control block */
static rt_thread_t tid1 = RT_NULL;
static rt_thread_t tid2 = RT_NULL;
static rt_thread_t tid3 = RT_NULL;
static rt_mutex_t mutex = RT_NULL;


#define THREAD_PRIORITY        10
#define THREAD_STACK_SIZE      512
#define THREAD_TIMESLICE    5

/* Thread 1 Entry */
static void thread1_entry(void *parameter)
{
    /* Let the low priority thread run first */
    rt_thread_mdelay(100);

    /* At this point, thread3 holds the mutex and thread2 is waiting to
hold the mutex */

    /* Check the priority level of thread2 and thread3 */
    if (tid2->current_priority != tid3->current_priority)
    {
        /* The priority is different, the test fails */
        rt_kprintf("the priority of thread2 is: %d\n",
tid2->current_priority);
        rt_kprintf("the priority of thread3 is: %d\n",
tid3->current_priority);
        rt_kprintf("test failed.\n");
        return;
    }
    else
    {
        rt_kprintf("the priority of thread2 is: %d\n",
tid2->current_priority);
        rt_kprintf("the priority of thread3 is: %d\n",
tid3->current_priority);
        rt_kprintf("test OK.\n");
    }
}

/* Thread 2 Entry */
static void thread2_entry(void *parameter)
{
    rt_err_t result;

    rt_kprintf("the priority of thread2 is: %d\n",
tid2->current_priority);

    /* Let the low-priority thread run first */
    rt_thread_mdelay(50);

    /*
     * Trying to hold a mutex lock. At this point, thread 3 has the mutex
lock, so the priority level of thread 3 should be raised
     * to the same level of priority as thread 2
     */
```

```
    result = rt_mutex_take(mutex, RT_WAITING_FOREVER);

    if (result == RT_EOK)
    {
        /* Release mutex lock */
        rt_mutex_release(mutex);
    }
}

/* Thread 3 Entry */
static void thread3_entry(void *parameter)
{
    rt_tick_t tick;
    rt_err_t result;

    rt_kprintf("the priority of thread3 is: %d\n",
tid3->current_priority);

    result = rt_mutex_take(mutex, RT_WAITING_FOREVER);
    if (result != RT_EOK)
    {
        rt_kprintf("thread3 take a mutex, failed.\n");
    }

    /* Operate a long cycle, 500ms */
    tick = rt_tick_get();
    while (rt_tick_get() - tick < (RT_TICK_PER_SECOND / 2)) ;

    rt_mutex_release(mutex);
}

int pri_inversion(void)
{
    /* Created a mutex lock */
    mutex = rt_mutex_create("mutex", RT_IPC_FLAG_FIFO);
    if (mutex == RT_NULL)
    {
        rt_kprintf("create dynamic mutex failed.\n");
        return -1;
    }

    /* Create thread 1*/
    tid1 = rt_thread_create("thread1",
                            thread1_entry,
                            RT_NULL,
                            THREAD_STACK_SIZE,
                            THREAD_PRIORITY - 1, THREAD_TIMESLICE);
    if (tid1 != RT_NULL)
        rt_thread_startup(tid1);

    /* Create thread 2 */
    tid2 = rt_thread_create("thread2",
                            thread2_entry,
                            RT_NULL,
                            THREAD_STACK_SIZE,
                            THREAD_PRIORITY, THREAD_TIMESLICE);
```

```
      if (tid2 != RT_NULL)
          rt_thread_startup(tid2);

      /* Create thread 3 */
      tid3 = rt_thread_create("thread3",
                              thread3_entry,
                              RT_NULL,
                              THREAD_STACK_SIZE,
                              THREAD_PRIORITY + 1, THREAD_TIMESLICE);
      if (tid3 != RT_NULL)
          rt_thread_startup(tid3);

      return 0;
}

/* Export to the msh command list */
MSH_CMD_EXPORT(pri_inversion, prio_inversion sample);
```

The results are as follows:

```
\ | /
- RT -      Thread Operating System
 / | \      3.1.0 build Aug 27 2018
 2006 - 2018 Copyright by rt-thread team

msh >pri_inversion
the priority of thread2 is: 10
the priority of thread3 is: 11
the priority of thread2 is: 10
the priority of thread3 is: 10
test OK.
```

The routine demonstrates how to use the mutex. Thread 3 holds the mutex first, and then thread 2 tries to hold the mutex, at which point thread 3's priority is raised to the same level as thread 2.

Notes: It is important to remember that mutex cannot be used in interrupt service routines.

Occasions to Use Mutex

The use of a mutex is relatively simple because it is a type of semaphore and it exists in the form of a lock. At the time of initialization, the mutex is always unlocked, and when it is held by the thread, it immediately becomes locked. Mutex is more suitable for:

1. When a thread holds a mutex multiple times. This can avoid the problem of deadlock caused by multiple recursive holdings of the same thread.
2. A situation in which priority inversion may occur due to multi-thread synchronization.

EVENT

An event set is also one of the mechanisms for synchronization between threads. An event set can contain multiple events. An event set can be used to complete one-to-many or many-to-many thread synchronization. Let's take the bus as an example to illustrate events. There may be the following situations when waiting for a bus at a bus stop:

1. P1 is taking a bus to a certain place, and only one type of bus can reach the destination. P1 can leave for the destination once that certain bus arrives.

2. P1 is taking a bus to a certain place, and three types of buses can reach the destination. P1 can leave for the destination once any one of the three types of buses arrives.
3. P1 is traveling with P2 to a certain place together; P1 can't leave for the destination unless two conditions are met. These two conditions are "P2 arrives at the bus stop" and "the bus arrives at the bus stop".

Here, P1 leaving for a certain place can be regarded as a thread, and "bus arrives at the bus stop" and "P2 arrives at the bus stop" are regarded as the occurrence of events. Situation ① is a specific event that wakes up the thread; situation ② is any single event to wake up the thread; situation ③ is when multiple events must occur simultaneously to wake up the thread.

EVENT SET WORKING MECHANISM

The event set is mainly used for synchronization between threads. Unlike the semaphore, it can achieve one-to-many or many-to-many synchronization. That is, the relationship between a thread and multiple events can be set as follows: any one of the events wakes up the thread, or several events wake up the thread for subsequent processing; likewise, the event can be multiple threads to synchronize multiple events. This collection of multiple events can be represented by a 32-bit unsigned integer variable, with each bit of the variable representing an event, and the thread associates one or more events by "logical AND" or "logical OR" to form an event combination. The "logical OR" of an event is also called independent synchronization, which means that the thread is synchronized with one of the events; the event "logical AND" is also called associative synchronization, which means that the thread is synchronized with several events.

The event set defined by RT-Thread has the following characteristics:

1. Events are related to threads only, and events are independent of each other: each thread can have 32 event flags, recorded with a 32-bit unsigned integer, and each bit representing an event.
2. An event is only used for synchronization and does not provide data transfer functionality.
3. Events are not queuing, that is, sending the same event to the thread multiple times (if the thread has not had time read it); the effect is equivalent to sending only once.

In RT-Thread, each thread has an event information tag with three attributes. They are RT_EVENT_FLAG_AND (logical AND), RT_EVENT_FLAG_OR (logical OR), and RT_EVENT_FLAG_CLEAR (clear flag). When the thread waits for event synchronization, it can determine whether the currently received event satisfies the synchronization condition by 32 event flags and this event information flag.

As shown in Figure 6.10, the 1st and 30th bits of the event flag of thread 1 are set. If the event information flag is set to logical AND, it means that thread 1 will be triggered to wake

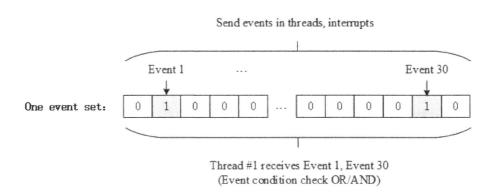

FIGURE 6.10 Event set work diagram.

up only after both event 1 and event 30 occur. If the event information flag is set to logical OR, the occurrence of either event 1 or event 30 will trigger to wake up thread 1. If the message flag also sets the clear flag bit, this means event 1 and event 30 will be automatically cleared to zero when thread 1 wakes up; otherwise, the event flag will still be present (set to 1).

EVENT SET CONTROL BLOCK

In RT-Thread, the event set control block is a data structure used by the operating system to manage events, represented by the structure struct rt_event. Another C type, rt_event_t, represents the handle of the event set, and the implementation in C language is a pointer to the event set control block. See the following code for a detailed definition of the event set control block structure:

```
struct rt_event
{
    struct rt_ipc_object parent;    /* Inherited from the ipc_object
class */

    /* The set of events, each bit represents 1 event, the value of the
bit can mark whether an event occurs */
    rt_uint32_t set;
};
/* rt_event_t is the pointer type pointing to the event structure */
typedef struct rt_event* rt_event_t;
```

The rt_event object is derived from the rt_ipc_object and is managed by the IPC container.

EVENT SET MANAGEMENT

The event set control block contains important parameters related to the event set and plays an important role in the implementation of the event set function. The event set–related interfaces are shown in Figure 6.11. The operations on an event set include create/initiate event sets, send events, receive events, and delete/detach event sets.

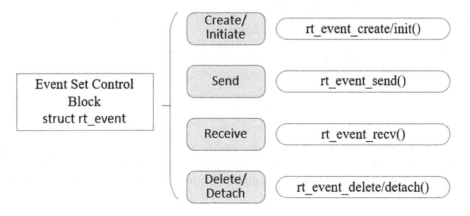

FIGURE 6.11 Event-related interface.

Create and Delete an Event Set

When creating an event set, the kernel first creates an event set control block and then performs basic initialization on the event set control block. The event set is created using the following function interface:

```
rt_event_t rt_event_create(const char* name, rt_uint8_t flag);
```

When the function interface is called, the system allocates the event set object from the object manager, initializes the object, and then initializes the parent class IPC object. Table 6.14 describes the input parameters and return values for this function.

TABLE 6.14
rt_event_create

Parameters	Description
name	Name of the event set
flag	The flag of the event set, which can take the following values: RT_IPC_FLAG_FIFO or RT_IPC_FLAG_PRIO
Return	——
RT_NULL	Creation failed
Handle of the event object	Creation successful

When the system no longer uses the event set object created by rt_event_create(), the system resource is released by deleting the event set object control block. To delete an event set, you can use the following function interface:

```
rt_err_t rt_event_delete(rt_event_t event);
```

When you call the rt_event_delete function to delete an event set object, you should ensure that the event set is no longer used. All threads that are suspended on the event set will be awakened before the deletion (the return value of the thread is -RT_ERROR), and then the memory block occupied by the event set object is released. Table 6.15 describes the input parameters and return values for this function.

TABLE 6.15
rt_event_delete

Parameters	Description
event	The handle of the event set object
Return	——
RT_EOK	Success

Initialize and Detach an Event Set

The memory of a static event set object is allocated by the compiler during system compilation, and is usually placed in a read-write data segment or an uninitialized data segment. Before using a static event set object, you need to initialize it first. The initialization event set uses the following function interface:

```
rt_err_t rt_event_init(rt_event_t event, const char* name, rt_uint8_t flag);
```

When the interface is called, you need to specify the handle of the static event set object (i.e., the pointer pointing to the event set control block), and then the system will initialize the event set object and add it to the system object container for management. Table 6.16 describes the input parameters and return values for this function.

TABLE 6.16
rt_event_init

Parameters	Description
event	The handle of the event set object
name	The name of the event set
flag	The flag of the event set, which can take the following values: RT_IPC_FLAG_FIFO or RT_IPC_FLAG_PRIO
Return	——
RT_EOK	Success

When the system no longer uses the event set object initialized by rt_event_init(), the system resources are released by detaching the event set object control block. Detaching an event set means to detach the event set object from the kernel object manager. To detach an event set, use the following function interface:

```
rt_err_t rt_event_detach(rt_event_t event);
```

When the developer calls this function, the system first wakes up all the threads suspended on the event set wait queue (the return value of the thread is -RT_ERROR) and then detaches the event set from the kernel object manager. Table 6.17 describes the input parameters and return values for this function.

TABLE 6.17
rt_event_detach

Parameters	Description
event	The handle of the event set object
Return	——
RT_EOK	Success

Send an Event

The send event function can send one or more events in the event set, as follows:

```
rt_err_t rt_event_send(rt_event_t event, rt_uint32_t set);
```

When using the function interface, the event flag value of the event set object is set by the event flag specified by the parameter set, and then the waiting thread linked list transversely waiting for the event set objects is used to determine whether a thread has the event activation requirement that matches the current event flag value. If there is, the thread is awakened. Table 6.18 describes the input parameters and return values for this function.

Receive an Event

The kernel uses a 32-bit unsigned integer to identify the event set. Each bit represents an event, so an event set object can wait to receive 32 events at the same time, and the kernel can decide how to

TABLE 6.18
rt_event_send

Parameters	Description
event	The handle of the event set object
set	The flag value of one or more events sent
Return	——
RT_EOK	Success

activate the thread by specifying the parameter "logical AND" or "logical OR." Using the "logical AND" parameter indicates that the thread is only activated when all waiting events occur, and using the "logical OR" parameter means that the thread is activated as soon as one waiting event occurs. To receive events, use the following function interface:

```
rt_err_t rt_event_recv(rt_event_t event,
                       rt_uint32_t set,
                       rt_uint8_t option,
                       rt_int32_t timeout,
                       rt_uint32_t* recved);
```

When the user calls the interface, the system first judges whether the event to be received is occurred according to the set parameters and receiving options. If it has already occurred, it determines whether to reset the corresponding flag of the event according to whether RT_EVENT_FLAG_CLEAR is set on the parameter option and then returns (where the received parameter returns to the received event); if it has not occurred, fill the waiting set and option parameters into the structure of the thread itself and then suspend the thread on this event until its waiting event satisfies the condition or until the specified timeout elapses. If the timeout is set to zero, this means that when the event to be accepted by the thread does not meet the requirements, it does not wait, but returns directly -RT_ETIMEOUT. Table 6.19 describes the input parameters and return values for this function.

TABLE 6.19
rt_event_recv

Parameters	Description
event	The handle of the event set object
set	Receive events of interest to the thread
option	Receive options
timeout	Timeout
recved	Point to the received event
Return	——
RT_EOK	Successful
-RT_ETIMEOUT	Timeout
-RT_ERROR	Error

The value of the option can be:

```
/* Select AND or OR to receive events */
RT_EVENT_FLAG_OR
RT_EVENT_FLAG_AND
```

```
/* Choose to clear reset event flag */
RT_EVENT_FLAG_CLEAR
```

Event Set Application Sample

This is the application routine for the event set, which initializes an event set and two threads. One thread waits for an event of interest to it, and another thread sends an event, as shown in Code Listing 6.5.

Code listing 6.5 Event Set Usage Routine

```c
#include <rtthread.h>

#define THREAD_PRIORITY      9
#define THREAD_TIMESLICE     5

#define EVENT_FLAG3 (1 << 3)
#define EVENT_FLAG5 (1 << 5)

/* Event control block */
static struct rt_event event;

ALIGN(RT_ALIGN_SIZE)
static char thread1_stack[1024];
static struct rt_thread thread1;

/* Thread 1 entry function*/
static void thread1_recv_event(void *param)
{
    rt_uint32_t e;

    /* The first time the event is received, either event 3 or event 5 can
trigger thread 1, clearing the event flag after receiving */
    if (rt_event_recv(&event, (EVENT_FLAG3 | EVENT_FLAG5),
                    RT_EVENT_FLAG_OR | RT_EVENT_FLAG_CLEAR,
                    RT_WAITING_FOREVER, &e) == RT_EOK)
    {
        rt_kprintf("thread1: OR recv event 0x%x\n", e);
    }

    rt_kprintf("thread1: delay 1s to prepare the second event\n");
    rt_thread_mdelay(1000);

    /* The second time the event is received, both event 3 and event 5 can
trigger thread 1, clearing the event flag after receiving */
    if (rt_event_recv(&event, (EVENT_FLAG3 | EVENT_FLAG5),
                    RT_EVENT_FLAG_AND | RT_EVENT_FLAG_CLEAR,
                    RT_WAITING_FOREVER, &e) == RT_EOK)
    {
        rt_kprintf("thread1: AND recv event 0x%x\n", e);
    }
    rt_kprintf("thread1 leave.\n");
}


ALIGN(RT_ALIGN_SIZE)
```

```
static char thread2_stack[1024];
static struct rt_thread thread2;

/* Thread 2 Entry */
static void thread2_send_event(void *param)
{
    rt_kprintf("thread2: send event3\n");
    rt_event_send(&event, EVENT_FLAG3);
    rt_thread_mdelay(200);

    rt_kprintf("thread2: send event5\n");
    rt_event_send(&event, EVENT_FLAG5);
    rt_thread_mdelay(200);

    rt_kprintf("thread2: send event3\n");
    rt_event_send(&event, EVENT_FLAG3);
    rt_kprintf("thread2 leave.\n");
}

int event_sample(void)
{
    rt_err_t result;

    /* Initialize event object */
    result = rt_event_init(&event, "event", RT_IPC_FLAG_FIFO);
    if (result != RT_EOK)
    {
        rt_kprintf("init event failed.\n");
        return -1;
    }

    rt_thread_init(&thread1,
                   "thread1",
                   thread1_recv_event,
                   RT_NULL,
                   &thread1_stack[0],
                   sizeof(thread1_stack),
                   THREAD_PRIORITY - 1, THREAD_TIMESLICE);
    rt_thread_startup(&thread1);

    rt_thread_init(&thread2,
                   "thread2",
                   thread2_send_event,
                   RT_NULL,
                   &thread2_stack[0],
                   sizeof(thread2_stack),
                   THREAD_PRIORITY, THREAD_TIMESLICE);
    rt_thread_startup(&thread2);

    return 0;
}

/* Export to the msh command list */
MSH_CMD_EXPORT(event_sample, event sample);
```

The results are as follows:

```
 \  |  /
- RT -       Thread Operating System
 /  |  \       3.1.0 build Aug 24 2018
 2006 - 2018 Copyright by rt-thread team

msh >event_sample
thread2: send event3
thread1: OR recv event 0x8
thread1: delay 1s to prepare the second event
msh >thread2: send event5
thread2: send event3
thread2 leave.
thread1: AND recv event 0x28
thread1 leave.
```

The routine demonstrates how to use the event set. Thread 1 receives events twice, before and after, using the "logical OR" and "logical AND," respectively.

OCCASIONS TO USE AN EVENT SET

Event set can be used in a variety of situations and can replace semaphores to some extent for inter-thread synchronization. A thread or interrupt service routine sends an event to the event set object, and the waiting thread awakens and the corresponding event is processed. However, unlike the semaphore, the event transmission operation is not cumulative until the event is cleared, and the release actions of semaphore are cumulative. Another feature of the event is that the receiving thread can wait for multiple events, meaning multiple events correspond to one thread or multiple threads. At the same time, according to thread waiting parameters, you can choose between a "logical OR" trigger or a "logical AND" trigger. This feature is not available for semaphores. The semaphore can only recognize a single release action and cannot wait for multiple types of release at the same time. Figure 6.12 shows the multi-event receiving diagram.

An event set contains 32 events, and a particular thread only waits for and receives events it is interested in. It can be a thread waiting for the arrival of multiple events (threads 1 and 2 are waiting for multiple events, logical "AND" or logical "OR" can be used to trigger the thread in events) or multiple threads waiting for an event to arrive (event 25). When events of interest to them occur, the thread will be awakened and subsequent processing actions will be taken.

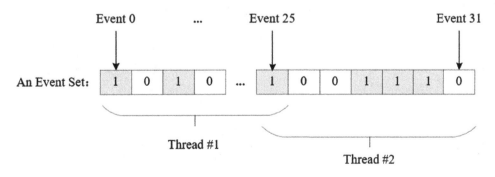

FIGURE 6.12   Multi-event receiving schematic.

CHAPTER SUMMARY

This chapter described inter-thread synchronization, and here's a look back at what to note:

1. The semaphore can be released in the interrupt, but cannot be obtained in the interrupt service program.
2. After obtaining the mutex, release it as soon as possible, and do not change the priority of the thread in the process of holding the mutex.
3. The mutex cannot be used in the interrupt service routine.

In order to deepen the understanding of the concepts of interrupt locks, scheduler locks, semaphore, mutex, and event sets, which are mentioned in this chapter, you'd better fully understand the routines or expand the functionality you want on routines.

7 Inter-Thread Communication

In the last chapter, we talked about inter-thread synchronization, and concepts such as semaphores, mutexes, and event sets were mentioned. To continue with the last chapter, this chapter is going to explain inter-thread communication. In bare-metal programming, global variables are often used for communication between functions. For example, some functions may change the value of a global variable due to some operations. Another function reads the global variable and will perform corresponding actions to achieve communication and collaboration according to the global variable values it reads. More tools are available in RT-Thread to help pass information between different threads. These tools are covered in more detail in this chapter. After reading this chapter, you will know how to use mailbox, message queue, and signal for communication between threads.

MAILBOX

A mailbox service is a typical inter-thread communication method in real-time operating systems. For example, there are two threads: thread 1 detects the state of the button and sends its state, and thread 2 reads the state of the button and turns the LED on or off according to the state of the button. Here, a mailbox can be used to communicate. Thread 1 sends the status of the button as a mail to the mailbox. Thread 2 reads the mail in the mailbox to get the button status and turn the LED on or off the accordingly.

Thread 1 here can also be extended to multiple threads. For example, there are three threads: thread 1 detects and sends the button state, thread 2 detects and sends the ADC information, and thread 3 performs different operations according to the type of information received.

Mailbox Working Mechanism

The mailbox of the RT-Thread operating system is used for inter-thread communication, which is characterized by low overhead and high efficiency. Each mail in the mailbox can only hold a fixed 4 bytes (for a 32-bit processing system, the pointer is 4 bytes in size, so a mail can hold only one pointer). A typical mailbox is also called a message exchange. As shown in Figure 7.1, a thread or an interrupt service routine sends a 4-byte message to a mailbox, and one or more threads can receive and process the message from the mailbox.

The sending process of the non-blocking mode mails can be safely used in ISR(Interrupt Service Routines). It is an effective way for the thread, the interrupt service, and the timer to send a mail to the thread. In general, the receiving of the mails can be a blocked process, depending on whether there is a mail in the mailbox and the timeout set when the mail was received. When there is no mail in the mailbox and the set timeout is not 0, the receiving of the mails will become blocked. In such cases, the mails can only be received by threads.

When a thread sends a mail to a mailbox, if the mailbox is not full, the mail will be copied to the mailbox. If the mailbox is full, the thread sending the mail can set a timeout and choose to wait and suspend or return directly -RT_EFULL. If the thread sending the mail chooses to suspend and wait, then when the mails in the mailbox are received and space is left open again, the thread sending the mail will be awakened and will continue to send.

When a thread receives a mail from a mailbox, if the mailbox is empty, the thread receiving the mail can choose whether to set a timeout or wait and suspend until a new mail is received to be awakened. When the set timeout is up and the mailbox still hasn't received the message, the thread that chose to wait for the timeout will be awakened and return -RT_ETIMEOUT. If there are mails in the mailbox, then the thread receiving the mail will copy the 4-byte mail in the mailbox to the receiving cache.

FIGURE 7.1   Mailbox working mechanism diagram.

MAILBOX CONTROL BLOCK

In RT-Thread, the mailbox control block is a data structure used by the operating system to manage mailboxes, represented by the structure `struct rt _ mailbox`. Another C expression, `rt _ mailbox _ t`, represents the handle of the mailbox, and the implementation in C language is a pointer to the mailbox control block. See the following code for a detailed definition of the mailbox control block structure:

```
struct rt_mailbox
{
    struct rt_ipc_object parent;

    rt_uint32_t* msg_pool;          /* the start address of the mailbox
buffer */
    rt_uint16_t size;               /* the size of the mailbox buffer    */

    rt_uint16_t entry;                /* the number of mail in the mailbox
*/
    rt_uint16_t in_offset, out_offset;  /* the entry and exit pointer of
the mailbox buffer    */
    rt_list_t suspend_sender_thread;   /* send the suspend and wait queue
of the thread */
};
typedef struct rt_mailbox* rt_mailbox_t;
```

The `rt _ mailbox` object is derived from `rt _ ipc _ object` and is managed by the IPC container.

MANAGEMENT OF MAILBOX

The mailbox control block is a structure that contains important parameters related to the mailbox, and it plays an important role in the function implementation of the mailbox. The relevant interfaces of the mailbox are shown in Figure 7.2. Mailbox operations include create/initiate a mailbox, send mail, receive mail, and delete/detach a mailbox.

Create and Delete a Mailbox

To create a mailbox object dynamically, call the following function interface:

```
rt_mailbox_t rt_mb_create (const char* name, rt_size_t size, rt_uint8_t
flag);
```

When a mailbox object is created, a mailbox object is first allocated from the object manager, and then a memory space is dynamically allocated for the mailbox to store the mail. The size of the memory is equal to the product of the mail size (4 bytes) and the mailbox capacity. Then initialize

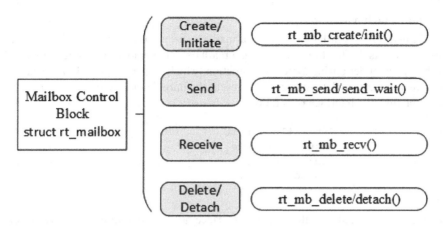

FIGURE 7.2 Mailbox-related interface.

the number of incoming mails and the offset of the outgoing mail in the mailbox. Table 7.1 describes the input parameters and return values for this function.

TABLE 7.1
rt_mb_create

Parameters	Description
name	The name of the mailbox
size	Mailbox capacity
flag	The mailbox flag, which can take the following values: RT_IPC_FLAG_FIFO or RT_IPC_FLAG_PRIO
Return	— —
RT_NULL	Creation failed
The handle of the mailbox object	Creation successful

When a mailbox created with rt_mb_create() is no longer used, it should be deleted to release the corresponding system resources. Once the operation is completed, the mailbox will be permanently deleted. The function interface for deleting a mailbox is as follows:

```
rt_err_t rt_mb_delete (rt_mailbox_t mb);
```

When deleting a mailbox, if a thread is suspended on the mailbox object, the kernel first wakes up all threads suspended on the mailbox (the thread return value is -RT_ERROR), then releases the memory used by the mailbox, and finally deletes the mailbox object. Table 7.2 describes the input parameters and return values for this function.

TABLE 7.2
rt_mb_delete

Parameters	Description
mb	The handle of the mailbox object
Return	— —
RT_EOK	Successful

Initialize and Detach a Mailbox

Initializing a mailbox is similar to creating a mailbox, except that the mailbox initialized is for static mailbox objects. Different from creating a mailbox, the memory of a static mailbox object is allocated by the compiler during system compilation, which is usually placed in a read-write data segment or an uninitialized data segment. The rest of the initialization is the same as the creation of a mailbox. The function interface is as follows:

```
rt_err_t rt_mb_init(rt_mailbox_t mb,
                    const char* name,
                    void* msgpool,
                    rt_size_t size,
                    rt_uint8_t flag)
```

When the mailbox is initialized, the function interface needs to obtain the mailbox object control block that the user has applied for, the pointer of the buffer, the mailbox name, and mailbox capacity (the number of mails that can be stored). Table 7.3 describes the input parameters and return values for this function.

TABLE 7.3
rt_mb_init

Parameters	Description
mb	The handle of the mailbox object
name	Mailbox name
msgpool	Buffer pointer
size	Mailbox capacity
flag	The mailbox flag, which can take the following values: RT_IPC_FLAG_FIFO or RT_IPC_FLAG_PRIO
Return	— —
RT_EOK	Successful

The size parameter here specifies the capacity of the mailbox. If the number of bytes in the buffer pointed to by msgpool is N, then the mailbox capacity should be $N/4$.

Detaching the mailbox means to detach the statically initialized mailbox objects from the kernel object manager. Use the following interface to detach the mailbox:

```
rt_err_t rt_mb_detach(rt_mailbox_t mb);
```

After using this function interface, the kernel wakes up all the threads suspended on the mailbox (the threads return -RT_ERROR) and then detaches the mailbox objects from the kernel object manager. Table 7.4 describes the input parameters and return values for this function.

TABLE 7.4
rt_mb_detach

Parameters	Description
mb	The handle of the mailbox object
Return	— —
RT_EOK	Successful

Send Mail

The thread or ISR can send mail to other threads through the mailbox. The function interface of sending mails is as follows:

```
rt_err_t rt_mb_send (rt_mailbox_t mb, rt_uint32_t value);
```

The mail sent can be any data that is 32-bit formatted, such as an integer value, or a pointer pointing to the buffer. When the mailbox is fully filled with mail, the thread or ISR sending the mail will receive a return value of -RT_EFULL. Table 7.5 describes the input parameters and return values for this function.

TABLE 7.5
rt_mb_send

Parameters	Description
mb	The handle of the mailbox object
value	Content of mail
Return	— —
RT_EOK	Sent successfully
-RT_EFULL	The mailbox is filled

Send Mails with Waiting

Users can also send mails to a specified mailbox through the following function interface:

```
rt_err_t rt_mb_send_wait (rt_mailbox_t mb,
                rt_uint32_t value,
                rt_int32_t timeout);
```

The difference between rt_mb_send_wait() and rt_mb_send() is that there is a waiting time. If the mailbox is full, the thread sending the mail will wait for the mailbox to release space as mails are received according to the set timeout parameter. If the timeout is expired and there is still no available space, then the thread sending the mail will wake up and return an error code. Table 7.6 describes the input parameters and return values for this function.

TABLE 7.6
rt_mb_send_wait

Parameters	Description
mb	The handle of the mailbox object
value	Mail content
timeout	Timeout
Return	— —
RT_EOK	Sent successfully
-RT_ETIMEOUT	Timeout
-RT_ERROR	Failed, return error

Receive Mails

Only when there is mail in the mailbox can the recipient receive the mail immediately and return RT_EOK. Otherwise, the thread receiving the mail will suspend on the waiting thread queue of the mailbox or return directly according to the set timeout. The receiving mail function interface is as follows:

```
rt_err_t rt_mb_recv (rt_mailbox_t mb, rt_uint32_t* value, rt_int32_t
timeout);
```

When receiving a mail, the recipient needs to specify the mailbox handle and specify the location to store the received mail and the maximum timeout that it can wait. If a timeout is set at the time of receiving, -RT_ETIMEOUT will be returned when the mail has not been received within the specified time. Table 7.7 describes the input parameters and return values for this function.

TABLE 7.7
rt_mb_recv

Parameters	Description
mb	The handle of the mailbox object
value	Mail content
timeout	Timeout
Return	— —
RT_EOK	Sent successfully
-RT_ETIMEOUT	Timeout
-RT_ERROR	Failed, return error

MAILBOX USAGE SAMPLE

This is a mailbox application routine that initializes two static threads, one static mailbox object, one of the threads sends mail to the mailbox, and one thread receives mail from the mailbox, as shown in Code Listing 7.1.

Code listing 7.1 Mailbox Usage Routine

```
#include <rtthread.h>

#define THREAD_PRIORITY      10
#define THREAD_TIMESLICE     5

/* Mailbox control block */
static struct rt_mailbox mb;
/* Memory pool for mails storage */
static char mb_pool[128];

static char mb_str1[] = "I'm a mail!";
static char mb_str2[] = "this is another mail!";
static char mb_str3[] = "over";

ALIGN(RT_ALIGN_SIZE)
static char thread1_stack[1024];
static struct rt_thread thread1;

/* Thread 1 entry */
static void thread1_entry(void *parameter)
{
    char *str;
```

```
    while (1)
    {
        rt_kprintf("thread1: try to recv a mail\n");

        /* Receive mail from the mailbox */
        if (rt_mb_recv(&mb, (rt_uint32_t *)&str, RT_WAITING_FOREVER) ==
RT_EOK)
        {
            rt_kprintf("thread1: get a mail from mailbox, the content:%s\n",
str);

            if (str == mb_str3)
                break;

            /* Delay 100ms */
            rt_thread_mdelay(100);
        }
    }
    /* Executing the mailbox object detachment */
    rt_mb_detach(&mb);
}

ALIGN(RT_ALIGN_SIZE)
static char thread2_stack[1024];
static struct rt_thread thread2;

/* Thread 2 entry*/
static void thread2_entry(void *parameter)
{
    rt_uint8_t count;

    count = 0;
    while (count < 10)
    {
        count ++;
        if (count & 0x1)
        {
            /* Send the mb_str1 address to the mailbox */
            rt_mb_send(&mb, (rt_uint32_t)&mb_str1);
        }
        else
        {
            /* Send the mb_str2 address to the mailbox */
            rt_mb_send(&mb, (rt_uint32_t)&mb_str2);
        }

        /* Delay 200ms */
        rt_thread_mdelay(200);
    }

    /* Send mail to inform thread 1 that thread 2 has finished running */
    rt_mb_send(&mb, (rt_uint32_t)&mb_str3);
}

int mailbox_sample(void)
{
    rt_err_t result;
```

```
    /* Initialize a mailbox */
    result = rt_mb_init(&mb,
                        "mbt",                    /* Name is mbt */
                        &mb_pool[0],              /* The memory pool used by
the mailbox is mb_pool */
                        sizeof(mb_pool) / 4,    /* The number of mails in
the mailbox, because a mail occupies 4 bytes */
                        RT_IPC_FLAG_FIFO);      /* Thread waiting in FIFO
approach */
    if (result != RT_EOK)
    {
        rt_kprintf("init mailbox failed.\n");
        return -1;
    }

    rt_thread_init(&thread1,
                   "thread1",
                   thread1_entry,
                   RT_NULL,
                   &thread1_stack[0],
                   sizeof(thread1_stack),
                   THREAD_PRIORITY, THREAD_TIMESLICE);
    rt_thread_startup(&thread1);

    rt_thread_init(&thread2,
                   "thread2",
                   thread2_entry,
                   RT_NULL,
                   &thread2_stack[0],
                   sizeof(thread2_stack),
                   THREAD_PRIORITY, THREAD_TIMESLICE);
    rt_thread_startup(&thread2);
    return 0;
}

/* Export to the msh command list */
MSH_CMD_EXPORT(mailbox_sample, mailbox sample);
```

The results are as follows:

```
 \ | /
- RT -     Thread Operating System
 / | \     3.1.0 build Aug 27 2018
 2006 - 2018 Copyright by rt-thread team

msh >mailbox_sample
thread1: try to recv a mail
thread1: get a mail from mailbox, the content:I'm a mail!
msh >thread1: try to recv a mail
thread1: get a mail from mailbox, the content:this is another mail!
…
thread1: try to recv a mail
thread1: get a mail from mailbox, the content:this is another mail!
thread1: try to recv a mail
thread1: get a mail from mailbox, the content:over
```

The routine demonstrates how to use the mailbox. Thread 2 sends the mails for a total of 11 times; thread 1 receives the mail for a total of 11 mails, prints the contents of the mails, and terminates.

Occasions to Use Mailbox

A mailbox is a simple inter-thread messaging method, which is characterized by low overhead and high efficiency. In the implementation of the RT-Thread operating system, a 4-byte message can be delivered at a time, and the mailbox has certain storage capabilities, which can cache a certain number of messages (the number of messages is determined by the capacity specified when creating and initializing the mailbox). The maximum length of a message in a mailbox is 4 bytes, so the mailbox can be used for messages less than 4 bytes. Since on the 32-bit system, 4 bytes can be placed right on a pointer, when a larger message needs to be transferred between threads, a pointer pointing to a buffer can be sent as a mail to the mailbox, which means the mailbox can also deliver a pointer. For example:

```
struct msg
{
    rt_uint8_t *data_ptr;
    rt_uint32_t data_size;
};
```

For such a message structure, it contains a pointer pointing to data data _ ptr and a variable data _ size of the length of the data block. When a thread needs to send this message to another thread, the following operations can be used:

```
struct msg* msg_ptr;

msg_ptr = (struct msg*)rt_malloc(sizeof(struct msg));
msg_ptr->data_ptr = ...;  /* Point to the corresponding data block address
*/
msg_ptr->data_size = len; /* Length of data block */
/* Send this message pointer to the mb mailbox */
rt_mb_send(mb, (rt_uint32_t)msg_ptr);
```

When receiving the thread, because the pointer is what's being received, and msg_ptr is a newly allocated memory block, after the thread receiving the message finishes processing, the corresponding memory block needs to be released:

```
struct msg* msg_ptr;
if (rt_mb_recv(mb, (rt_uint32_t*)&msg_ptr) == RT_EOK)
{
    /* After the thread receiving the message finishes processing, the
corresponding memory block needs to be released: */
    rt_free(msg_ptr);
}
```

MESSAGE QUEUE

Message queue is another commonly used inter-thread communication method, which is an extension of the mailbox. It can be used on a variety of occasions, like message exchange between threads, use the serial port to receive variable-length data, etc.

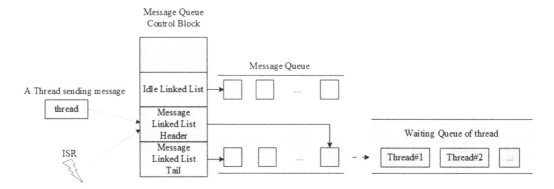

FIGURE 7.3   Message queue working mechanism diagram.

MESSAGE QUEUE WORKING MECHANISM

The message queue can receive messages with an unfixed length from threads or ISR and cache messages in their own memory space. Other threads can also read the corresponding message from the message queue, and when the message queue is empty, the thread reading the messages can be suspended. When a new message arrives, the suspended thread will be awakened to receive and process the message. A message queue is an asynchronous way of communication.

As shown in Figure 7.3, a thread or interrupt service routine can put one or more messages into the message queue. Similarly, one or more threads can also get messages from the message queue. When multiple messages are sent to the message queue, the message that enters the message queue first is passed to the thread. That is, the thread always gets the message first in the message queue—in other words, it follows the first-in/first-out principle (FIFO).

The message queue object of the RT-Thread operating system consists of multiple elements. When a message queue is created, it is assigned a message queue control block: name, memory buffer, message size, and queue length. At the same time, each message queue object contains multiple message boxes, and each message box can store one message. The first and last message boxes in the message queue are, respectively, called the message linked list header and the message linked list tail, corresponding to msg _ queue _ head and msg _ queue _ tail in the queue control block. Some message boxes may be empty; they form a linked list of idle message boxes via msg _ queue _ free. The total number of message boxes in all message queues is the length of the message queue, which can be specified when creating the message queue.

MESSAGE QUEUE CONTROL BLOCK

In RT-Thread, a message queue control block is a data structure used by the operating system to manage message queues, represented by the structure struct rt _ messagequeue. Another C expression, rt _ mq _ t, represents the handle of the message queue. The implementation of this handle in C language is a pointer to the message queue control block. See the following code for a detailed definition of the message queue control block structure:

```
struct rt_messagequeue
{
    struct rt_ipc_object parent;

    void* msg_pool;          /* Pointer pointing to the buffer storing the
messages */
```

```
    rt_uint16_t msg_size;      /* The length of each message */
    rt_uint16_t max_msgs;      /* Maximum number of messages that can be
stored */

    rt_uint16_t entry;         /* Number of messages already in the queue */

    void* msg_queue_head;      /* Message linked list header  */
    void* msg_queue_tail;      /* Message linked list tail   */
    void* msg_queue_free;      /* Idle message linked list */
};
typedef struct rt_messagequeue* rt_mq_t;
```

The rt_messagequeue object is derived from rt_ipc_object and is managed by the IPC container.

MANAGEMENT OF MESSAGE QUEUE

The message queue control block is a structure that contains important parameters related to the message queue and plays an important role in the implementation of functions of the message queue. The relevant interfaces of the message queue are shown in Figure 7.4. Operations on a message queue include create a message queue, send a message, receive a message, and delete a message queue.

Create and Delete a Message Queue

The message queue should be created before it is used, or the existing static message queue objects should be initialized. The function interface for creating the message queue is as follows:

```
rt_mq_t rt_mq_create(const char* name, rt_size_t msg_size,
                rt_size_t max_msgs, rt_uint8_t flag);
```

When creating a message queue, first allocate a message queue object from the object manager, then allocate memory space to the message queue object forming an idle message linked list. The size of this memory = [message size + message header (for linked list connection) size] × the number of messages in the message queue. Then initialize the message queue, at which point the message queue is empty. Table 7.8 describes the input parameters and return values for this function.

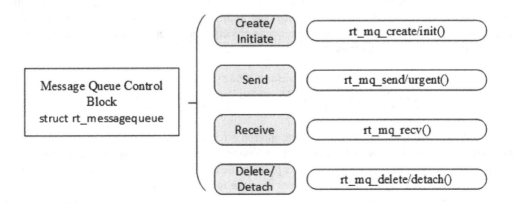

FIGURE 7.4   Message queue–related interfaces.

TABLE 7.8
rt_mq_create

Parameters	Description
name	The name of the message queue
msg_size	The maximum length of a message in the message queue, in bytes
max_msgs	The number of messages in the message queue
flag	The waiting method used by the message queue, which can take the following values: RT_IPC_FLAG_FIFO or RT_IPC_FLAG_PRIO
Return	— —
The handle of the message queue object	Successful
RT_NULL	Fail

When the message queue is no longer in use, it should be deleted to free up system resources, and once the operation is complete, the message queue will be permanently deleted. The function interface for deleting the message queue is as follows:

```
rt_err_t rt_mq_delete(rt_mq_t mq);
```

When deleting a message queue, if a thread is suspended on the message queue's waiting queue, the kernel wakes up all the threads suspended on the message waiting queue (the thread returns the value -RT_ERROR) and then releases the memory used by the message queue. Finally, delete the message queue object. Table 7.9 describes the input parameters and return values for this function.

TABLE 7.9
rt_mq_delete

Parameters	Description
mq	The handle of the message queue object
Return	— —
RT_EOK	Successful

Initialize and Detach a Message Queue

Initializing a static message queue object is similar to creating a message queue object, except that the memory of a static message queue object is allocated by the compiler during system compilation and is typically placed in a read data segment or an uninitialized data segment. Initialization is required before using such static message queue objects. The function interface for initializing a message queue object is as follows:

```
rt_err_t rt_mq_init(rt_mq_t mq, const char* name,
               void *msgpool, rt_size_t msg_size,
               rt_size_t pool_size, rt_uint8_t flag);
```

When the message queue is initialized, the interface requires the handle of the message queue object that the user has requested (i.e., a pointer pointing to the message queue object control block), the message queue name, the message buffer pointer, the message size, and the message queue buffer size. As shown in Figure 7.3, all messages after the message queue is initialized are suspended on the idle message list, and the message queue is empty. Table 7.10 describes the input parameters and return values for this function.

TABLE 7.10
rt_mq_init

Parameters	Description
mq	The handle of the message queue object
name	The name of the message queue
msgpool	Pointer pointing to the buffer storing the messages
msg_size	The maximum length of a message in the message queue, in bytes
pool_size	The buffer size for storing messages
flag	The waiting method used by the message queue, which can take the following values: RT_IPC_FLAG_FIFO or RT_IPC_FLAG_PRIO
Return	— —
RT_EOK	Successful

Detaching the message queue will cause the message queue object to be detached from the kernel object manager. The following interface is used to detach from the message queue:

```
rt_err_t rt_mq_detach(rt_mq_t mq);
```

After using this function interface, the kernel wakes up all threads suspended on the message waiting queue object (the thread return value is -RT_ERROR) and then detaches the message queue object from the kernel object manager. Table 7.11 describes the input parameters and return values for this function.

TABLE 7.11
rt_mq_detach

Parameters	Description
mq	The handle of the message queue object
Return	— —
RT_EOK	Successful

Send a Message

A thread or ISR can send a message to the message queue. When sending a message, the message queue object first takes an idle message block from the idle message list, then copies the content of the message sent by the thread or the interrupt service program to the message block, and then suspends the message block to the end of the message queue. The sender can successfully send a message if and only if there is an idle message block available on the idle message list; when there is no message block available on the idle message list, it means that the message queue is full; at this time, the thread or interrupt program that sent the message will receive an error code (-RT_EFULL). The function interface for sending messages is as follows:

```
rt_err_t rt_mq_send (rt_mq_t mq, void* buffer, rt_size_t size);
```

When sending a message, the sender specifies the object handle of the sent message queue (i.e., a pointer pointing to the message queue control block) and specifies the content of the message being sent and the size of the message. As shown in Figure 7.3, after sending a normal message, the first message on the idle message list is transferred to the end of the message queue. Table 7.12 describes the input parameters and return values for this function.

TABLE 7.12
rt_mq_send

Parameter	Description
mq	The handle of the message queue object
buffer	Message content
size	Message size
Return	— —
RT_EOK	Successful
-RT_EFULL	Message queue is full
-RT_ERROR	Failed, indicating that the length of the sent message is greater than the maximum length of the message in the message queue

Send an Emergency Message

The process of sending an emergency message is almost the same as sending a message. The only difference is that when an emergency message is sent, the message block taken from the idle message list is not put at the end of the message queue, but at the head of the queue. The recipient can receive the emergency message preferentially, so that the message can be processed in time. The function interface for sending an emergency message is as follows:

```
rt_err_t rt_mq_urgent(rt_mq_t mq, void* buffer, rt_size_t size);
```

Table 7.13 describes the input parameters and return values for this function.

TABLE 7.13
rt_mq_urgent

Parameters	Description
mq	The handle of the message queue object
buffer	Message content
size	Message size
Return	— —
RT_EOK	Successful
-RT_EFULL	Message queue is full
-RT_ERROR	Fail

Receive a Message

Only when there is a message in the message queue can the receiver receive the message; otherwise, the receiver will suspend on the waiting queue of the message queue or return directly—it depends on the timeout setting. The function interface to receive a message is as follows:

```
rt_err_t rt_mq_recv (rt_mq_t mq, void* buffer,
                rt_size_t size, rt_int32_t timeout);
```

When receiving a message, the receiver needs to specify the message queue object handle that stores the message and specify a memory buffer into which the contents of the received message will be copied. Besides, the timeout when the message is not received in time needs to be set. As shown in Figure 7.3, the message on the front of the message queue is transferred to the end of the idle message linked list after receiving a message. Table 7.14 describes the input parameters and return values for this function.

TABLE 7.14
rt_mq_recv

Parameters	Description
Mq	The handle of the message queue object
buffer	Message content
size	Message size
timeout	Specified timeout
Return	— —
RT_EOK	Received successfully
-RT_ETIMEOUT	Timeout
-RT_ERROR	Fail, return error

MESSAGE QUEUE APPLICATION EXAMPLE

This is a message queue application routine. Two static threads are initialized in the routine: one thread will receive messages from the message queue, and another thread will periodically send regular messages and emergency messages to the message queue, as shown in Code Listing 7.2.

Code listing 7.2 Message Queue Usage Routine

```
#include <rtthread.h>

/* Message queue control block */
static struct rt_messagequeue mq;
/* The memory pool used to place messages in the message queue */
static rt_uint8_t msg_pool[2048];

ALIGN(RT_ALIGN_SIZE)
static char thread1_stack[1024];
static struct rt_thread thread1;
/* Thread 1 entry function */
static void thread1_entry(void *parameter)
{
    char buf = 0;
    rt_uint8_t cnt = 0;

    while (1)
    {
        /* Receive messages from the message queue */
        if (rt_mq_recv(&mq, &buf, sizeof(buf), RT_WAITING_FOREVER) ==
RT_EOK)
        {
            rt_kprintf("thread1: recv msg from msg queue, the content:%c\n",
buf);
            if (cnt == 19)
            {
                break;
            }
        }
        /* Delay 50ms */
        cnt++;
        rt_thread_mdelay(50);
    }
```

```
    rt_kprintf("thread1: detach mq \n");
    rt_mq_detach(&mq);
}


ALIGN(RT_ALIGN_SIZE)
static char thread2_stack[1024];
static struct rt_thread thread2;
/* Thread 2 entry */
static void thread2_entry(void *parameter)
{
    int result;
    char buf = 'A';
    rt_uint8_t cnt = 0;

    while (1)
    {
        if (cnt == 8)
        {
            /* Send emergency message to the message queue */
            result = rt_mq_urgent(&mq, &buf, 1);
            if (result != RT_EOK)
            {
                rt_kprintf("rt_mq_urgent ERR\n");
            }
            else
            {
                rt_kprintf("thread2: send urgent message - %c\n", buf);
            }
        }
        else if (cnt>= 20)/* Exit after sending 20 messages */
        {
            rt_kprintf("message queue stop send, thread2 quit\n");
            break;
        }
        else
        {
            /* Send a message to the message queue */
            result = rt_mq_send(&mq, &buf, 1);
            if (result != RT_EOK)
            {
                rt_kprintf("rt_mq_send ERR\n");
            }

            rt_kprintf("thread2: send message - %c\n", buf);
        }
        buf++;
        cnt++;
        /* Delay 5ms */
        rt_thread_mdelay(5);
    }
}

/* Initialization of the message queue example */
int msgq_sample(void)
{
    rt_err_t result;
```

```
    /* Initialize the message queue */
    result = rt_mq_init(&mq,
                     "mqt",
                     &msg_pool[0],          /* Memory pool points to
msg_pool */
                     1,                     /* The size of each message
is 1 byte */
                     sizeof(msg_pool),      /* The size of the memory
pool is the size of msg_pool */
                     RT_IPC_FLAG_FIFO);   /* If there are multiple
threads waiting, assign messages in first come first get mode. */

    if (result != RT_EOK)
    {
        rt_kprintf("init message queue failed.\n");
        return -1;
    }

    rt_thread_init(&thread1,
                   "thread1",
                   thread1_entry,
                   RT_NULL,
                   &thread1_stack[0],
                   sizeof(thread1_stack), 25, 5);
    rt_thread_startup(&thread1);

    rt_thread_init(&thread2,
                   "thread2",
                   thread2_entry,
                   RT_NULL,
                   &thread2_stack[0],
                   sizeof(thread2_stack), 25, 5);
    rt_thread_startup(&thread2);

    return 0;
}

/* Export to the msh command list */
MSH_CMD_EXPORT(msgq_sample, msgq sample);
```

The results are as follows:

```
\ | /
- RT -     Thread Operating System
 / | \     3.1.0 build Aug 24 2018
 2006 - 2018 Copyright by rt-thread team

msh > msgq_sample
msh >thread2: send message - A
thread1: recv msg from msg queue, the content:A
thread2: send message - B
thread2: send message - C
thread2: send message - D
thread2: send message - E
thread1: recv msg from msg queue, the content:B
thread2: send message - F
thread2: send message - G
```

```
thread2: send message - H
thread2: send urgent message - I
thread2: send message - J
thread1: recv msg from msg queue, the content:I
thread2: send message - K
thread2: send message - L
thread2: send message - M
thread2: send message - N
thread2: send message - O
thread1: recv msg from msg queue, the content:C
thread2: send message - P
thread2: send message - Q
thread2: send message - R
thread2: send message - S
thread2: send message - T
thread1: recv msg from msg queue, the content:D
message queue stop send, thread2 quit
thread1: recv msg from msg queue, the content:E
thread1: recv msg from msg queue, the content:F
thread1: recv msg from msg queue, the content:G
...
thread1: recv msg from msg queue, the content:T
thread1: detach mq
```

The routine demonstrates how to use the message queue. Thread 1 receives messages from the message queue; thread 2 periodically sends regular and emergency messages to the message queue. Since the message "I" that thread 2 sent is an emergency message, it will be inserted directly into the front of the message queue. So after receiving the message "B," thread 1 receives the emergency message and then receives the message "C."

OCCASIONS TO USE MESSAGE QUEUE

A message queue can be used where occasional long messages are sent, including exchanging message between threads and sending messages to threads in interrupt service routines (interrupt service routines cannot receive messages). The following sections describe the use of message queues from two perspectives: sending messages and synchronizing messages.

Sending Messages

The obvious difference between a message queue and a mailbox is that the length of the message is not limited to 4 bytes. In addition, the message queue includes a function interface for sending emergency messages. But when you create a message queue with a maximum length of 4 bytes for all messages, the message queue object will be reduced to a mailbox. This unlimited length message is reflected in the code, as follows:

```
struct msg
{
    rt_uint8_t *data_ptr;     /* Data block starting address */
    rt_uint32_t data_size;    /* Data block size   */
};
```

Similar to the mailbox example is the message structure definition. Now, let's assume that such a message needs to be sent to a thread receiving messages. In the mailbox example, this structure can only send pointers pointing to this structure (after the function pointer is sent, the thread receiving messages can access the content pointing to this address; usually this piece of data needs to be left to the thread receiving the messages to release). How message queues are used is quite different:

```
void send_op(void *data, rt_size_t length)
{
    struct msg msg_ptr;

    msg_ptr.data_ptr = data;  /* Point to the corresponding data block
address */
    msg_ptr.data_size = length; /* Datablock length */

    /* Send this message pointer to the mq message queue */
    rt_mq_send(mq, (void*)&msg_ptr, sizeof(struct msg));
}
```

Note that in the above code, the data content of a local variable is sent to the message queue. In the thread that receives the message, the same structure that receives messages using local variables is used:

```
void message_handler()
{
    struct msg msg_ptr; /* Local variable used to place the message */

    /* Receive messages from the message queue into msg_ptr */
    if (rt_mq_recv(mq, (void*)&msg_ptr, sizeof(struct msg)) == RT_EOK)
    {
        /* Successfully received the message, corresponding data processing
is performed */
    }
}
```

Because the message queue is a direct copy of data content, in the earlier example, the local structure is used to save the message structure, which eliminates the trouble of dynamic memory allocation (and no need to worry, because message memory space has already been released when the thread receives the message).

Synchronizing Messages

In general system design, the problem of sending synchronous messages is often encountered. At this time, corresponding implementations can be selected according to the state of the time: two threads can be implemented in the form of [message queue + semaphore or mailbox]. The thread sending messages sends the corresponding message to the message queue in the form of message sending. After the message is sent, it is expecting to receive the confirmation from the threads receiving messages. The working diagram is shown in Figure 7.5.

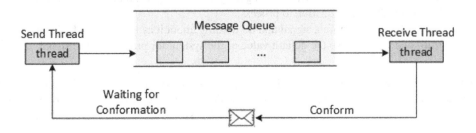

FIGURE 7.5 Synchronizing messages diagram.

Depending on the message confirmation, the message structure can be defined as:

```
struct msg
{
    /* Other members of the message structure */
    struct rt_mailbox ack;
};
/* or */
struct msg
{
    /* Other members of the message structure */
    struct rt_semaphore ack;
};
```

The first type of message uses a mailbox as a confirmation flag, while the second type of message uses a semaphore as a confirmation flag. The mailbox is used as a confirmation flag, which means that the receiving thread can notify some status values to the thread sending messages; the semaphore is used as a confirmation flag that can only notify the thread sending messages in a single way, and the message has been confirmed to be received.

SIGNAL

A signal (also known as a soft interrupt signal), from a software perspective, is a simulation of an interrupt mechanism. In principle, a thread receiving a signal is similar to the processor receiving an interrupt request.

SIGNAL WORKING MECHANISM

Signals are used for asynchronous communication in RT-Thread. The POSIX standard defines that sigset_t type defines a signal set. However, the sigset_t type may be defined differently in different systems. In RT-Thread, sigset_t is defined as unsigned long and named rt_sigset_t; the signals that the application can use are SIGUSR1(10) and SIGUSR2(12).

The essence of the signal is a soft interrupt, which is used to notify the thread that an asynchronous event has occurred. It is used to notify of an abnormality and handle emergencies between threads. No operation is needed for a thread to wait for the signal's arrival. In fact, the thread does not know when the signal will arrive. Threads can send soft interrupt signals by calling rt _ thread _ kill() between each other.

The thread that receives the signals has different processing methods for various signals, and these methods can be divided into three categories:

1. The first one is an interrupt-like processing program. For signals that need to be processed, the thread can specify a function to process.
2. The second one is to ignore a signal and do nothing about it as if it had not happened.
3. The third one is to reserve the default value of the system for processing the signal.

As shown in Figure 7.6, suppose that thread 1 needs to process the signal. First, thread 1 installs a signal and unmasks it. At the same time, it sets the approach of how to process abnormality of signals. Then other threads can send a signal to thread 1, triggering thread 1 to process the signal.

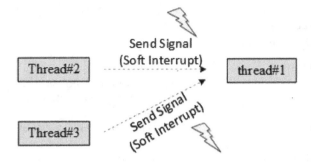

FIGURE 7.6 Signal working mechanism.

When the signal is passed to thread 1, if thread 1 is suspended, thread 1 will be switched to a ready state to process the corresponding signal. If thread 1 is running, thread 1 will create a new stack frame space on its current thread stack to process the corresponding signal. It should be noted that the thread stack size used will also increase accordingly.

MANAGEMENT OF SIGNAL

Operations on signals include install a signal, block a signal, unblock a signal, send a signal, and wait for a signal. The interfaces of the signal are shown in Figure 7.7.

Install a Signal

If the thread is to process a signal, then the signal needs to be installed in the thread. Signal installation is primarily used to determine the mapping relation between the signal value and the actions of the thread on that signal value—that is, what signal is to be processed and what actions will be taken when the signal is passed to the thread. See the following code for a detailed definition:

```
rt_sighandler_t rt_signal_install(int signo, rt_sighandler_t[] handler);
```

rt _ sighandler _ t is the type of function pointer that defines the signal process function. Table 7.15 describes the input parameters and return values for this function.

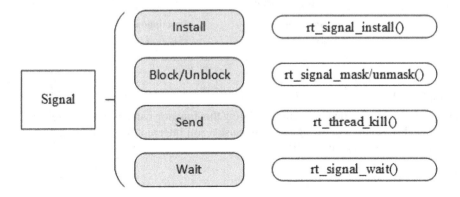

FIGURE 7.7 Signal related interface.

TABLE 7.15
rt_signal_install

Parameters	Description
signo	Signal value (only SIGUSR1 and SIGUSR2 are open to the user, the same applies below)
handler	Set the approach to process signal values
Return	— —
SIG_ERR	Wrong signal
The handler value before the signal is installed	Successful

Setting the handler parameter during signal installation determines the different processing methods for this signal. Processing methods can be divided into three types:

1. Similar to the interrupt processing method, the parameter points to the user-defined processing function when the signal occurs and is processed by the function.
2. The parameter is set to SIG_IGN. Ignore a signal, and do nothing on the signal as if it had not happened.
3. The parameter is set to SIG_DFL, and the system will call the default function _signal_default_handler() to process.

Block a Signal

Blocking a signal can also be understood as shielding signals. If the signal is blocked, the signal will not be delivered to the thread that installed the signal and will not cause soft interrupt processing. Call rt_signal_mask() to block the signal:

```
void rt_signal_mask(int signo);
```

Table 7.16 describes the input parameters for this function.

TABLE 7.16
rt_signal_mask

Parameters	Description
signo	Signal value

Unblock a Signal

Several signals can be installed in the thread. Using this function can give some "attention" to some of the signals; then sending these signals will cause a soft interrupt for the thread. Call rt_signal_unmask() to unblock the signal as follows:

```
void rt_signal_unmask(int signo);
```

Table 7.17 describes the input parameters for this function.

TABLE 7.17
rt_signal_unmask

Parameters	Description
signo	Signal value

Send Signals

When we need to process an abnormality, a signal can be sent to the relevant thread. Call rt_thread_kill() to send a signal to thread:

```
int rt_thread_kill(rt_thread_t tid, int sig);
```

Table 7.18 describes the input parameters and return values for this function.

TABLE 7.18
rt_thread_kill

Parameters	Description
tid	The thread receiving the signal
sig	Signal value
Return	— —
RT_EOK	Sent successfully
-RT_EINVAL	Parameter error

Wait for a Signal

When waiting for a signal, if it does not arrive, suspend the thread until the signal does arrive or wait until the specified timeout is exceeded. If the signal arrived, the pointer pointing to the signal body is stored in si. The function to wait for the signal is as follows:

```
int rt_signal_wait(const rt_sigset_t *set,
                   rt_siginfo_t[] *si, rt_int32_t timeout);
```

rt_siginfo_t is the data type that defines the signal information. Table 7.19 describes the input parameters and return values of the function.

TABLE 7.19
rt_signal_wait

Parameters	Description
set	Specify the signal to wait
si	Pointer pointing to signal information
timeout	Waiting time
Return	— —
RT_EOK	Signal arrives
-RT_ETIMEOUT	Timeout
-RT_EINVAL	Parameter error

SIGNAL APPLICATION EXAMPLE

Here is a signal application routine, as shown in Code Listing 7.3. This routine creates one thread. When the signal is installed, the signal processing mode is set to custom processing. The processing function of the signal is defined to be thread1_signal_handler(). After the thread is running and the signal is installed, a signal is sent to this thread. This thread will receive the signal and print the message.

Code listing 7.3 Signal Usage Routine

```
#include <rtthread.h>

#define THREAD_PRIORITY         25
#define THREAD_STACK_SIZE       512
#define THREAD_TIMESLICE        5

static rt_thread_t tid1 = RT_NULL;

/* Signal process function for thread 1 signal handler */
void thread1_signal_handler(int sig)
{
    rt_kprintf("thread1 received signal %d\n", sig);
}

/* Entry function for thread 1 */
static void thread1_entry(void *parameter)
{
    int cnt = 0;

    /* Install signal */
    rt_signal_install(SIGUSR1, thread1_signal_handler);
    rt_signal_unmask(SIGUSR1);

    /* Run for 10 times */
    while (cnt < 10)
    {
        /* Thread 1 runs with low-priority and prints the count value all
through*/
        rt_kprintf("thread1 count : %d\n", cnt);

        cnt++;
        rt_thread_mdelay(100);
    }
}

/* Initialization of the signal example */
int signal_sample(void)
{
    /* Create thread 1 */
    tid1 = rt_thread_create("thread1",
                            thread1_entry, RT_NULL,
                            THREAD_STACK_SIZE,
                            THREAD_PRIORITY, THREAD_TIMESLICE);

    if (tid1 != RT_NULL)
        rt_thread_startup(tid1);
```

```
        rt_thread_mdelay(300);

        /* Send signal SIGUSR1 to thread 1 */
        rt_thread_kill(tid1, SIGUSR1);

        return 0;
}

/* Export to the msh command list */
MSH_CMD_EXPORT(signal_sample, signal sample);
```

The results are as follows:

```
 \ | /
- RT -      Thread Operating System
 / | \      3.1.0 build Aug 24 2018
 2006 - 2018 Copyright by rt-thread team

msh >signal_sample
thread1 count : 0
thread1 count : 1
thread1 count : 2
msh >thread1 received signal 10
thread1 count : 3
thread1 count : 4
thread1 count : 5
thread1 count : 6
thread1 count : 7
thread1 count : 8
thread1 count : 9
```

In the routine, the thread first installs the signal and unblocks it and then sends a signal to the thread. The thread receives the signal and prints out the signal received: SIGUSR1 (10).

CHAPTER SUMMARY

This chapter described inter-thread communication. Here's a look back at what to note:

1. You can't receive mail in an interrupt, and you can't send mail in an interrupt with a waiting way.
2. You can't receive a message from a message queue in an interrupt.
3. When the signal is installed, the exception handling process of the signal is also set, such as ignoring the signal, processing the system in default, or user-defined functions.

Here's a summary of a few differences between the message queue and the mailbox:

1. The size of each mail is a fixed 4 bytes, but a message queue does not limit the length of each message.
2. The message queue uses the linked list mechanism to send the emergency message (like jump the queue, inserted into the header of the linked list), so that the receiving thread receives this emergency message first.

3. When sending a message pointer with mailbox, the buffer that stores the message should be defined as a global or static form, because the mailbox does not copy the contents of the message. The message queue copies the contents of the message and can send local variables.

4. A message queue is an asynchronous way of communication. If the receiver needs to tell the sender that the message has been received, this can be achieved by the mailbox or semaphore.

To deepen your understanding of the concepts of mailboxes, message queues, and signals, you should understand the routines or expand the functionality you want on your routines. Hands-on experience is valuable in this regard.

8 Memory Management

In a computing system, there are usually two types of memory space: internal memory space and external memory space. The internal memory is quick to access and can be accessed randomly according to the variable address. It is what we usually called RAM (random-access memory) and can be understood as the computer's memory. In the external memory, the content stored is relatively fixed, and the data will not be lost even after the power is turned off. It is what we usually called ROM (read-only memory) and can be understood as the hard disk of the computer.

In a computer system, variables and intermediate data are generally stored in RAM, and they are only transferred from RAM to the CPU for calculation when needed. The memory size required by some data needs to be determined according to the actual situation during the running of the program, which requires the system to have the ability to dynamically manage the memory space. The user asks the system to allocate a block of memory, and then the system selects a suitable memory space to allocate for the user. After the user finishes using it, the memory space is released back to the system, which enables the system to recycle the memory space.

This chapter introduces two kinds of memory management methods in RT-Thread, namely dynamic memory heap management and static memory pool management. After reading this chapter, readers will understand the memory management principle and usage of RT-Thread.

MEMORY MANAGEMENT FEATURES

Since timing is critical in real-time systems, the requirements to the memory management are much higher than those in general-purpose operating systems:

1. Time for allocating memory must be deterministic. The general memory management algorithm is to find a free memory block that fits the size of the data to be stored and then store the data therein. The time it takes to find such a free block of memory is uncertain. But for real-time systems, non-deterministic timing in memory allocation is unacceptable because a real-time system, as the name implies, requires deterministic timing of its task responses and it can't be done without a deterministic memory allocator.
2. As memory is constantly being allocated and released, it will accumulate fragments in some of the memory regions. This is because some of the allocated memory blocks sit in between other free memory blocks and prevent a bigger consecutive memory block from being allocated. Even though there is enough free memory in the system, the blocks are not consecutive and hence cannot form a continuous block of memory for an allocation request. For general-purpose systems, this inappropriate memory allocation algorithm can be solved by a system reboot once in a while. But it is unacceptable for embedded systems that need to work continuously and reliably in the field all the year round.
3. The resource environment of embedded system is also different. Some systems have relatively tight resources; only tens of kilobytes of memory are available for allocation, while some systems have several megabytes of memory. This makes choosing an efficient memory allocation algorithm for these different systems more complicated.

The RT-Thread operating system provides different memory allocation algorithms for memory management according to different upper-layer applications and system resources. There are two categories in general: memory heap and memory pool. A memory heap also has three cases according to specific memory devices:

1. Allocation management for small memory blocks (small memory management algorithm);
2. Allocation management for large memory blocks (slab management algorithm);
3. Allocation management for multiple memory heaps (memheap management algorithm).

MEMORY HEAP

A memory heap is used for managing a contiguous memory space. We introduced the memory distribution of RT-Thread in the chapter "Kernel Basics." As shown in Figure 8.1, RT-Thread uses the space at "the end of the ZI segment" to the end of the memory as the memory heap.

If the current resource allows, memory heap can allocate memory blocks of any size according to the needs of users. When these memory blocks are not needed anymore, they can be released back to the heap for other applications. In order to meet different needs, the RT-Thread system provides different memory management algorithms, including a small memory management algorithm, slab management algorithm, and memheap management algorithm.

The small memory management algorithm is mainly for systems with fewer resources and with less than 2MB of memory. The slab memory management algorithm mainly provides a fast algorithm similar to multiple memory pool management algorithms when the system resources are rich. In addition to the above, RT-Thread has a management algorithm for a multi-memory heap, namely the memheap management algorithm. The memheap management algorithm is suitable for the scenario where there are multiple memory heaps in the system and the user wants to use them as a unified memory heap. It provides a very convenient way for the user to make good use of all the memory space and in the meantime doesn't need to take much care about the details of how the allocations are handled in different heaps. However, RT-Thread doesn't support having more than one heap management algorithm running simultaneously. The interface for these different algorithms is unified, which makes it much easier for the user to fine-tune their application performance by switching between different algorithms.

Notes: The heap memory manager is thread-safe; thus, usually it will require locks internally and will lead to unexpected thread suspension or even deadlock if used in an interrupt service routine (ISR). So please do not use heap memory in ISR.

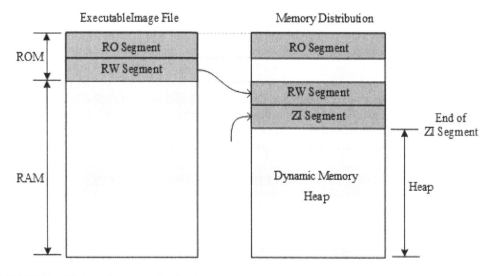

FIGURE 8.1 RT-thread memory distribution.

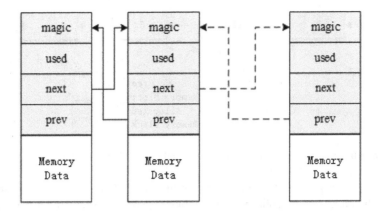

FIGURE 8.2   Small memory management working mechanism diagram.

SMALL MEMORY MANAGEMENT ALGORITHM

The small memory management algorithm is a simple memory allocation algorithm. It is initialized with a continuous trunk of memory, and when a block needs to be allocated, the matching memory block is allocated from a large memory block, and the remaining part is returned to the heap management system. Each memory block has a header containing pointers to the previous and next memory block so that all blocks form a doubly linked list, as shown in Figure 8.2.

Each memory block (whether it is an allocated memory block or a free memory block) contains a data head, including:

1. **magic**: Variable (also called the magic number). It will be initialized to 0x1ea0 (represents the word heap), which is used to mark this memory block as a memory data block for memory management. The variable is not only used to identify that the block is a memory data block for memory management. It is also a memory protection mechanism: if this value is changed, it means that the memory block is illegally overridden (normally only the memory manager will operate on this memory).
2. **used**: Indicates whether the current memory block is in use.

The performance of a memory management algorithm mainly relies on the performance of allocation and release operations. An example of a small memory management algorithm is as follows.

As shown in Figure 8.3, the free list pointer lfree initially points to a 32-byte block of memory. When the user thread needs to allocate a 64-byte memory block, since the memory block pointed to by this lfree pointer is only 32 bytes and does not meet the requirements, the memory manager will continue to search from the next memory block. When the next memory block with 128 bytes

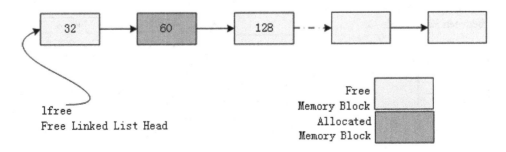

FIGURE 8.3   Small memory management algorithm linked list structure diagram 1.

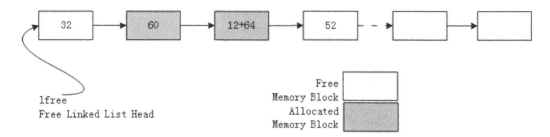

FIGURE 8.4   Small memory management algorithm linked list structure diagram 2.

is found, it meets the requirements of the allocation. Because this memory block is larger than the required size, the allocator will split the memory block, leaving the remaining memory block (52 bytes) in the lfree linked list, as shown in Figure 8.4, which is after 64 bytes is allocated.

In addition, a 12-byte data head is reserved for magic, used information, and linked list nodes before each memory block is allocated. The address returned to the application is offset by 12 bytes, and the 12-byte data header is something the user should never touch. (Note: The header might not always be 12 bytes; it depends on a compiler's alignment configuration.)

Memory deallocation is the reverse process. The allocator will also check if the adjacent memory blocks are free, and if so, the allocator will merge them into one large free memory block.

SLAB MANAGEMENT ALGORITHM

RT-Thread's slab allocator is based on the slab allocator implemented by DragonFly BSD founder Matthew Dillon and optimized for embedded systems. The most primitive slab algorithm is Jeff Bonwick's efficient kernel memory allocation algorithm introduced for the Solaris operating system.

RT-Thread's slab allocator implementation mainly removes the object construction and destruction process, leaving only the pure buffered memory pool algorithm. The slab allocator will divide memory into multiple zones according to the size of the object. It can also be seen as a combination of multiple memory pools, where each pool holds a certain type of object, as shown in Figure 8.5.

A zone is between 32K and 128K bytes in size, and the allocator will automatically adjust this size based on the total heap size during initialization. The zone in the system includes up to 72 objects, which can allocate up to 16K of memory at a time. If the allocation exceeds 16K bytes, it will directly allocate from the page allocator instead. The size of the memory block allocated on each zone is fixed. Zones that can allocate blocks of the same size are linked in a linked list. The zone linked lists of the 72 objects are managed in an array (zone_array[]).

Here are the two main operations for the memory allocator.

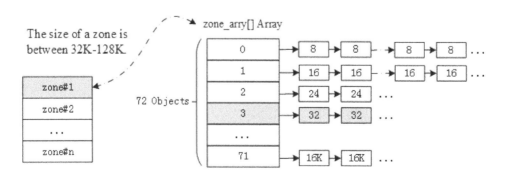

FIGURE 8.5   Slab memory allocation structure.

Memory Allocation

Assuming a 32-byte memory is being allocated, the slab memory allocator first finds the corresponding zone linked list from the linked list head of zone array in accordance with the 32-byte value. If the linked list is empty, a new zone will be assigned from the page allocator and return the first free block of memory from the zone. If the linked list is not empty, a free block must exist in the first zone node in the zone linked list (otherwise, it would not have been placed in the linked list) and then take the corresponding free block. If all free memory blocks in the zone are used after the allocation, the allocator needs to remove this zone node from the linked list.

Memory Release

The allocator needs to find the zone node where the memory block is located and then link the memory block to the zone's free memory block linked list. If the free linked list of the zone indicates that all the memory blocks of the zone have been released, which means that the zone is completely free, the system will release the full free zone to the page allocator when the number of free zones in the zone linked list reaches a certain number.

MEMHEAP MANAGEMENT ALGORITHM

The memheap management algorithm is suitable for systems with multiple memory heaps that are not contiguous. Using memheap memory management can simplify the use of multiple memory heaps in the system: when there are multiple memory heaps in the system, the user only needs to initialize multiple needed memheaps during system initialization and turn on the memheap function to glue multiple memheaps (addresses can be discontinuous) for the system's heap allocation.

Notes: When using memheap, the original heap function will be turned off. You can choose whether to use memheap by enabling or disabling RT_USING_MEMHEAP_AS_HEAP.

The working mechanism of memheap is shown in Figure 8.6. First, add multiple blocks of memory to the memheap_item linked list to glue. The allocation of a memory block starts by allocating memory from the default memory heap. When it cannot be allocated, the allocator will look up the

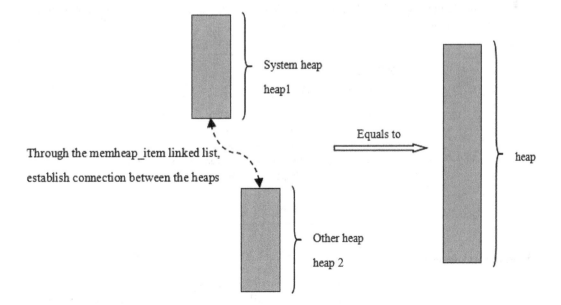

FIGURE 8.6 The memheap function handles multiple memory heaps.

memheap_item linked list and try to allocate a memory block from another one. The combination of multiple heaps is completely transparent to the application. It works like the application is allocating memory from one single heap.

MEMORY HEAP CONFIGURATION AND INITIALIZATION

When using the memory heap, the initialization must be done at system initialization by calling the following function:

```
void rt_system_heap_init(void* begin_addr, void* end_addr);
```

This function will use the memory space between begin_addr and end_addr parameters as a memory heap. Table 8.1 describes the input parameters for this function.

TABLE 8.1
rt_system_heap_init

Parameters	Description
begin_addr	Start address for heap memory area
end_addr	End address for heap memory area

When using the "memheap" heap memory, you must initialize the heap memory at system initialization, which can be done through the following function interface:

```
rt_err_t rt_memheap_init(struct rt_memheap  *memheap,
                const char  *name,
                void        *start_addr,
                rt_uint32_t size)
```

If there are multiple non-contiguous memheaps, the function can be called multiple times to initialize it and join the memheap_item linked list. Table 8.2 describes the input parameters and return values for this function.

TABLE 8.2
rt_memheap_init

Parameters	Description
memheap	memheap control block
name	The name of the memory heap
start_addr	Heap memory area start address
size	Heap memory size
Return	— —
RT_EOK	Successful

MEMORY HEAP MANAGEMENT

Operations of the memory heap are as shown in Figure 8.7, including initialization, allocation of memory blocks, and deallocation of memory blocks. All dynamic memory should be released after use for future use by other programs.

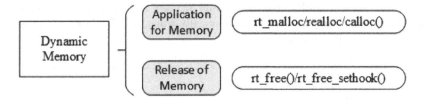

FIGURE 8.7   Operations of the memory heap.

Allocate and Release a Memory Block

Allocate a memory block of a user-specified size from the memory heap. The function interface is as follows:

```
void *rt_malloc(rt_size_t nbytes);
```

The rt_malloc function finds a memory block of the appropriate size from the system heap space and returns the available address of the memory block to the user. Table 8.3 describes the input parameters and return values for this function.

TABLE 8.3
rt_malloc

Parameters	Description
nbytes	The size of the memory block to be allocated, in bytes
Return	— —
Allocated memory block address	Successful
RT_NULL	Fail

After the application uses the memory applied from the memory allocator, it must be released in time; otherwise, it will cause a memory leak. The function interface for releasing the memory block is as follows:

```
void rt_free (void *ptr);
```

The rt_free function will return the to-be-released memory back to the heap manager. When calling this function, user needs to pass the to-be-released pointer of the memory block. If it is a null pointer, it returns directly. Table 8.4 describes the input parameters for this function.

TABLE 8.4
rt_free

Parameters	Description
ptr	To-be-released memory block pointer

Reallocate a Memory Block

Reallocating the size of the memory block (increase or decrease) based on the allocated memory block can be done through the following function interface:

```
void *rt_realloc(void *rmem, rt_size_t newsize);
```

When the memory block is reallocated, the original memory block data remains the same (in the case of reduction, the subsequent data is automatically truncated). Table 8.5 describes the input parameters and return values for this function.

TABLE 8.5
rt_realloc

Parameters	Description
rmem	Point to the allocated memory block
newsize	Reallocated memory size
Return	— —
Reallocated memory block address	Successful

Allocate Multiple Memory Blocks

Allocating multiple memory blocks with contiguous memory addresses from the memory heap can be done through the following function interface:

```
void *rt_calloc(rt_size_t count, rt_size_t size);
```

Table 8.6 describes the input parameters and return values for this function.

TABLE 8.6
rt_calloc

Parameters	Description
count	Number of memory block
size	Size of memory block
Return	— —
Pointer pointing to the first memory block address	Successful, all allocated memory blocks are initialized to zero
RT_NULL	Allocation failed

Set the Memory Hook Function

When allocating memory blocks, the user can set a hook function. The function interface called is as follows:

```
void rt_malloc_sethook(void (*hook)(void *ptr, rt_size_t size));
```

The hook function set will call back after the memory allocation. During the callback, the allocated memory block address and size are passed as input parameters. Table 8.7 describes the input parameters for this function.

TABLE 8.7
rt_malloc_sethook

Parameters	Description
hook	Hook function pointer

The hook function interface is as follows:

```
void hook(void *ptr, rt_size_t size);
```

Table 8.8 describes the input parameters for the hook function.

TABLE 8.8
Hook Function

Parameters	Description
ptr	The allocated memory block pointer
size	The size of the allocated memory block

When releasing memory, the user can set a hook function; the function interface called is as follows:

```
void rt_free_sethook(void (*hook)(void *ptr));
```

The hook function set will call back before the memory release is completed. During the callback, the released memory block address is passed in as the entry parameter (the memory block is not released at this time). Table 8.9 describes the input parameters for this function.

TABLE 8.9
rt_free_sethook

Parameters	Description
hook	Hook function pointer

The hook function interface is as follows:

```
void hook(void *ptr);
```

Table 8.10 describes the input parameters for the hook function.

TABLE 8.10
Hook Function

Parameters	Description
ptr	Memory block pointer to be released

MEMORY HEAP MANAGEMENT APPLICATION EXAMPLE

This is an example of a memory heap application. This program creates a dynamic thread that dynamically requests memory and releases it. Each time it applies for more memory, it ends when it can't apply for it, as shown in Code Listing 8.1.

Code listing 8.1 Memory Heap Management

```
#include <rtthread.h>

#define THREAD_PRIORITY      25
```

```
#define THREAD_STACK_SIZE     512
#define THREAD_TIMESLICE      5

/* thread entry */
void thread1_entry(void *parameter)
{
    int i;
    char *ptr = RT_NULL; /* memory block pointer */

    for (i = 0; ; i++)
    {
        /* memory space for allocating (1 << i) bytes each time */
        ptr = rt_malloc(1 << i);

        /* if allocated successfully */
        if (ptr != RT_NULL)
        {
            rt_kprintf("get memory :%d byte\n", (1 << i));
            /* release memory block */
            rt_free(ptr);
            rt_kprintf("free memory :%d byte\n", (1 << i));
            ptr = RT_NULL;
        }
        else
        {
            rt_kprintf("try to get %d byte memory failed!\n", (1 << i));
            return;
        }
    }
}
int dynmem_sample(void)
{
    rt_thread_t tid = RT_NULL;

    /* create thread 1 */
    tid = rt_thread_create("thread1",
                           thread1_entry, RT_NULL,
                           THREAD_STACK_SIZE,
                           THREAD_PRIORITY,
                           THREAD_TIMESLICE);
    if (tid != RT_NULL)
        rt_thread_startup(tid);

    return 0;
}
/* Export to the msh command list */
MSH_CMD_EXPORT(dynmem_sample, dynmem sample);
```

The results are as follows:

```
 \ | /
- RT -     Thread Operating System
 / | \     3.1.0 build Aug 24 2018
 2006 - 2018 Copyright by rt-thread team

msh >dynmem_sample
msh >get memory :1 byte
```

```
free memory :1 byte
get memory :2 byte
free memory :2 byte
...
get memory :16384 byte
free memory :16384 byte
get memory :32768 byte
free memory :32768 byte
try to get 65536 byte memory failed!
```

The memory is successfully allocated in the routine and the information is printed; when trying to allocate 65,536 bytes, 64KB, of memory, the allocation fails because the total RAM size is only 64K and the available RAM is less than 64K.

MEMORY POOL

The memory heap manager can allocate blocks of any size, which is very flexible and convenient. But it also has obvious shortcomings. First, the allocation efficiency is not high because the free memory block needs to be looked up for each allocation. Second, it is easy to generate memory fragmentation. In order to improve memory allocation efficiency and avoid memory fragmentation, the RT-Thread provides another method of memory management: the memory pool.

Memory pool is a memory allocation method for allocating a large number of small memory blocks of the same size. It can greatly speed up memory allocation and release, and can avoid memory fragmentation as much as possible. In addition, RT-Thread's memory pool supports the thread suspension feature. The idea is that when there is no free memory block in the memory pool, the application thread will be suspended until a new memory block is available. Then the suspended application thread will be awakened.

The thread suspension feature of the memory pool is very suitable for situations when applications need to be synchronized by memory resources. For example, when playing music, the player thread decodes a music file and then sends it to the sound card driver to drive the hardware to play music.

As shown in Figure 8.8, when the player thread needs to decode the data, it will request the memory block from the memory pool. If there is no memory block available, the thread will be suspended; otherwise, it will obtain the memory block to place the decoded data.

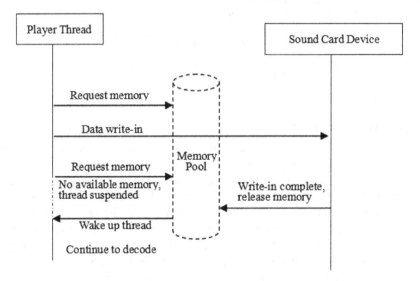

FIGURE 8.8 Player thread and sound card driver relationship.

The player thread then writes the memory block containing the decoded data to the sound card abstraction device (the thread will return immediately and continue to decode more data).

After the sound card device is written, the callback function set by the player thread is called to release the written memory block. If the player thread was suspended because there was no memory block in the memory pool available, it will be awakened to continue to decode.

How a Memory Pool Works

Memory Pool Control Block

The memory pool control block is a data structure used by the operating system to manage the memory pool. It stores some information about the memory pool, such as the start address of the data area in the memory pool, memory block size, and memory block list. It also includes the memory blocks, linked list structure used for the connection between memory blocks, event set of the thread suspended due to the memory block being unavailable, and so on.

In the RT-Thread real-time operating system, the memory pool control block is represented by the structure `struct rt_mempool`. Another C expression, `rt_mp_t`, represents the memory block handle. The implementation in C language is a pointer pointing to the memory pool control block. For details, see the following code:

```
struct rt_mempool
{
    struct rt_object parent;

    void    *start_address;  /* start address of memory pool data area */
    rt_size_t    size;          /* size of memory pool data area */

    rt_size_t    block_size;   /* size of memory block   */
    rt_uint8_t   *block_list;  /* list of memory block   */

    /* maximum number of memory blocks that can be accommodated in the
memory pool data area  */
    rt_size_t    block_total_count;
    /* number of free memory blocks in the memory pool  */
    rt_size_t    block_free_count;
    /* list of threads suspended because memory blocks are unavailable */
    rt_list_t    suspend_thread;
    /* number of threads suspended because memory blocks are unavailable
*/
    rt_size_t    suspend_thread_count;
};
typedef struct rt_mempool* rt_mp_t;
```

Memory Block Allocation Mechanism

When the memory pool is being created, it first applies a large amount of memory from the system. Then it divides the memory into multiple small memory blocks of the same size. The small memory blocks are directly connected by a linked list (this linked list is also called a free linked list). At each allocation, the first memory block is taken from the head of the free linked list and provided to the applicant. In Figure 8.9, there are multiple memory pools of different sizes allowed in physical memory. Each memory pool is composed of multiple free memory blocks, which are used by the kernel for memory management. When a memory pool object is created, the memory pool object is assigned to a memory pool control block. The parameters of the memory control block include the memory pool name, memory buffer, memory block size, number of blocks, and a queue of threads waiting.

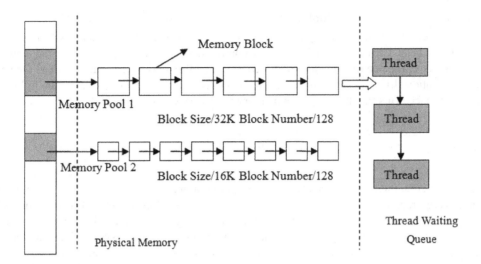

FIGURE 8.9 Memory pool working mechanism diagram.

The kernel is responsible for allocating memory pool control blocks to the memory pool. It also receives the requests for allocation of memory blocks from the user thread. When this information is obtained, the kernel can allocate memory from the memory pool. Once the memory pool is initialized, the size of the memory blocks inside will no longer be available for adjustment.

Each memory pool object consists of the above structure, where suspend_thread forms a list for the thread waiting for memory blocks, that is, when there is no memory block available in the memory pool, and the request thread allows waiting, the thread applying for the memory block will be suspended on the suspend_thread linked list.

MEMORY POOL MANAGEMENT

The memory pool control block is a structure that contains important parameters related to the memory pool and acts as a link between various states of the memory pool. The related interfaces of the memory pool are as shown in Figure 8.10. The operation of the memory pool includes create/initialize a memory pool, apply for a memory block, release a memory block, and delete/detach a memory pool. Note that not all memory pools will be deleted. The deletion is related to the needs of the designer, but the used memory blocks should all be released.

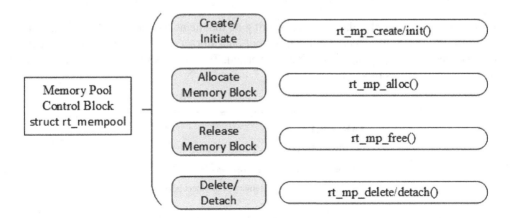

FIGURE 8.10 Related interfaces of a memory pool.

Create and Delete a Memory Pool

To create a memory pool, a memory pool object is created first and then a memory heap is allocated from the heap. Creating a memory pool is a prerequisite for allocating and releasing memory blocks from the corresponding memory pool. After the memory pool is created, the thread then can perform operations like allocation, release, and so on. To create a memory pool, use the following function interface. This function returns a memory pool object that has been created.

```
rt_mp_t rt_mp_create(const char* name,
                     rt_size_t block_count,
                     rt_size_t block_size);
```

Using this function interface can create a memory pool that matches the size and number of memory blocks required. The creation will be successful if system resources allow it (most importantly, memory heap memory resources). When you create a memory pool, you need to give the memory pool a name. The kernel then applies for a memory pool object from the system. Next, a memory buffer calculated from the number and sizes of blocks will be allocated from the memory heap. Then the memory pool object is initialized. Afterward, the successfully applied memory block buffer is organized into an idle linked list used for allocation. Table 8.11 describes the input parameters and return values for this function.

TABLE 8.11
rt_mp_create

Parameters	Description
name	Name of the memory pool
block_count	Number of memory blocks
block_size	Size of memory block
Return	— —
Handle of memory pool	Creation of memory pool object successful
RT_NULL	Creation of memory pool failed

Deleting a memory pool will delete the memory pool object and release the applied memory. Use the following function interface:

```
rt_err_t rt_mp_delete(rt_mp_t mp);
```

When a memory pool is deleted, all threads waiting on the memory pool object will be first awakened (return -RT_ERROR). Then the memory pool data storage area allocated from the memory heap is released, and the memory pool object is deleted. Table 8.12 describes the input parameters and return values for this function.

TABLE 8.12
rt_mp_delete

Parameters	Description
mp	Memory pool object handle
Return	— —
RT_EOK	Deletion successful

Initialize and Detach a Memory Pool

Memory pool initialization is similar to memory pool creation, except that the memory pool initialization is for static memory management mode, and the memory pool control block is derived from static objects that the user applies in the system. In addition, unlike memory pool creation, the memory space used by the memory pool object here is a buffer space specified by user. The user passes the pointer of the buffer to the memory pool control block; the rest of the initialization is the same as the creation of the memory pool. The function interface is as follows:

```
rt_err_t rt_mp_init(rt_mp_t mp,
                    const char* name,
                    void *start,
                    rt_size_t size,
                    rt_size_t block size);
```

When initializing the memory pool, pass the memory pool object that needs to be initialized to the kernel, as well as the memory space used by the memory pool, with the number and sizes of memory blocks managed by the memory pool, and assign a name to the memory pool. This way, the kernel can initialize the memory pool and organize the memory space used by the memory pool into a free block linked list for allocation. Table 8.13 describes the input parameters and return values for this function.

TABLE 8.13
rt_mp_init

Parameters	Description
mp	Memory pool object
name	Memory pool name
start	Starting address of memory pool
size	Memory pool data area size
block_size	Memory pool size
Return	— —
RT_EOK	Initialization successful
-RT_ERROR	Fail

The number of memory pool blocks = size/(block_size + 4-byte, linked list pointer size), and the calculation result needs to be rounded (an integer).

For example, the size of the memory pool data area is set to 4096 bytes, and the memory block size block_size is set to 80 bytes; then the number of memory blocks applied is 4096 / (80 + 4) = 48.

Detaching the memory pool means the memory pool object will be detached from the kernel object manager. Use the following function interface to detach the memory pool:

```
rt_err_t rt_mp_detach(rt_mp_t mp);
```

After using this function interface, the kernel wakes up all threads waiting on the memory pool object and then detaches the memory pool object from the kernel object manager. Table 8.14 describes the input parameters and return values for this function.

Allocate and Release a Memory Block

To allocate a memory block from the specified memory pool, use the following interface:

```
void *rt_mp_alloc (rt_mp_t mp, rt_int32_t time);
```

TABLE 8.14
rt_mp_detach

Parameters	Description
mp	Memory pool object
Return	— —
RT_EOK	Successful

The time parameter means the timeout period for applying for allocation of memory blocks. If there is a memory block available in the memory pool, remove a memory block from the free linked list of the memory pool, reduce the number of free blocks, and return this memory block; if there is no free memory block in the memory pool, determine the timeout time setting: if the timeout period is set to zero, the empty memory block is immediately returned; if the waiting time is greater than zero, the current thread is suspended on the memory pool object until there is free memory block available in the memory pool, or the waiting time elapses. Table 8.15 describes the input parameters and return values for this function.

TABLE 8.15
rt_mp_alloc

Parameters	Description
mp	Memory pool object
time	Timeout
Return	— —
Allocated memory block address	Successful
RT_NULL	Fail

Any memory block must be released after it has been used. Otherwise, memory leaks will occur. The memory block is released using the following interface:

```
void rt_mp_free (void *block);
```

When using the function interface, first, the memory pool object of (or belongs to) the memory block will be calculated by the pointer of the memory block that needs to be released. Second, the number of available memory blocks of the memory pool object will be increased. Third, the released memory block to the linked list of free memory blocks will be added. Then, it will be determined whether there is a suspended thread on the memory pool object; if so, the first thread on the suspended thread linked list will be awakened. Table 8.16 describes the input parameters for this function.

TABLE 8.16
rt_mp_free

Parameters	Description
block	Memory block pointer

MEMORY POOL APPLICATION EXAMPLE

This is a static internal memory pool application routine that creates a static memory pool object and two dynamic threads. One thread will try to get the memory block from the memory pool, and the other thread will release the memory block, as shown in Code Listing 8.2.

Code listing 8.2 Memory Pool Usage Example

```c
#include <rtthread.h>

static rt_uint8_t *ptr[50];
static rt_uint8_t mempool[4096];
static struct rt_mempool mp;

#define THREAD_PRIORITY      25
#define THREAD_STACK_SIZE     512
#define THREAD_TIMESLICE      5

/* pointer pointing to the thread control block */
static rt_thread_t tid1 = RT_NULL;
static rt_thread_t tid2 = RT_NULL;

/* thread 1 entry */
static void thread1_mp_alloc(void *parameter)
{
    int i;
    for (i = 0 ; i < 50 ; i++)
    {
        if (ptr[i] == RT_NULL)
        {
            /* Trying to apply for a memory block 50 times, when no
memory block is available,
                thread 1 suspends, thread 2 runs */
            ptr[i] = rt_mp_alloc(&mp, RT_WAITING_FOREVER);
            if (ptr[i] != RT_NULL)
                rt_kprintf("allocate No.%d\n", i);
        }
    }
}

/* thread 2 entry, thread 2 has a lower priority than Thread 1, so thread
1 should be executed first. */
static void thread2_mp_release(void *parameter)
{

    int i;

    rt_kprintf("thread2 try to release block\n");
    for (i = 0; i < 50 ; i++)
    {
        /* release all successfully allocated memory blocks */
        if (ptr[i] != RT_NULL)
        {
            rt_kprintf("release block %d\n", i);
            rt_mp_free(ptr[i]);
            ptr[i] = RT_NULL;
        }
    }
```

```
    }
}

int mempool_sample(void)
{
    int i;
    for (i = 0; i < 50; i ++) ptr[i] = RT_NULL;

    /* initialize the memory pool object */
    rt_mp_init(&mp, "mp1", &mempool[0], sizeof(mempool), 80);

    /* create thread 1: applying for memory pool */
    tid1 = rt_thread_create("thread1", thread1_mp_alloc, RT_NULL,
                            THREAD_STACK_SIZE,
                            THREAD_PRIORITY, THREAD_TIMESLICE);
    if (tid1 != RT_NULL)
        rt_thread_startup(tid1);


    /* create thread 2: release memory pool */
    tid2 = rt_thread_create("thread2", thread2_mp_release, RT_NULL,
                            THREAD_STACK_SIZE,
                            THREAD_PRIORITY + 1, THREAD_TIMESLICE);
    if (tid2 != RT_NULL)
        rt_thread_startup(tid2);

    return 0;
}

/* export to the msh command list */
MSH_CMD_EXPORT(mempool_sample, mempool sample);
```

The results are as follows:

```
 \ | /
- RT -       Thread Operating System
 / | \       3.1.0 build Aug 24 2018
 2006 - 2018 Copyright by rt-thread team

msh >mempool_sample
msh >allocate No.0
allocate No.1
allocate No.2
allocate No.3
allocate No.4
...
allocate No.46
allocate No.47
thread2 try to release block
release block 0
allocate No.48
release block 1
allocate No.49
release block 2
release block 3
release block 4
release block 5
```

```
...
release block 47
release block 48
release block 49
```

This routine initializes 4096 / (80 + 4) = 48 memory blocks when initializing the memory pool object.

1. After thread 1 applies for 48 memory blocks, the memory block has been used up and needs to be released elsewhere to be applied again; but at this time, thread 1 has applied for another one in the same waiting way because it cannot be allocated, so thread 1 suspends.
2. Thread 2 starts to execute the operation of releasing the memory. When thread 2 releases a memory block, it means there is a memory block that is free. Wake up thread 1 to apply for memory, and then apply again after the application is successful; thread 1 suspends again and repeats ②.
3. Thread 2 continues to release the remaining memory blocks until the release is complete.

CHAPTER SUMMARY

In this chapter, we learned the functional characteristics, working mechanism, and some methods using several memory management algorithms provided by RT-Thread. Let's review some points that need to be noted:

1. Memory pools can greatly speed up memory allocation and release, but once initialization is complete, the internal memory block size can no longer be adjusted.
2. Whether it is memory pool or memory heap management, the requested memory must be released in a timely manner after the use of the requested memory.

9 Interrupt Management

What is an interrupt? Simply speaking, when the system is processing a normal event, it is suddenly interrupted by an urgent event, which needs to be handled immediately. The system suspends the current event to process an urgent event. When it is done, the system turns back to carry out the interrupted event. In our lives, this kind of scenario happens all the time.

When you are reading a book, the phone rings; then you mark down the page number and go to pick up the phone. When you are done with the phone, you go on with the book from where you stopped. This is a classic procedure of interrupt.

If the phone was from your teacher, urging you to finish your homework, which you consider is of higher priority than reading books. So after the phone call, you won't continue to read the book until your homework is completed. This is a classic procedure of switching tasks in interrupts.

This kind of scenario is also common in embedded systems. When the CPU is processing a normal task, an external urgent event occurs; the CPU has to suspend the current task to handle the asynchronous event. After the external event has been handled, the CPU then returns to the previous position to continue the work that was interrupted. The system that implements this function is called the interrupt system, and the source of the asynchronous event requesting the CPU handling is called the interrupt source. An interrupt is a kind of exception. An exception is any event that causes the processor to move away from the normal process and execute special code. If the exception is not processed in time, the system may either encounter an error or face a complete breakdown. So appropriately handling exceptions to avoid errors is a very important part of improving software robustness (stability). Figure 9.1 is a simple interrupt diagram.

Interrupt processing is closely related to the CPU architecture. Therefore, this chapter briefly introduces the ARM Cortex-M CPU architecture first and then introduces the RT-Thread interrupt management mechanism in conjunction with the Cortex-M CPU architecture. After reading this chapter, you will learn more about the interrupt handling process of RT-Thread, how to add an interrupt service routine (ISR), and other matters related to this.

CORTEX-M CPU ARCHITECTURE FOUNDATION

Unlike older classic ARM processors (like ARM7 and ARM9), the ARM Cortex-M processor has a very different architecture. Cortex-M is a series of models that contains Cortex M0/M3/M4/M7 and so on. There are some differences between each model. For example, the Cortex-M4 is equipped with more floating-point calculation functions than the Cortex-M3, but their programming models are basically the same, so the parts of the book that describe interrupt management and porting are not going to be too finely differentiated from Cortex M0/M3/M4/M7. This section focuses on the architectural aspects related to RT-Thread interrupt management.

INTRODUCTION TO REGISTER

The register set of the Cortex-M series has 16 general registers from R0 to R15 and several special function registers, as shown in Figure 9.2.

R13 is used as the stack pointer register (SP); R14 is used as the link register (LR), which is used to store the return address when the subroutine is called; R15 is used as the program counter (PC). Note that the stack pointer register can be either the main stack pointer (MSP) or the process stack pointer (PSP).

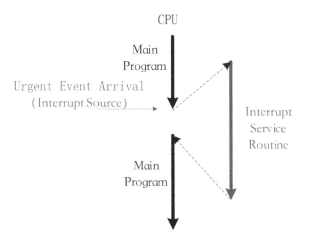

FIGURE 9.1 Interrupt diagram.

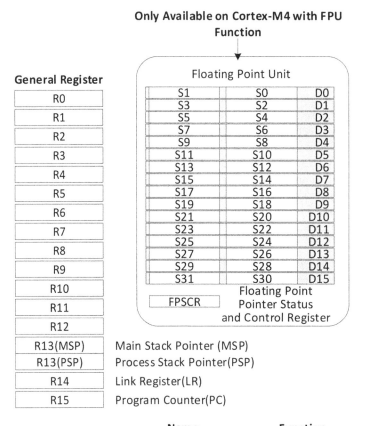

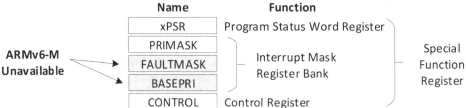

FIGURE 9.2 Register schematic.

Special function registers include the program status word register bank (PSRs), interrupt mask register banks (PRIMASK, FAULTMASK, BASEPRI), and control registers (CONTROL). Special function registers can be accessed through MSR/MRS instructions, such as:

```
MRS R0, CONTROL ; Read CONTROL to R0
MSR CONTROL, R0 ; Write R0 to the CONTROL register
```

The program status word registers store arithmetic and logic flags, such as negative flags, zero flags, overflow flags, and so on. The interrupt mask register bank controls when to disable the Cortex-M interrupts. The control registers are used to define the privilege level and decide which stack pointer is to be used.

In the case of Cortex-M4 or Cortex-M7 with a floating-point unit, the control register is also used to indicate whether the floating-point unit is currently in use. The floating-point unit contains 32 general-purpose floating-point registers S0–S31 and a special FPSCR register (floating-point status and control register).

OPERATING SCHEME AND PRIVILEGE LEVEL

Cortex-M introduces the concept of operating modes: Thread mode and Handler mode. If the CPU starts exception or interrupt processing, it enters Handler mode; otherwise, it is in Thread mode.

On the other side, Cortex-M has two running levels: privilege-level and user-level, which is controlled by the CONTROL special register. Thread mode can work at either the privilege-level or user-level, while Handler mode always works at the privilege-level. The switching of different working modes is shown in Figure 9.3.

Cortex-M's stack register SP corresponds to two physical registers: MSP and PSP, MSP is the main stack; PSP is the process stack. Handler mode always uses MSP as the stack; Thread mode can choose to use MSP or PSP as the stack, which is also controlled through the special register CONTROL. After reset, Cortex-M enters Thread mode, privilege-level, and uses the MSP stack by default.

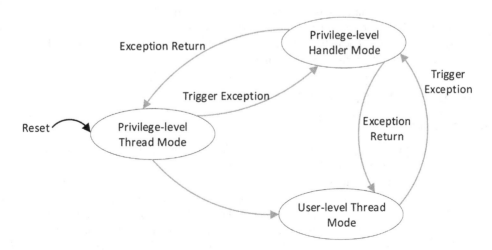

FIGURE 9.3   Cortex-M working mode switching diagram.

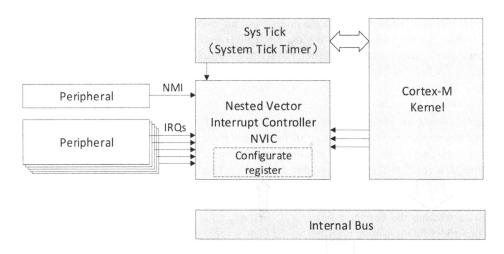

FIGURE 9.4 Relationship between cortex-M kernel and NVIC diagram.

Nested Vectored Interrupt Controller

The Cortex-M interrupt controller is called NVIC (nested vectored interrupt controller) and supports interrupt nesting. When an interrupt is triggered and the system responds, the processor hardware automatically pushes the current context into the interrupt stack. As shown in Figure 9.4, the registers in this section include the PSR, PC, LR, R12, and R3–R0 registers.

When the system is servicing an interrupt, if a higher-priority interrupt occurs, then the processor will also interrupt the currently running ISR and then save the context of the interrupt service program register PSR, PC, LR, R12, and R3–R0 to the interrupt stack.

PendSV System Call

PendSV, also known as the suspendable system call, is an exception that can be suspended like a normal interrupt. It is specifically designed to assist the operating system with context switching. PendSV exceptions are initialized as the lowest-priority exceptions. Each time a context switch is required, the PendSV exception is triggered manually, and the context switch is performed in the PendSV exception handler. The detailed process of operating system context switching through the PendSV mechanism will be illustrated in the next chapter.

RT-THREAD INTERRUPTION MECHANISM

Interrupt Vector Table

The interrupt vector table is the entry point for all interrupt handlers. Figure 9.5 shows the interrupt handlers for the Cortex-M series: a function (user ISR) is linked to each interrupt in a virtual interrupt vector table. Once an interrupt is triggered, the corresponding ISR will be executed.

On the Cortex-M cores, all interrupts are processed using the interrupt vector table, which means that when an interrupt is triggered, the processor will directly determine the interrupt source and then jump directly to the corresponding fixed location for processing. The ISRs must be placed together, starting from a uniform address (this address must be set to the NVIC interrupt vector offset register). The interrupt vector table is generally defined in the form of an array or stored in the startup code by default:

```
__Vectors       DCD      __initial_sp           ; Top of Stack
                DCD      Reset_Handler          ; Reset processing function
                DCD      NMI_Handler            ; NMI processing function
                DCD      HardFault_Handler      ; Hard Fault processing function
                DCD      MemManage_Handler      ; MPU Fault processing function
                DCD      BusFault_Handler       ; Bus Fault processing function
                DCD      UsageFault_Handler     ; Usage Fault processing
function
                DCD      0                      ; reserve
                DCD      0                      ; reserve
                DCD      0                      ; reserve
                DCD      0                      ; reserved
                DCD      SVC_Handler            ; SVCall processing function
                DCD      DebugMon_Handler       ; Debug Monitor processing
function
                DCD      0                      ; reserve
                DCD      PendSV_Handler         ; PendSV processing function
                DCD      SysTick_Handler        ; SysTick processing function

... ...

NMI_Handler              PROC
                EXPORT NMI_Handler                         [WEAK]
                B        .
                ENDP
HardFault_Handler        PROC
                EXPORT HardFault_Handler                   [WEAK]
                B        .
                ENDP

... ...
```

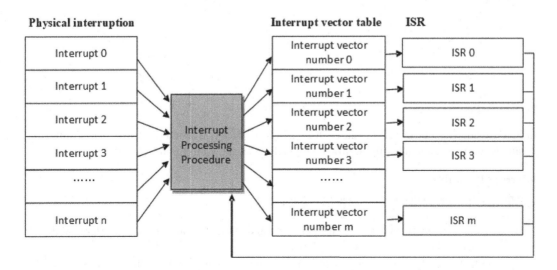

FIGURE 9.5 Interrupt processing.

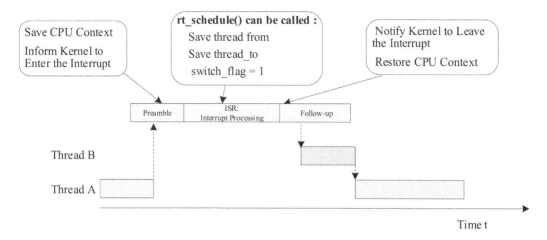

FIGURE 9.6   Three parts of the interrupt handler.

Notes: The [WEAK] after the code is a weak-symbol identifier. The symbols labeled by [WEAK], like NMI_Handler and HardFault_Handler, are weak symbols, which means that if the same name exists in another place and is not decorated by [WEAK], the alternative definition of the symbol is used instead of the weak ones, and the code associated with the weak symbols will be automatically discarded.

Take the SysTick interrupt as an example. In the system startup code, the weak symbol SysTick_Handler has been used as the handler for SysTick interrupt; users only need to implement the function of the same name to respond to the SysTick interrupt. The interrupt handler function sample program is shown as follows:

```
void SysTick_Handler(void)
{
    /* enter interrupt */
    rt_interrupt_enter();

    rt_tick_increase();

    /* leave interrupt */
    rt_interrupt_leave();
}
```

Interrupt Processing

In RT-Thread interrupt management, each interrupt handler is divided into three parts: interrupt preamble, user ISR, and interrupt follow-up procedure, as shown in Figure 9.6.

Interrupt Preamble

The main job of the interrupt preamble is as follows:

1. Save the CPU interrupt context. This part is related to the CPU architecture. Different CPU architectures are implemented differently.

   For Cortex-M, this part of the work is done automatically by hardware. When an interrupt is triggered and the system responds, the processor hardware automatically pushes the registers that constitute the current context into the interrupt stack. The registers in this section include the PSR, PC, LR, R12, and R3–R0 registers.

2. Inform the kernel to enter the interrupt state by invoking the rt_interrupt_enter() function, which adds 1 to the global variable rt_interrupt_nest to record the level of interrupt nesting. The code is as follows:

```
void rt_interrupt_enter(void)
{
    rt_base_t level;

    level = rt_hw_interrupt_disable();
    rt_interrupt_nest ++;
    rt_hw_interrupt_enable(level);
}
```

User Interrupt Service Routine

In the user ISR, two cases need to be distinguished. The first case is that no thread switching is required. In this case, after the user ISR and interrupt subsequent program finish running, it exits and returns to the interrupted thread.

In the other case, thread switching becomes necessary during interrupt processing. In this case, the rt_hw_context_switch_interrupt() function is called for context switching. This function is related to the CPU architecture, and different CPU architectures are implemented differently.

In Cortex-M architecture, the implementation of the function rt_hw_context_switch_interrupt() is shown in Figure 9.7. It sets the thread rt_interrupt_to_thread variable to the thread that needs to be switched to and then triggers the PendSV exception (PendSV exception is specifically used to assist context switching and is initialized to the lowest level). After the PendSV exception is triggered, the PendSV exception interrupt handler will not be executed immediately, as the interrupt processing is still in progress; instead, it will be entered only when the interrupt subsequent program completes and is about to exit the interrupt procedure.

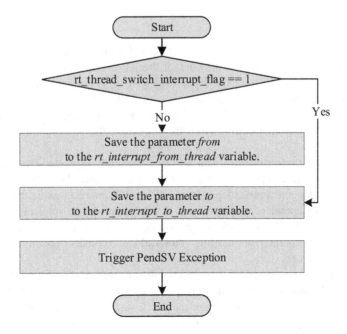

FIGURE 9.7 Function rt_hw_context_switch_interrupt() implementation process.

Interrupt Follow-up Procedure

The main work done by the interrupt follow-up procedure is as follows:

1. Inform the kernel to leave the interrupt state and reduce the global variable rt_interrupt_ nest by 1 through calling the rt_interrupt_leave() function. The code is as follows:

```
void rt_interrupt_leave(void)
{
    rt_base_t level;

    level = rt_hw_interrupt_disable();
    rt_interrupt_nest --;
    rt_hw_interrupt_enable(level);
}
```

2. Restore the CPU context prior to the interrupt. If thread switching did not happen during the interrupt processing, the CPU context of the *from* thread is restored. If thread switching did happen, the CPU context of the *to* thread is restored. This part of the implementation is related to the CPU architecture. Different CPU architectures are implemented differently. The implementation process in the Cortex-M architecture is shown in Figure 9.8.

INTERRUPT NESTING

When interrupt nesting is enabled, during the execution of the ISR, if an interrupt with a higher priority occurs, the CPU suspends the current ISR and starts to execute a new one. When the new interrupt with the higher priority has been served, the previous ISR resumes. If thread switching is required, the thread context switch will occur after all interrupt handlers are completed, as shown in Figure 9.9.

INTERRUPT STACK

When the interrupt occurs and before the system responds to the interrupt, the software code (or processor) needs to save the context of the current thread (usually stored on the thread stack of the current thread) and then call the ISR. When the interrupt is being served (essentially calling the user's ISR function), the interrupt handler function is likely to have its local variables, which also require the corresponding stack space. The new interrupt data can be saved on the stack of the interrupted thread. When exiting from the interrupt, the corresponding thread resumes execution.

On the other hand, the interrupt stack can also be completely separated from the thread stack; that is, when entering the interrupt each time, after the interrupt thread context is saved, it switches to the new interrupt stack and runs independently. When the interrupt exits, the previous context is restored. Using an independent interrupt stack is relatively easy to implement, and it is easier to understand and grasp the thread stack usage (otherwise, it must reserve space for the interrupt stack. If the system supports interrupt nesting, you should also consider how much space should be reserved for nested interrupt).

RT-Thread chooses the implementation with separate interrupt stacks. When an interrupt occurs, the preprocessor of the interrupt will place the user's stack pointer into the interrupt stack space reserved by the system in advance and restore the user's stack when the interrupt exits. This way, the interrupt does not occupy the stack space of the thread, thereby improving the utilization of the

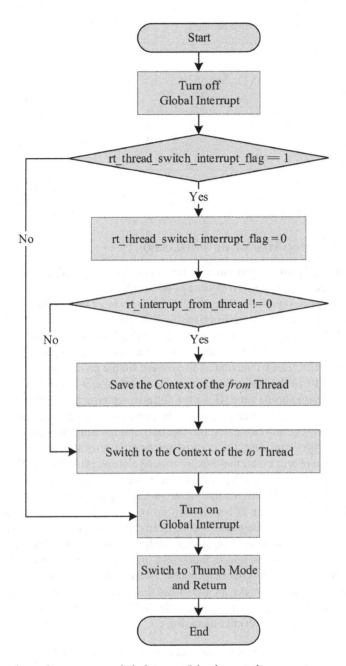

FIGURE 9.8 Function rt_hw_context_switch_interrupt() implementation process.

memory space, and as the number of threads increases, it becomes more effective to reduce the memory consumption.

There are two stack pointers in the Cortex-M processor core. One is the MSP, which is the stack pointer by default. It is used before the first thread is created or in the interrupt and exception handlers. The other is the thread stack pointer (PSP) used by threads. When the interrupt and exception service routine exits, the value of the second bit of the LR register is modified to 1, and the SP of the thread is switched from MSP to PSP.

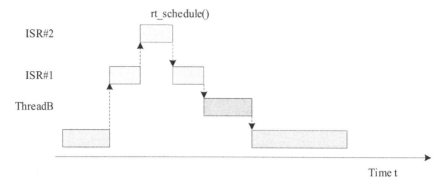

FIGURE 9.9   Thread switching during interrupt.

PROCESSING OF THE BOTTOM HALF OF THE INTERRUPTION

RT-Thread does not make any assumptions or restrictions on the processing time required by the ISR, but like other operating systems (realtime or non-realtime), users need to ensure that all ISRs are completed in the shortest possible time (the ISR is like to be of the highest priority in the system and will preempt all threads to execute first). In this way, during interrupt nesting or being masked, the handling of other interrupts or of the same interrupt that happens again will not be postponed.

When an interrupt occurs, the ISR needs to obtain the corresponding hardware state or data. If the ISR, such as for a CPU clock interrupt, is to perform simple processing on the state or data, the ISR only needs to add 1 to the system clock variable and then exit the ISR. Such interrupts often require a relatively short running time. However, for other interrupts, the ISR needs to perform a series of more time-consuming processing after obtaining the hardware state or data. Usually, the interrupt handler is divided into two parts: the **top half** and the **bottom half**. In the top half, after getting the hardware state and data, the masked interrupt is re-enabled and a notification is sent (which can be the semaphore, event, mailbox, or message queue provided by RT-Thread) and then the interrupt handler ends. On the other side, the relevant thread, after receiving the notification, further processes the state or data—this part of the processing is called bottom-half processing.

To illustrate the implementation of bottom-half processing in RT-Thread, we take a virtual network device receiving network data packets as an example, as shown in Code Listing 9.1. Assume that after receiving the data message, the system analyzes and processes the message, which is a relatively time-consuming process and is much less important than an external interrupt source signal. It can also be processed without masking the interrupt source signal.

The program in this example creates an nwt thread that is blocked on the nw_bh_sem signal after it starts to run. Once this semaphore is released, the following nw_packet_parser process will be executed to begin the bottom-half event processing.

Code listing 9.1 Interrupt Bottom-Half Processing Example

```
/*
 * program list: interrupt bottom half processing example
 */

/* semaphore used to wake up threads */
rt_sem_t nw_bh_sem;

/* thread for data reading and analysis */
void demo_nw_thread(void *param)
{
    /*First, perform the necessary initialization work on the device. */
```

```
    device_init_setting();

    /*.. other operations..*/

    /* create a semaphore to respond to Bottom Half events */
    nw_bh_sem = rt_sem_create("bh_sem", 0, RT_IPC_FLAG_FIFO);

    while(1)
    {
        /* Finally, let demo_nw_thread wait on nw_bh_sem. */
        rt_sem_take(nw_bh_sem, RT_WAITING_FOREVER);

        /* After receiving the semaphore signal, start the real Bottom Half
processing. */
        nw_packet_parser (packet_buffer);
        nw_packet_process(packet_buffer);
    }
}

int main(void)
{
    rt_thread_t thread;

    /* create processing thread */
    thread = rt_thread_create("nwt",demo_nw_thread, RT_NULL, 1024, 20, 5);

    if (thread != RT_NULL)
        rt_thread_startup(thread);
}
```

Let's take a look at how the top half is handled in demo_nw_isr and how the bottom half is started, as in the following example:

```
void demo_nw_isr(int vector, void *param)
{
    /* When the network device receives the data, it is met with an
interrupt exception and starts executing this ISR. */
    /* Start the processing of the Top Half, such as reading the status of
the hardware device to determine what kind of interruption occurred. */
    nw_device_status_read();

    /*.. Some other data operations, etc. ..*/

    /* Release nw_bh_sem, send a signal to demo_nw_thread, ready to start
Bottom Half */
    rt_sem_release(nw_bh_sem);

    /* Then exit the interrupted Top Half section and end the device's ISR */
}
```

As showed in the two code snippets earlier, the ISR is done by waiting and releasing a semaphore at the start and the end of the bottom half, respectively. Since the interrupt processing is divided into two parts, top and bottom, the interrupt processing becomes an asynchronous process. With the system overhead introduced by the bottom half, the user needs to consider seriously whether the interrupt handling time is longer than the time cost by sending a notification and switching to the bottom-half thread to handle it in RT-Thread.

RT-THREAD INTERRUPT MANAGEMENT INTERFACE

To isolate the operating system from the underlying exception and interrupt hardwares, RT-Thread encapsulates interrupt and exception handlers into a set of abstract interfaces, as shown in Figure 9.10.

MOUNT INTERRUPT SERVICE ROUTINE

The system associates the user's interrupt handler with the specified interrupt number. You can call the following interface to register a new ISR:

```
rt_isr_handler_t rt_hw_interrupt_install(int vector,
                                         rt_isr_handler_t  handler,
                                         void *param,
                                         char *name);
```

After calling rt_hw_interrupt_install(), when the interrupt source generates an interrupt, the system will automatically call the registered ISR. Table 9.1 describes the input parameters and return values for this function.

TABLE 9.1
rt_hw_interrupt_install

Parameters	Description
vector	Vector is the mounted interrupt number
handler	Newly mounted interrupt service routine
param	Param will be passed as a parameter to the interrupt service routine
name	Name of the interrupt
Return	——
return	The handle of the interrupt service routine mounted before the interrupt service routine was mounted

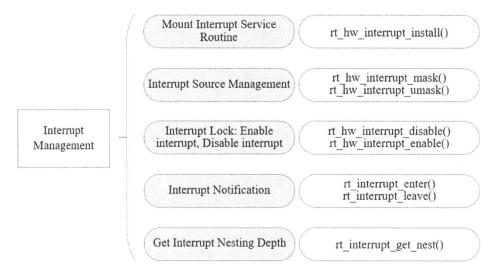

FIGURE 9.10 Interrupt-related interfaces.

Notes: This interface does not appear in every migration branch. For example, there is usually no such interface in the migration branch of Cortex-M0/M3/M4.

The ISR is a kind of runtime environment that requires special attention. It runs in a non-threaded execution environment (generally a special operating mode of the chip (privileged mode)). In this runtime environment, the current execution flow cannot be suspended because it is not a thread at all. Invocation of thread-related functions causes such prompt information to appear, "Function [abc_func] shall not be used in ISR," meaning a function that should not be called in the ISR.

INTERRUPT SOURCE MANAGEMENT

Before the ISR is invoked to handle an interrupt, its source is masked until the state and data are handled; then the masked interrupt source should be enabled in time. Masking the interrupt source ensures that the hardware state or data will not be disturbed during the next process. This is done through the following function:

```
void rt_hw_interrupt_mask(int vector);
```

After the rt_hw_interrupt_mask function interface is called, the corresponding interrupt will be masked (usually when the same interrupt appears again, the interrupt status register will change accordingly, but the CPU will not be informed to handle it). Table 9.2 describes the input parameters for this function.

TABLE 9.2
rt_hw_interrupt_mask

Parameters	Description
vector	Interrupt number to be masked

Notes: This interface does not appear in every migration branch. For example, there is usually no such interface in the migration branch of Cortex-M0/M3/M4.

To avoid losing the hardware interrupt signals as much as possible, the following function interface can be called to enable the blocked interrupt source:

```
void rt_hw_interrupt_umask(int vector);
```

After the rt_hw_interrupt_umask function interface is called, if the interrupt (and corresponding peripheral) is configured correctly after the interrupt is triggered, the processor will be informed to handle it. Table 9.3 describes the input parameters for this function.

TABLE 9.3
rt_hw_interrupt_umask

Parameters	Description
vector	Enable the blocked interrupt number

Notes: This interface does not appear in every migration branch. For example, there is usually no such interface in the migration branch of Cortex-M0/M3/M4.

ENABLE AND DISABLE GLOBAL INTERRUPT

Enable and disable global interrupts, also known as the interrupt lock, are the easiest way of preventing several threads from accessing the critical sections at the same time. By disabling interrupts, the current thread is assured it will not be interrupted by other events (because the entire system no longer responds to those external events that could trigger a thread rescheduling); that is, the current thread will not be preempted unless the thread voluntarily relinquishes the control of the processor. When you need to disable all the interrupts of the system, you can call the following function interface:

```
rt_base_t rt_hw_interrupt_disable(void);
```

Table 9.4 describes the return values for this function.

TABLE 9.4
rt_hw_interrupt_disable

Return	Description
Interrupt Status	Interrupt status before the function rt_hw_interrupt_disable runs

To restore interrupts is also known as to enable interrupts. The rt_hw_interrupt_enable() function is used to "enable" interrupts, which restores the previously interrupted state before the rt_hw_interrupt_disable() function is called. If the interrupts are disabled before the invocation of rt_hw_interrupt_disable(), they are still disabled after that. The functions to enable and disable interrupts are often used in pairs. The function interface called is as follows:

```
void rt_hw_interrupt_enable(rt_base_t level);
```

Table 9.5 describes the input parameters for this function.

TABLE 9.5
rt_hw_interrupt_enable

Parameters	Description
level	The interrupt status returned by the previous rt_hw_interrupt_disable

The method of accessing critical sections through the interrupt lock can be applied on any occasion, and other types of synchronization methods are implemented based on the interrupt lock. Generally speaking, the interrupt lock is the most powerful and efficient synchronization method. The main problem of using interrupt locks is that the system will no longer respond to any interrupts during the lock time, so the external events cannot be handled. Therefore, the impact of the interrupt lock on the real-time system is very large. When used improperly, the system will be completely non-real-time (may cause the system to severely violate the time requirement); when used properly, it will become a fast, efficient synchronization procedure.

For example, to ensure that a line of code (such as assignments) run exclusively, the quickest way is to use interrupt locks instead of semaphores or mutexes:

```
/* turn off the interrupt */
level = rt_hw_interrupt_disable();
```

```
    a = a + value;
    /* resume interrupt */
    rt_hw_interrupt_enable(level);
```

When using an interrupt lock, you need to ensure that the interrupt is locked for a very short time, such as a = a + value in the above code; you can also switch to another method, such as using semaphores:

```
    /* get semaphore lock */
    rt_sem_take(sem_lock, RT_WAITING_FOREVER);
    a = a + value;
    /* release the semaphore lock */
    rt_sem_release(sem_lock);
```

In the implementation of rt_sem_take and rt_sem_release, interrupt locks are actually used to protect the semaphore's internal variable, so for operations such as a = a + value; it is more concise and faster to use interrupt locks directly.

Notes: The function rt_base_t rt_hw_interrupt_disable(void) and the function void rt_hw_interrupt_enable(rt_base_t level) generally need to be used in pairs to ensure the correct interrupt status.

In RT-Thread, the interfaces to enable and disable global interrupts support multi-level nesting. The code for simple nested interrupts is shown in Code Listing 9.2.

Code listing 9.2 A Simple Example of Nested Interrupts

```
#include <rthw.h>

void global_interrupt_demo(void)
{
    rt_base_t level0;
    rt_base_t level1;

    /* The global interrupt is turned off for the first time. The global
interrupt status before being turned off may be turned on or off. */
    level0 = rt_hw_interrupt_disable();
    /* The global interrupt is turned off for the second time. The global
interrupt status before being turned off may be turned on or off. */
    level1 = rt_hw_interrupt_disable();

    do_something();

    /* Resume the global interrupt to the state before the second turn-off,
so the global interrupt is still turned off after this enable. */
    rt_hw_interrupt_enable(level1);
    /* Resume the global interrupt to the state before the first turn-off,
so the global interrupt status can be on or off. */
    rt_hw_interrupt_enable(level0);
}
```

This feature can bring great convenience to software development. For example, a function disables interrupts, invokes some sub-functions, and then enables interrupts again. However, the sub-functions themselves may also disable or enable interrupts. Since the interface for global interrupts allows interrupt nesting, users do not need to do special processing for this code.

INTERRUPT NOTIFICATION

When the entire system is interrupted by an interrupt and the interrupt handler starts to take care of it, the kernel needs to be informed that the processor is in the interrupt state. In this case, the following interfaces can be used:

```
void rt_interrupt_enter(void);
void rt_interrupt_leave(void);
```

These two interfaces are used, respectively, in the interrupt preamble and interrupt postamble procedures and will both modify the value of rt_interrupt_nest (interrupt nesting depth).

Whenever an interrupt is entered, the rt_interrupt_enter() function can be called to notify the kernel that it has entered the interrupt state and increased the interrupt nesting depth (execute rt_interrupt_nest++).

Whenever an interrupt is exited, the rt_interrupt_leave() function can be called to notify the kernel that it has exited the interrupt state and reduced the interrupt nesting depth (execute rt_interrupt_nest--). Be careful not to call these two interface functions in user applications.

The role of using rt_interrupt_enter/leave() is that in the ISR, if a kernel-related function (such as releasing a semaphore) is called, the kernel can be adjusted in time according to the current interrupt status. For example, if a semaphore is released in the interrupt and a thread is awakened but the kernel is aware that it is still in the interrupt context, the thread switching should be implemented through the mechanism of scheduling in interrupts and not switched immediately.

However, if the ISR does not call kernel-related functions (release semaphores, etc.), you may not call the rt_interrupt_enter/leave() function at this time.

In the upper application, the rt_interrupt_get_nest() interface is called when the kernel needs to know whether it has entered the interrupt state or what the currently nested interrupt depth is. It will return the value of rt_interrupt_nest, as follows. Table 9.6 describes the return value.

```
rt_uint8_t rt_interrupt_get_nest(void);
```

TABLE 9.6
rt_interrupt_get_nest

Return	Description
0	Current system is not in an interrupt context
1	Current system is in an interrupt context
Bigger Than 1	Current interrupt nesting level

INTERRUPT AND POLLING

When the peripheral is working, how to get the changed state informed, by interrupts or by polling, is often the first problem to be considered by the driver developer, and there is a difference between the real-time operating system and the time-sharing operating system on how to solve this problem. Because the polling mode itself executes sequentially, check the event and deal with the state. Therefore, the polling mode is relatively simple and clear to implement. For example, to write data to the serial port, the program code writes the next data only when the serial controller has completed the previous write operation (otherwise, the data is discarded). The corresponding code may look like this:

```
/* polling mode writes data to the serial port */
    while (size)
```

```
{
    /* Determine if the data in the UART peripheral is sent. */
    while (!(uart->uart_device->SR & USART_FLAG_TXE));
    /* Send the next data when all data has been sent. */
    uart->uart_device->DR = (*ptr & 0x1FF);

    ++ptr; --size;
}
```

In the real-time system, the polling mode may be very problematic. Since the program keeps running (when polling) in a real-time operating system, the thread never stops and the threads with lower priority will never get the chance to run. In a time-sharing system, the opposite happens. There is almost no difference between different priorities. You can run this program in a one-time slice and then run another program in another slice.

So generally, in real-time systems, interrupt mode is mostly used to drive peripherals. When the data has arrived, the relevant processing threads are awakened by the interrupt and then the subsequent actions are executed. For example, some serial peripherals with FIFOs (FIFO queue with a certain amount of data) can be implemented, as shown in Figure 9.11.

The thread first writes data to the serial port's FIFO. When the FIFO is full, the thread actively suspends. The serial controller continuously fetches data from the FIFO and sends it out at a configured baud rate (e.g., 115,200 bps). When all data in the FIFO is sent, an interrupt is triggered to inform the processor; then the ISR is executed and the thread is awakened. Here is an example of a FIFO-type device. In reality, there are also DMA-type devices used in similar principles.

For low-speed devices, this mode is perfect, since the processor can run other threads before the serial peripheral fetches data from the FIFO, which improves the overall operating efficiency of the system. (Even for time-sharing systems, such a mode is often necessary.) But for some high-speed devices, such as when the transmission speed reaches 10Mbps, assuming that the amount of data sent at one time is 32 bytes, we can calculate the time required to send such a piece of data: $(32 \times 8) \times 1/10\text{Mbps} = 25\text{us}$. When data needs to be transmitted continuously, the system will trigger an interrupt after 25us to wake up the upper thread to continue the next transmission. Suppose the system's thread switching time is 8us (usually, the real-time operating system's thread context switching only takes a few us); then, when the entire system is running, the data bandwidth utilization rate will be only $25/(25 + 8) = 75.8\%$. However, with polling mode, the data bandwidth utilization rate may reach 100%. This is also why people generally think that the data throughput in the real-time system is insufficient. The system overhead is consumed in the thread switching (some real-time systems may even use the bottom half processing and hierarchical interrupt processing, as described earlier in this chapter, which lengthens the time consumption from interrupts to the activation of threads that actually send data, and the efficiency will be further reduced).

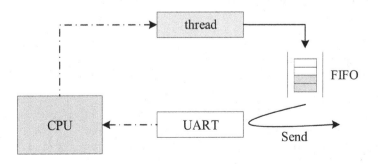

FIGURE 9.11 Interrupt mode drive peripheral.

Through the said calculation process, we can see some of the key factors: the smaller the amount of transmitted data is, the faster the transmission speed will be, and the greater the impact on data throughput it will cause. Ultimately, it depends on how often the system generates interrupts. When a real-time system wants to increase data throughput, several ways can be considered:

1. Increase the length of each data volume for each transmission and try to let the peripherals send as much data as possible every time.
2. Change the interrupt mode to polling mode, if necessary. At the same time, in order to solve the problem that the processor is always occupied in polling mode and other low-priority threads cannot run, the priority of the polling thread can be lowered accordingly.

GLOBAL INTERRUPT USAGE EXAMPLE

This is an interrupted application routine: when multiple threads access the same variable, the variable is protected by enabling/disabling global interrupts, as shown in the following Code Listing 9.3.

Code listing 9.3 Protect the Variable by Enabling/Disabling Global Interrupt

```
#include <rthw.h>
#include <rtthread.h>

#define THREAD_PRIORITY       20
#define THREAD_STACK_SIZE     512
#define THREAD_TIMESLICE      5

/* global variables accessed simultaneously */
static rt_uint32_t cnt;
void thread_entry(void *parameter)
{
    rt_uint32_t no;
    rt_uint32_t level;

    no = (rt_uint32_t) parameter;
    while (1)
    {
        /* turn off glocal interrupt */
        level = rt_hw_interrupt_disable();
        cnt += no;
        /* resume glocal interrupt */
        rt_hw_interrupt_enable(level);

        rt_kprintf("protect thread[%d]'s counter is %d\n", no, cnt);
        rt_thread_mdelay(no * 10);
    }
}

/* user application entry */
int interrupt_sample(void)
{
    rt_thread_t thread;

    /* create t1 thread */
    thread = rt_thread_create("thread1", thread_entry, (void *)10,
                      THREAD_STACK_SIZE,
                      THREAD_PRIORITY, THREAD_TIMESLICE);
```

```
    if (thread != RT_NULL)
        rt_thread_startup(thread);


    /* create t2 thread */
    thread = rt_thread_create("thread2", thread_entry, (void *)20,
                                THREAD_STACK_SIZE,
                                THREAD_PRIORITY, THREAD_TIMESLICE);
    if (thread != RT_NULL)
        rt_thread_startup(thread);

    return 0;
}

/* export to the msh command list */
MSH_CMD_EXPORT(interrupt_sample, interrupt sample);
```

The results are as follows:

```
 \ | /
- RT -      Thread Operating System
 / | \      3.1.0 build Aug 27 2018
 2006 - 2018 Copyright by rt-thread team

msh >interrupt_sample
msh >protect thread[10]'s counter is 10
protect thread[20]'s counter is 30
protect thread[10]'s counter is 40
protect thread[20]'s counter is 60
protect thread[10]'s counter is 70
protect thread[10]'s counter is 80
protect thread[20]'s counter is 100
protect thread[10]'s counter is 110
protect thread[10]'s counter is 120
protect thread[20]'s counter is 140
...
```

Notes: Since disabling the global interrupts will cause the entire system to fail to respond to any new interrupts, when using the global interrupt to implement exclusive accesses to the critical section, it is necessary to ensure that the closure of global interrupts is very short, such as the time to run several machine instructions.

CHAPTER SUMMARY

We have highlighted the interrupt mechanism and management method of RT-Thread. Let's review a few points to note in this chapter:

1. The interrupt service program should be as streamlined as possible, and the time-consuming data processing can be put into threads. When considering whether bottom-half processing should be used in an interrupt, the user needs to consider whether the interrupt service processing time is greater than the time to send notifications to the bottom half and the time to process.
2. All interrupts in the ARM Cortex-M series are handled in the form of an interrupt vector table.
3. It's recommended to use the interrupt model in real-time systems. If the polling model is applied, the thread must be set to a low priority.

10 Kernel Porting

After reading the previous chapters, everyone has a better understanding of RT-Thread, but many people are not familiar with how to port the RT-Thread kernel to different hardware platforms. Kernel porting refers to the RT-Thread kernel running on different chip architectures and different boards. It has functions such as thread management and scheduling, memory management, inter-thread synchronization and communication, and timer management. Porting can be divided into two parts: CPU architecture porting and BSP (board support package) porting.

This chapter will introduce CPU architecture porting and BSP porting. The CPU architecture porting part will be introduced in conjunction with the Cortex-M CPU architecture. Therefore, it is necessary to review the "Cortex-M CPU Architecture Foundation" section of the previous chapter: "Interrupt Management." After reading this chapter, you will know how to complete the RT-Thread kernel porting.

CPU ARCHITECTURE PORTING

There are many different CPU architectures in the embedded world, for example, Cortex-M, ARM920T, MIPS32, RISC-V, etc. In order to enable RT-Thread to run on different CPU architecture chips, RT-Thread provides a libcpu abstraction layer to adapt to different CPU architectures. The libcpu layer provides unified interfaces to the kernel, including global interrupt switches, thread stack initialization, context switching, and more.

RT-Thread's libcpu abstraction layer provides a unified set of CPU architecture porting interfaces downward. This part of the interface includes global interrupt switch functions, thread context switch functions, OS Tick configuration and interrupt functions, cache, and so on. Table 10.1 shows the interfaces and variables that the CPU architecture migration needs to implement.

IMPLEMENT GLOBAL INTERRUPT ENABLE/DISABLE

Regardless of kernel code or user code, there may be some variables that need to be used in multiple threads or interrupts. If there is no corresponding protection mechanism, it may lead to critical section problems. To solve this problem, RT-Thread provides a series of inter-thread synchronization and communication mechanisms. But these mechanisms require the global interrupt enable/disable function provided in libcpu. They are, respectively:

```
/* disable global interrupt */
rt_base_t rt_hw_interrupt_disable(void);

/* enable global interrupt */
void rt_hw_interrupt_enable(rt_base_t level);
```

The following describes how to implement these two functions on the Cortex-M architecture. As mentioned earlier, the Cortex-M implements the CPS instruction in order to achieve fast switch interrupts, which can be used here.

```
CPSID I; PRIMASK=1, ; disable global interrupt
CPSIE I; PRIMASK=0, ; enable global interrupt
```

Disable Global Interrupt

The functions that need to be done in order in the rt_hw_interrupt_disable() function are:

1. Save the current global interrupt status and use the status as the return value of the function.
2. Disable the global interrupt.

TABLE 10.1
libcpu Porting-Related API

Functions and Variables	Description
rt_base_t rt_hw_interrupt_disable(void);	Disable global interrupt
void rt_hw_interrupt_enable(rt_base_t level);	Enable global interrupt
rt_uint8_t *rt_hw_stack_init(void *tentry, void *parameter, rt_uint8_t *stack_addr, void *texit);	The initialization of the thread stack; the kernel will call this function during thread creation and thread initialization
void rt_hw_context_switch_to(rt_uint32 to);	Context switch without source thread, which is called when the scheduler starts the first thread, and is called in the signal
void rt_hw_context_switch(rt_uint32 from, rt_uint32 to);	Switch from *from* thread to *to* thread; used for switching between threads
void rt_hw_context_switch_interrupt(rt_uint32 from, rt_uint32 to);	Switch from *from* thread to *to* thread; used for switch in interrupt
rt_uint32_t rt_thread_switch_interrupt_flag;	A flag indicating that a switch is needed in the interrupt
rt_uint32_t rt_interrupt_from_thread, rt_interrupt_to_thread;	Used to save *from* and *to* threads when the thread is context switching

Based on MDK, the global interrupt disabled function on the Cortex-M core is shown in Code Listing 10.1.

Code listing 10.1 Disable Global Interrupt

```
;/*
; * rt_base_t rt_hw_interrupt_disable(void);
; */
rt_hw_interrupt_disable     PROC        ;PROC pseudoinstruction definition
function
    EXPORT  rt_hw_interrupt_disable ;EXPORT output defined function,
similar to C language extern
    MRS     r0, PRIMASK             ;read the value of the PRIMASK
register to the r0 register
    CPSID   I                       ;disable global interrupt
    BX      LR                      ;function renturn
    ENDP                            ;ENDP end of function
```

The said code first uses the MRS instruction to save the value of the PRIMASK register to the r0 register, then disables the global interrupt with the "CPSID I" instruction, and finally returns with the BX instruction. The data stored by r0 is the return value of the function. Interrupts can occur between the "MRS r0, PRIMASK" instruction and "CPSID I," which do not cause a global interrupt status disorder.

There are different conventions for different CPU architectures regarding how registers are managed during function calls and in interrupt handlers. A more detailed introduction to the use of registers for Cortex-M can be found in the official ARM manual, *Procedure Call Standard for the ARM ® Architecture.*"

Enable Global Interrupt

In rt_hw_interrupt_enable(rt_base_t level), the variable *level* is used as the state to be restored, overriding the global interrupt status of the chip.

Based on MDK, implementation on the Cortex-M core enables a global interrupt, as shown in Code Listing 10.2.

Code listing 10.2 Enable Global Interrupt

```
;/*
; * void rt_hw_interrupt_enable(rt_base_t level);
```

```
; */
rt_hw_interrupt_enable    PROC        ; PROC pseudoinstruction definition
function
    EXPORT  rt_hw_interrupt_enable   ; EXPORT output defined function,
similar to "extern" in C language
    MSR     PRIMASK, r0              ; write the value of the r0 register to
the PRIMASK register
    BX      LR                      ; function renturn
    ENDP                            ; ENDP end of function
```

The said code first uses the MSR instruction to write the value register of r0 to the PRIMASK register, thus restoring the previous interrupt status.

IMPLEMENT THREAD STACK INITIALIZATION

When dynamically creating threads and initializing threads, the internal thread initialization function _rt_thread_init() is used. The _rt_thread_init() function calls the stack initialization function *rt_hw_stack_init()*, which manually constructs a context in the stack initialization function. The context will be used as the initial value for each thread's first execution. The layout of the context on the stack is shown in Figure 10.1.

Code Listing 10.3 is the stack initialization code.

Code listing 10.3 Build a Context on the Stack

```
rt_uint8_t *rt_hw_stack_init(void        *tentry,
                             void        *parameter,
                             rt_uint8_t  *stack_addr,
                             void        *texit)
{
    struct stack_frame *stack_frame;
    rt_uint8_t          *stk;
    unsigned long       i;
```

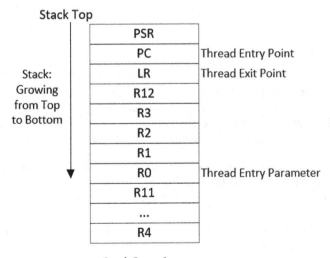

FIGURE 10.1   Context information in the stack.

```
    /* align the incoming stack pointer */
    stk  = stack_addr + sizeof(rt_uint32_t);
    stk  = (rt_uint8_t *)RT_ALIGN_DOWN((rt_uint32_t)stk, 8);
    stk -= sizeof(struct stack_frame);

    /* obtain the pointer to the stack frame of the context */
    stack_frame = (struct stack_frame *)stk;

    /* set the default value of all registers to 0xdeadbeef */
    for (i = 0; i < sizeof(struct stack_frame) / sizeof(rt_uint32_t); i ++)
    {
        ((rt_uint32_t *)stack_frame)[i] = 0xdeadbeef;
    }

    /* save the first parameter in the r0 register according to the ARM
APCS calling standard */
    stack_frame->exception_stack_frame.r0  = (unsigned long)parameter;
    /* set the remaining parameter registers to 0 */
    stack_frame->exception_stack_frame.r1  = 0;             /* r1 register
*/
    stack_frame->exception_stack_frame.r2  = 0;             /* r2 register
*/
    stack_frame->exception_stack_frame.r3  = 0;             /* r3 register
*/
    /* set IP (Intra-Procedure-call scratch register.) to 0 */
    stack_frame->exception_stack_frame.r12 = 0;             /* r12
register */
    /* save the address of the thread exit function in the lr register */
    stack_frame->exception_stack_frame.lr  = (unsigned long)texit;
    /* save the address of the thread entry function in the pc register */
    stack_frame->exception_stack_frame.pc  = (unsigned long)tentry;
    /* Set the value of psr to 0x01000000L, which means that the default
switch is Thumb mode. */
    stack_frame->exception_stack_frame.psr = 0x01000000L;

    /* return the stack address of the current thread        */
    return stk;
}
```

IMPLEMENT CONTEXT SWITCHING

In different CPU architectures, context switches between threads, and context switches from inter-rupts to contexts, the register portion of the context may be different or the same. In Cortex-M, context switching is done uniformly using PendSV exceptions, and there is no difference in the switching parts. However, to adapt to different CPU architectures, RT-Thread's libcpu abstraction layer still needs to implement three thread switching–related functions:

1. rt_hw_context_switch_to(): No source thread, switching to the target thread, which is called when the scheduler starts the first thread.
2. rt_hw_context_switch(): In a threaded environment, switches from the current thread to the target thread.
3. rt_hw_context_switch_interrupt(): In the interrupt environment, switches from the current thread to the target thread.

There are differences between switching in a thread environment and switching in an interrupt envi-ronment. In the thread environment, if the rt_hw_context_switch() function is called, the context

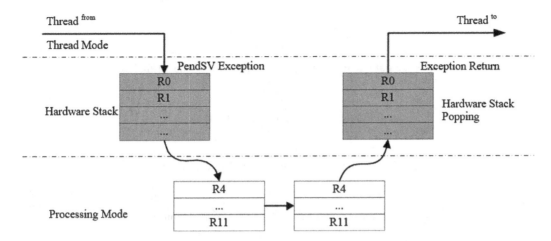

FIGURE 10.2 Context switch between threads.

switch can be performed immediately; in the interrupt environment, it needs to wait for the interrupt handler to complete processing the functions before switching.

Due to this difference, the implementation of rt_hw_context_switch() and rt_hw_context_switch_interrupt() is not the same on platforms such as ARM9. If the thread's schedule is triggered in the interrupt handler, rt_hw_context_switch_interrupt() is called in the dispatch function to trigger the context switching. It will check the rt_thread_switch_interrupt_flag variable before the interrupt exits after the interrupt handler processed the interrupt. If the value of the variable is 1, the context switch of the thread is completed according to the rt_interrupt_from_thread variable and the rt_interrupt_to_thread variable.

In the Cortex-M processor architecture, context switching can be made more compact based on the features of automatic partial push and PendSV.

Context switching between threads is shown in Figure 10.2.

The hardware automatically saves the PSR, PC, LR, R12, and R3–R0 registers of the *from* thread before entering the PendSV interrupt, then saves the R11–R4 registers of the *from* thread in PendSV, and restores the R4~R11 registers of the *to* thread, and finally the hardware automatically restores the R0~R3, R12, LR, PC, and PSR registers of the *to* thread after exiting the PendSV interrupt.

The context switch from interrupt to thread can be represented by Figure 10.3.

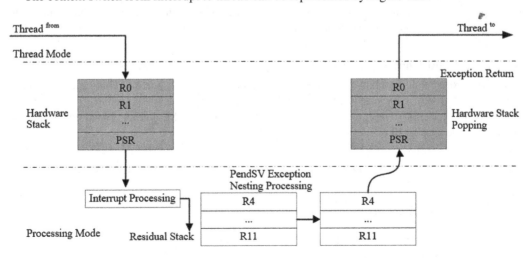

FIGURE 10.3 Interrupt to thread switching.

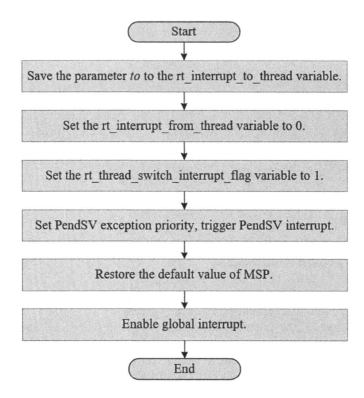

FIGURE 10.4   rt_hw_context_switch_to() flowchart.

The hardware automatically saves the PSR, PC, LR, R12, and R3–R0 registers of the *from* thread before entering the interrupt and then triggers a PendSV exception. R11~R4 registers of the *from* thread are saved, and R4~R11 registers of the *to* thread are restored in the PendSV exception handler. Finally, the hardware automatically restores the R0~R3, R12, PSR, PC, and LR registers of the *to* thread after exiting the PendSV interrupt.

Obviously, in the Cortex-M kernel, the rt_hw_context_switch() and rt_hw_context_switch_interrupt() have same functions, which is finishing saving and replying to the remaining contexts in PendSV. So we just need to implement a piece of code to simplify the porting.

Implement rt_hw_context_switch_to()

rt_hw_context_switch_to() has only the target thread and no source thread. This function implements the function of switching to the specified thread. Figure 10.4 is the flowchart.

The rt_hw_context_switch_to() implementation on the Cortex-M3 kernel (based on MDK), as shown in Code Listing 10.4.

Code listing 10.4 MDK Version rt_hw_context_switch_to() Implementation

```
;/*
; * void rt_hw_context_switch_to(rt_uint32 to);
; * r0 --> to
; * this fucntion is used to perform the first thread switch
; */
rt_hw_context_switch_to    PROC
```

```
        EXPORT rt_hw_context_switch_to
        ; r0 is a pointer pointing to the SP member of the thread control block
of the to thread
        ; save the value of the r0 register to the rt_interrupt_to_thread
variable
        LDR     r1, =rt_interrupt_to_thread
        STR     r0, [r1]


        ; set the from thread to empty, indicating that no context is needed to
save from
        LDR     r1, =rt_interrupt_from_thread
        MOV     r0, #0x0
        STR     r0, [r1]


        ; set the flag to 1, indicating that switching is required, this
variable will be cleared when switching in the PendSV exception handler
        LDR     r1, =rt_thread_switch_interrupt_flag
        MOV     r0, #1
        STR     r0, [r1]


        ; set PendSV exception priority to lowest priority
        LDR     r0, =NVIC_SYSPRI2
        LDR     r1, =NVIC_PENDSV_PRI
        LDR.W   r2, [r0,#0x00]          ; read
        ORR     r1,r1,r2                ; modify
        STR     r1, [r0]                ; write-back


        ; trigger PendSV exception (PendSV exception handler will be executed)
        LDR     r0, =NVIC_INT_CTRL
        LDR     r1, =NVIC_PENDSVSET
        STR     r1, [r0]


        ; abandon the stack from chip startup to before the first context
switch, set the value of the MSP as when it is started
        LDR     r0, =SCB_VTOR
        LDR     r0, [r0]
        LDR     r0, [r0]
        MSR     msp, r0


        ; enable global interrupts and global exceptions. After enabling, the
PendSV exception handler will be entered.
        CPSIE   F
        CPSIE   I


        ; will not execute to here
        ENDP
```

Implement rt_hw_context_switch()/rt_hw_context_switch_interrupt()

The function rt_hw_context_switch() and the function rt_hw_context_switch_interrupt() have two parameters: the *from* thread and the *to* thread. They implement the function to switch from the *from* thread to the *to* thread. Figure 10.5 is a specific flowchart.

The rt_hw_context_switch() and rt_hw_context_switch_interrupt() implementations on the Cortex-M3 kernel (based on MDK) are shown in Code Listing 10.5.

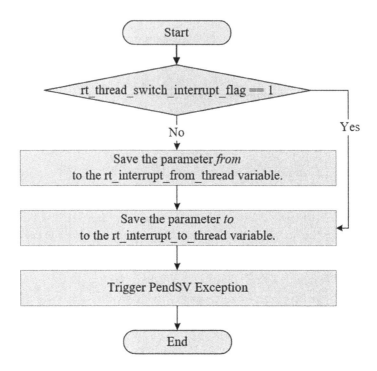

FIGURE 10.5   rt_hw_context_switch()/rt_hw_context_switch_interrupt() flowchart.

Code listing 10.5 Implement rt_hw_context_switch()/
rt_hw_context_switch_interrupt()

```
;/*
; * void rt_hw_context_switch(rt_uint32 from, rt_uint32 to);
; * r0 --> from
; * r1 --> to
; */
rt_hw_context_switch_interrupt
    EXPORT rt_hw_context_switch_interrupt
rt_hw_context_switch      PROC
    EXPORT rt_hw_context_switch

    ; check if the rt_thread_switch_interrupt_flag variable is 1
    ; skip updating the contents of the thread from if the variable is 1
    LDR     r2, =rt_thread_switch_interrupt_flag
    LDR     r3, [r2]
    CMP     r3, #1
    BEQ     _reswitch
    ; set the rt_thread_switch_interrupt_flag variable to 1
    MOV     r3, #1
    STR     r3, [r2]

    ; update the rt_interrupt_from_thread variable from parameter r0
    LDR     r2, =rt_interrupt_from_thread
    STR     r0, [r2]

_reswitch
    ; update the rt_interrupt_to_thread variable from parameter r1
    LDR     r2, =rt_interrupt_to_thread
    STR     r1, [r2]
```

```
    ; trigger PendSV exception, will enter the PendSV exception handler to
complete the context switch
    LDR      r0, =NVIC_INT_CTRL
    LDR      r1, =NVIC_PENDSVSET
    STR      r1, [r0]
    BX       LR
```

Implement PendSV Interrupt

In Cortex-M3, the PendSV interrupt handler is PendSV_Handler(). The actual thread switching is done in PendSV_Handler(). Figure 10.6 is a specific flowchart.

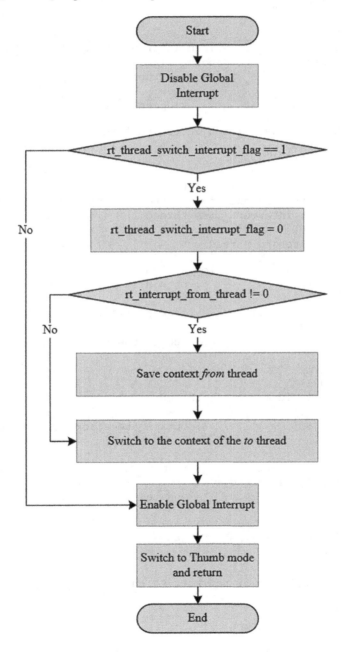

FIGURE 10.6 PendSV interrupt handling.

Code Listing 10.6 is a PendSV_Handler implementation.

Code listing 10.6 PendSV_Handler

```
; r0 --> switch from thread stack
; r1 --> switch to thread stack
; psr, pc, lr, r12, r3, r2, r1, r0 are pushed into [from] stack
PendSV_Handler    PROC
    EXPORT PendSV_Handler

    ; disable global interrupt
    MRS     r2, PRIMASK
    CPSID   I

    ; check if the rt_thread_switch_interrupt_flag variable is 0
    ; if it is zero, jump to pendsv_exit
    LDR     r0, =rt_thread_switch_interrupt_flag
    LDR     r1, [r0]
    CBZ     r1, pendsv_exit           ; pendsv already handled

    ; clear the rt_thread_switch_interrupt_flag variable
    MOV     r1, #0x00
    STR     r1, [r0]

    ; check the rt_thread_switch_interrupt_flag variable
    ; if it is 0, the context save of the from thread is not performed.
    LDR     r0, =rt_interrupt_from_thread
    LDR     r1, [r0]
    CBZ     r1, switch_to_thread

    ; save the context of the from thread
    MRS     r1, psp               ; obtain the stack pointer of the thread
from
    STMFD   r1!, {r4 - r11}       ; save r4~r11 to the thread's stack
    LDR     r0, [r0]
    STR     r1, [r0]              ; update the SP pointer of the thread's
control block

switch_to_thread
    LDR     r1, =rt_interrupt_to_thread
    LDR     r1, [r1]
    LDR     r1, [r1]              ; obtain the stack pointer of the thread
to

    LDMFD   r1!, {r4 - r11}       ; restore the register value of the thread
to in the stack of the thread
    MSR     psp, r1               ; update the value of r1 to psp

pendsv_exit
    ; restore global interrupt status
    MSR     PRIMASK, r2

    ; modify bit 2 of the lr register to ensure that the process uses the
PSP stack pointer
    ORR     lr, lr, #0x04
    ; exit interrupt function
    BX      lr
    ENDP
```

IMPLEMENT OS TICK

With the basics of switching global interrupts and context switching understood, RTOS can perform functions such as creating, running, and scheduling threads. With OS Tick, RT-Thread can do time slice round-robin scheduling for threads of the same priority, implementing timer functions, implementing rt_thread_delay() delay functions, and so on.

To finish the porting of libcpu, we need to ensure that the rt_tick_increase() function is called periodically during the clock tick interrupt. The call cycle depends on the value of the RT_TICK_PER_SECOND macro of rtconfig.h.

In Cortex-M, the OS Tick can be implemented by SysTick's interrupt handler.

```
void SysTick_Handler(void)
{
    /* enter interrupt */
    rt_interrupt_enter();

    rt_tick_increase();

    /* leave interrupt */
    rt_interrupt_leave();
}
```

BSP PORTING

In a practical project, for the same CPU architecture, different boards may use the same CPU architecture, carry different peripheral resources, and complete different products, so we also need to adapt to the board. RT-Thread provides a BSP abstraction layer to fit boards commonly seen. If you want to use the RT-Thread kernel on a board, in addition to the need to have the corresponding chip architecture porting, you need to port the corresponding board, that is, implement a basic BSP. The job is to establish a basic environment for the operating system to run. The main tasks that need to be completed are:

1. Initialize the CPU internal registers and set the RAM operation timing.
2. Implement clock driver and interrupt controller driver and interrupt management.
3. Implement UART and GPIO driver.
4. Initialize the dynamic memory heap to implement dynamic heap memory management.

KERNEL PORTING SAMPLE

This section takes the example of RT-Thread IoT Board (STM32L475, Cortex-M4 kernel) and describes from scratch how to complete RT-Thread kernel porting, with the goal that RT-Thread kernel will function properly on the IoT Board. This IoT board is shown in Figure 10.7.

PREPARE FOR THE BARE METAL PROJECT

Starting with the preparation of a basic bare metal MDK project, we can find it in the chapter10-bare-metal directory from the accompanying materials, which contains a project.uvprojx file and opens the project file, as shown in Figure 10.8.

The code is divided into three groups: Application, Drivers, and STM32L4xx_HAL_Driver.

1. The Application directory is the app code.
2. The Drivers directory is the driving code, including various hardware-related code.
3. The STM32L4xx_HAL_Driver directory is the firmware library code for STM32L4xx.

FIGURE 10.7   IoT board.

LED blink and UART output functions are implemented in this project.

Using a USB cable to connect the IoT board and computer, in the MDK project, click on the "Flash-> Download" to download the program to the development board. After successfully downloading, the program runs, and you can see the LED blinking, as shown in Figure 10.9.

Using the PuTTY serial tool to connect to the carrier serial port, the baud is set to 115200/N-8-1, and you can see the continuous output "01" sequence, as shown in Figure 10.10.

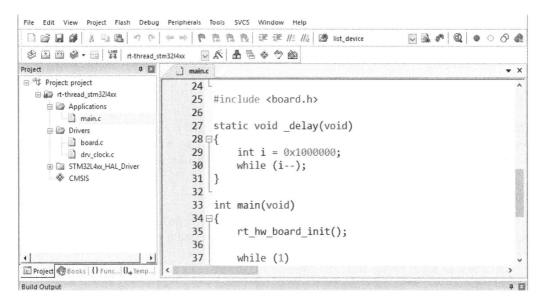

FIGURE 10.8   Project file.

FIGURE 10.9   IoT board LED blinking.

BUILD THE RT-THREAD PROJECT

Next, build the RT-Thread project on the basis of the previous 0-bare-metal project, add the core source code of RT-Thread, the CPU architecture porting code for Cortex-M4 and the configuration header file of RT-Thread, and add the corresponding header file search path.

Find the accompanying materials of ①src folder under rt-thread-3.1.0, ②libcpu-m4 folder under rt-thread-3.1.0, ③include folder under rt-thread-3.1.0, and ④ rtconfig.h files under the chapter10 folder and copy them to the chapter10-bare-metal directory.

The src folder is the kernel source code of RT-Thread. We need to add the following files in the folder to the Kernel group in the project (create your own Kernel group in the project):

1. .\rt-thread\src\clock.c
2. .\rt-thread\src\components.c
3. .\rt-thread\src\idle.c
4. .\rt-thread\src\ipc.c
5. .\rt-thread\src\irq.c
6. .\rt-thread\src\kservice.c
7. .\rt-thread\src\mem.c
8. .\rt-thread\src\object.c
9. .\rt-thread\src\scheduler.c
10. .\rt-thread\src\signal.c
11. .\rt-thread\src\thread.c
12. .\rt-thread\src\timer.c

FIGURE 10.10   Output.

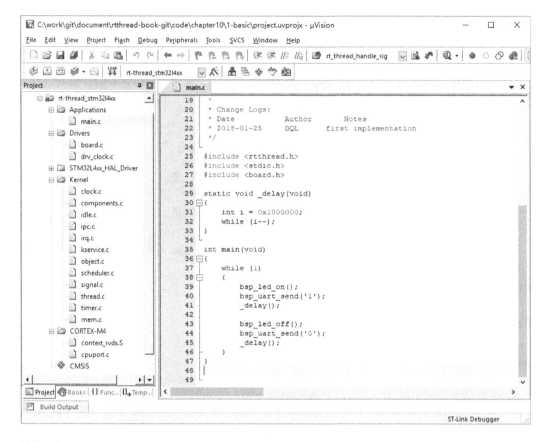

FIGURE 10.11   Project.

The libcpu file is a chip-related file. Cortex-M4's porting code includes the following files, which can be found under the directory libcpu-arm-cortex-m4. We add them to the CORTEX-M4 group in the MDK project (create your own CORTEX-M4 group in the project):

1. .\rt-thread\libcpu\arm\cortex-m4\cpuport.c
2. .\rt-thread\libcpu\arm\cortex-m4\context_rvds.S

Notes: Double-click on the newly created engineering group in Keil to add files to this group. By default, you add a file of the .c file type, at which point you can change the file type to All files (*.*) to display all the files.

The result is as shown in Figure 10.11.

In addition to the corresponding source code, we need to add the relevant header file search path to the project options. Click on "Project -> Options for Target 'rt-thread_stm32l4xx' -> C/C++ -> Include Paths" in turn to add all of the following paths:

1. .\rt-thread\include
2. .\rt-thread\libcpu\arm\cortex-m4
3. .\rt-thread\libcpu\arm\common

This is shown in Figure 10.12.

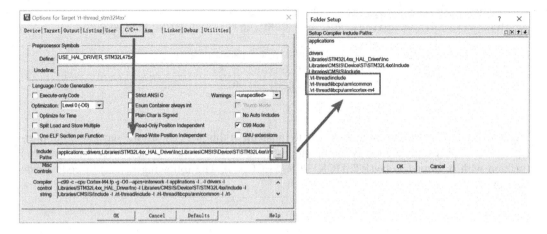

FIGURE 10.12   Set include paths for project.

Since the rt_hw_board_init() function has been called in the RT-Thread kernel components.c, the rt_hw_board_init() function called in the main() function needs to be removed.

Once we have all the said changes done, we can compile and download the modified project to get RT-Thread up on the IoT board. The complete project can refer to the accompanying material under the chapter10-basic directory of the project.

It is important to notice that the delay function _delay() used in the main() function at this time is still achieving by the while loop method. It's because the system has not yet implemented clock management functions and cannot use rt_thread_delay() to delay.

IMPLEMENT CLOCK MANAGEMENT

The ARM Cortex-M kernel offers a system tick timer: SysTick. SysTick's clock source is derived from the system clock, so it is also known as a periodic overflow time-based timer. After the SysTick count overflows, it enters the SysTick interrupt handler function, and RTOS typically uses SysTick as the OS Tick.

The specific work of SysTick porting is to:

1. Configure the interrupt frequency of SysTick.
2. Timed call rt_tick_increase() in the SysTick interrupt handler function of SysTick_Handler().

Add the function as shown in Code Listing 10.7 in board.c to complete the configuration of the OS Tick.

Code listing 10.7 System Tick Porting Modifications as Shown in Board.c

```
void SysTick_Handler(void)
{
    rt_interrupt_enter();
    rt_tick_increase();
    rt_interrupt_leave();
}

void rt_hw_board_init(void)
{
    /* ......Keep previous code...... */
```

```
    /* Configure the Systick interrupt time */
    HAL_SYSTICK_Config(HAL_RCC_GetHCLKFreq() / RT_TICK_PER_SECOND);
    /* Configure the Systick */
    HAL_SYSTICK_CLKSourceConfig(SYSTICK_CLKSOURCE_HCLK);
    /* SysTick_IRQn interrupt configuration */
    HAL_NVIC_SetPriority(SysTick_IRQn, 0, 0);
}
```

That's all SysTick's configuration work, and to finish all of this, we can test the success of our porting with a simple example: modify the main() function in main.c and replace the original _delay() function with rt_thread_delay(RT_TICK_PER_SECOND). The following is the test code inside main.c:

```
int main(void)
{
    while (1)
    {
        bsp_led_on();
        bsp_uart_send('1');
        rt_thread_delay(RT_TICK_PER_SECOND);

        bsp_led_off();
        bsp_uart_send('0');
        rt_thread_delay(RT_TICK_PER_SECOND);
    }
}
```

The complete project can refer to the accompanying materials under the chapter10/2-os-tick-porting directory. When you download the program into the IoT board and run it, you can see the LED lights keep flashing, and you can also receive 10 data through the serial port, as shown in Figure 10.13.

IMPLEMENT CONSOLE OUTPUT

RT-Thread provides the rt_kprintf() function that can be used to output debugging information. The rt_kprintf() function can be output based on the character device of the device framework or by the rt_hw_console_output() function. In general, when there is no character device, you can use

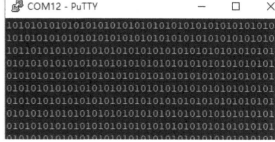

FIGURE 10.13   LED and console output.

rt_hw_console_output() to support debugging information output. This part of porting is modified in board.c, adding the function as shown in Code Listing 10.8.

Code listing 10.8 Board.c Modified Console Porting

```
void rt_hw_console_output(const char *str)
{
    RT_ASSERT(str != RT_NULL);

    while (*str != '\0')
    {
        if (*str == '\n')
        {
            bsp_uart_send('\r');
        }
        bsp_uart_send(*str++);
    }
}
```

That's the initialization code for the console, and then we can test the success of our porting with a simple example: modify main() function in main.c, and replace bsp_led_xx() function with rt_kprintf(). The test code inside the modified main.c is shown in Code Listing 10.9.

Code listing 10.9 Console Test Code

```
int main(void)
{
    while (1)
    {
        rt_kprintf("led on\n");
        rt_thread_delay(RT_TICK_PER_SECOND);

        rt_kprintf("led off\n");
        rt_thread_delay(RT_TICK_PER_SECOND);
    }
}
```

The complete project can refer to the accompanying materials under the chapter10/3console-porting directory. When you download the program into the IoT board and run it, you can see the output information in PuTTY, as shown in Figure 10.14.

IMPLEMENT DYNAMIC HEAP MEMORY MANAGEMENT

Turning on dynamic memory management requires turning on the following macro definitions in the rtconfig.h configuration file:

```
#define RT_USING_HEAP
#define RT_USING_SMALL_MEM
```

The initialization of dynamic memory management is done through the rt_system_heap_init() function:

```
void rt_system_heap_init(void *begin_addr, void *end_addr);
```

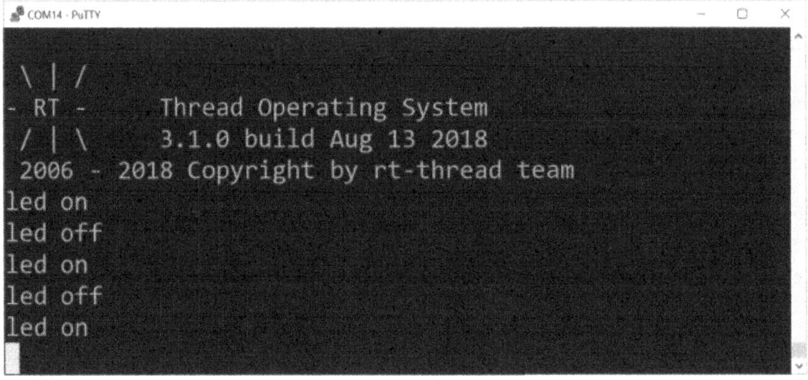

FIGURE 10.14 Console output.

Add rt_system_heap_init() to the rt_hw_board_init() function of board.c, and define a large static array to manage as dynamic memory, as shown in Code Listing 10.10.

Code listing 10.10 Dynamic Memory Initialization

```
void rt_hw_board_init(void)
{
    static uint8_t heap_buf[10 * 1024];
    ......
    /* init heap*/
rt_system_heap_init(heap_buf, heap_buf + sizeof(heap_buf) - 1);
}
```

That's the initialization code for dynamic memory, which we can modify in main.c, and add the creation of dynamic threads to verify the porting, as shown in Code Listing 10.11.

Code listing 10.11 Dynamic Memory Test Code

```
#define THREAD_PRIORITY         25
#define THREAD_STACK_SIZE       512
#define THREAD_TIMESLICE        5

void led_thread_entry(void *parameter)
{
    while (1)
    {
        rt_kprintf("enter test thread\n");
        rt_thread_delay(RT_TICK_PER_SECOND);
    }
}

int main(void)
{
    rt_thread_t tid;

    tid = rt_thread_create("led",
                           led_thread_entry, RT_NULL,
                           THREAD_STACK_SIZE, THREAD_PRIORITY,
                THREAD_TIMESLICE);
```

```
    if (tid != RT_NULL)
    {
        rt_thread_startup(tid);
        return 0;
    }
    else
    {
        return -1;
    }
}
```

The complete project can refer to the accompanying materials under the chapter10/4-heap-init directory. Download the program to the IoT board, and you can see the output log information of the IoT board in the PuTTY tool interface, as shown in Figure 10.15. At this point, RT-Thread is running smoothly, and the porting RT-Thread succeeds!

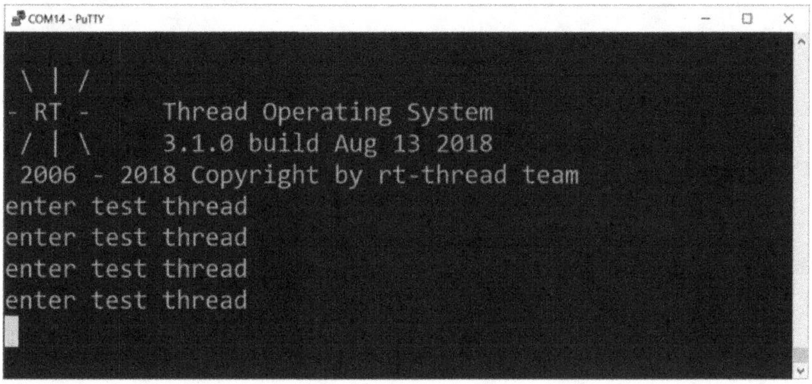

FIGURE 10.15 Console output.

PORT TO MORE DEVELOPMENT BOARDS

According to the earlier methods, the RT-Thread kernel can be easily ported to other development boards.

CHAPTER SUMMARY

This chapter described what porting is, what CPU architecture porting is, and what BSP porting is. Here's a review of this chapter to summarize the following key points:

1. Enable or disable the global interrupts and ensure that the rt_hw_interrupt_disable() and rt_hw_interrupt_enable() appear in pairs.
2. The porting of the CPU architecture includes the implementation of global interrupt-related, context-related interfaces, and OS Tick.
3. The porting of the OS Tick needs to ensure that the number of times its interrupt handler function executes per second is RT_TICK_PER_SECOND (this macro is defined in rtconfig.h).
4. The rt_kprintf() function provided by console to the upper layer can be implemented on the basis of the device or on the basis of rt_hw_console_output().

11 Env-Assisted Development Environment

In the previous chapter, we experienced the RT-Thread kernel porting process in a way of building a project from scratch, which deepening our understanding of RT-Thread. In fact, during project development, we rarely need to port the RT-Thread kernel to a board, because RT-Thread has been ported to many mainstream chips and boards. Dozens of BSPs(Board Support Packages) have been supported in RT-Thread version 3.1.0. What we need to do is find the specific BSP suited to our own development board and learn how to use the Env-assisted development environment to build a project framework, configure the kernel, and component functions, download software packages required by the project, and automatically generate the MDK(Microcontroller Development Kit) project for creating applications on top of this basis.

This chapter begins with an example of the features and the usage of the Env-assisted development environment. Then it shows an example of the RT-Thread application development process.

ENV INTRODUCTION

RT-Thread supports dozens of BSPs, multiple compilers, and IDEs(Integrated Development Environment), as well as a wide range of underlying components and a growing number of software packages. For project development, however, only one or a limited number of MCUs(Microcontroller Unit) are required, using a familiar IDE development environment and using limited peripherals and components. Env is an auxiliary development environment for the development scenario with RT-Thread, which helps developers build a project that fits their target application based on a full-featured version of the RT-Thread source code. The Env environment consists of SCons compiling and construction tools, menuconfig graphical configuration tools, and packages software package management tools, QEMU(Quick EMUlator) simulators, etc. The following sections will explain how these tools are mainly used.

Env is portable software; the Env version in this book is 1.0.0. The Env interface is shown in Figure 11.1.

ENV FEATURES

COMPILING AND CONSTRUCTION

In Env, we use the compiling construction tool SCons for compiling and constructing the source code of the RT-Thread projects. The SCons's main functions are:

1. Create the RT-Thread project framework. Extract suitable source code from the RT-Thread source code repository, which is suited for a particular board.
2. Generate the mainstream IDE (such as MDK, IAR, etc.) projects with RT-Thread ported according to the rtconfig.h configuration file.
3. It also provides a command-line compiling method that supports RT-Thread project compiling using different compilers, such as GCC, ARMCC, etc.

After entering the BSP root directory with the Env tool, you can manage the BSP with some of the commands provided by SCons. The following is an introduction for the commonly used SCons command.

FIGURE 11.1 Env working interface.

BUILD PROJECT FRAMEWORK

The RT-Thread framework can be generated using the "scons --dist" command. First, select a corresponding BSP from the RT-Thread source code, and then use the "scons --dist" command in the BSP root directory to generate a dist directory, which is the new project framework, containing the RT-Thread source code and BSP-related project. Unrelated BSP folders and libcpus are removed and can be copied to any directory at will.

GENERATE A NEW PROJECT

If you are using an IDE, such as MDK/IAR for project development, you can use one of the following commands to rebuild the project after the configuration so that the source code associated with the configuration options is automatically added to the new project. Later, you can use the IDE to open the project, compile, and download to your board.

The following command is used to generate the Keil MDK5 project:

```
scons --target=mdk5
```

The following command is used to generate the Keil MDK4 project:

```
scons --target=mdk4
```

The following command is used to generate the IAR project:

```
scons --target=iar
```

COMPILE PROJECT

In the BSP directory, enter the "scons" command to compile the project using the default ARM_GCC toolchain. Most of the BSP of the ARM platform's chip supports this command.

GRAPHICAL SYSTEM CONFIGURATION

Menuconfig is a Kconfig-based graphical configuration tool that RT-Thread uses to configure, tailor the entire system, and eventually generate the rtconfig.h configuration file required by the project. The main functions of menuconfig are as follows:

1. The RT-Thread kernel, components, and software packages can be graphically configured and tailored to automatically generate the rtconfig.h configuration file.
2. The ability to automatically process the dependency between configuration items.

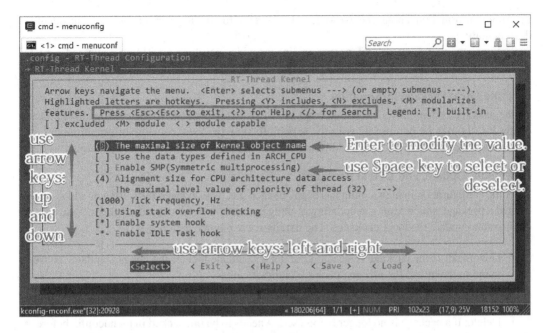

FIGURE 11.2   Introduction to shortcut keys of menuconfig.

Use the Env tool access to the BSP root directory, and enter the menuconfig command to open the configuration interface. The configuration menu is divided into the following three categories:

```
RT-Thread Kernel    --->              [Kernel Configuration]

RT-Thread Components   --->           [Component Configuration]

RT-Thread online packages   --->     [Online Software Package Configuration]
```

The common shortcut keys for menuconfig are shown in Figure 11.2.

There are many types of configuration items in the menuconfig. The modification method is also different. The common types are shown as follows:

1. Bool type: Use the space key to check or uncheck. The * in parentheses means it is selected.
2. Numeric, string type: A dialog box appears when the Enter key is pressed, and the configuration items are modified in the dialog box.

After the configuration is finished, press ESC key to exit, and select Save to automatically generate the rtconfig.h configuration file.

PACKAGE MANAGEMENT

RT-Thread provides an online software package management platform where it stores official or community packages. The platform offers developers a wide range of reusable packages. This is one of the highlights in the RT-Thread ecosystem.

The official packages provided by RT-Threads are hosted on GitHub. Most packages are provided with detailed documentation and examples of use. The current number of packages is 200+.

For more information about the package, please refer to the RT-Thread package introduction documentation.

The package management tool, as part of Env, provides developers with the download, update, and delete management functions of the package.

Upgrade Local Package Information

As the package system grows, more and more packages will be added. So the package list in menuconfig locally may be out-of-date with the server. To sync the local package information with the server, use "`pkgs --upgrade`." Additionally, this command will update Env's functional scripts.

Download, Update, and Delete Packages

After the configuration in manuconfig has been changed (i.e., you changed the setting or added and/or deleted a package, which results in updating rtconfig.h), you will need to use the "`pkgs --update`" command to update the local packages. The details are explained as follows:

- Download: If the package is selected but not downloaded, enter the "pkgs --update" command to download the missing package automatically.
- Update: If the selected package has been updated on the server side and the version number selects the "latest", the local package will be updated.
- Delete: If a package is no longer being used, it needs to be unchecked in menuconfig before executing the pkgs --update command. Packages that have been downloaded locally but are not selected will be deleted.

Notes: When using the "pkgs --update" command, the "git clone" feature is used, so you need to install the git tool first; otherwise, you will be prompted to download or the package will fail to update.

QEMU Simulator-Assisted Development

In the absence of a hardware environment, QEMU simulators can be used to virtualize the hardware environment. QEMU is a virtual machine that supports cross-platform virtualization and can virtualize many hardware environments. RT-Thread provides an ARM vexpress A9 BSP simulated by QEMU. Developers can run RT-Thread based on this BSP. Virtual hardware environments can assist in application development and debugging, reduce development costs, and improve efficiency. To learn how to use QEMU please refer to the appendix "Getting Started with QEMU (Ubuntu)."

EXAMPLE OF BUILDING A PROJECT WITH Env

The main features of the Env tool were described earlier. This section will demonstrate a practical example of how to build an MDK project with FinSH functionality based on the stm32l475-atk-pandora BSP. (Note: The code is located in the chapter12 directory with the accompanying materials.) The built project is compiled to run on the Internet of Things (IoT) board.

The tools and source codes required for building the project are under the root directory of the accompanying code package.

REGISTER Env to the Right-Click Menu

Find the Env directory and double-click env.exe or env.bat to open the Env tool with the Env console, interface as shown here. Register Env to the right-click menu, as per Figures 11.3 and 11.4.

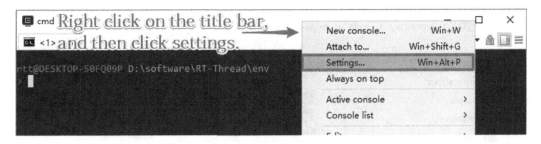

FIGURE 11.3 Open Env for settings.

After registration, right-click under a BSP path (e.g., "code-rt-thread-3.1.0\bsp\stm32l475-atk-pandora" directory). The "ConEmu Here" option will show up on the right-click menu. Click "ConEmu Here" to configure the BSP using Env, as shown in Figure 11.5.

Click "ConEmu Here" to open Env; therefore, the path in Env is located to the BSP directory, so that you can start configuring easily. See Figure 11.6 for details.

Note 1: Because of the need to set the environmental variables of the Env process, the first start-up may indicate false-positive anti-virus results. Allow Env-related programs to run if this happens, and then add the relevant programs to the whitelist.

Note 2: In Env working environments, Chinese characters or spaces are not allowed in any paths.

Note 3: If you do not want to register Env to the right-click menu, you can open env.exe or env.bat, and use the command "cd+space+a BSP path" to navigate to your specific BSP root directory.

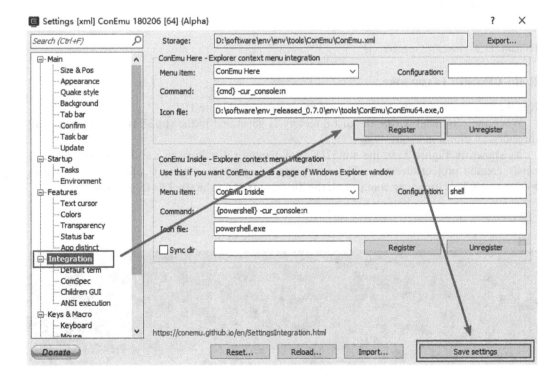

FIGURE 11.4 Register Env to the right-click menu.

FIGURE 11.5 Result of register.

BUILD A PROJECT FRAMEWORK

Run the "`scons --dist`" command in Env, as shown in Figure 11.7.

When the command is run, the dist directory is generated in the stm32l475-atk-pandora BSP directory, as shown in Figure 11.8.

As shown in Figure 11.9, the stm32l475-atk-pandora directory in the dist directory is the newly created project directory. The new project framework directory structure is shown in Figure 11.10. This directory contains all the files required for the stm32l475-atk-pandora project,

FIGURE 11.6 Click ConEmu here to open Env tool.

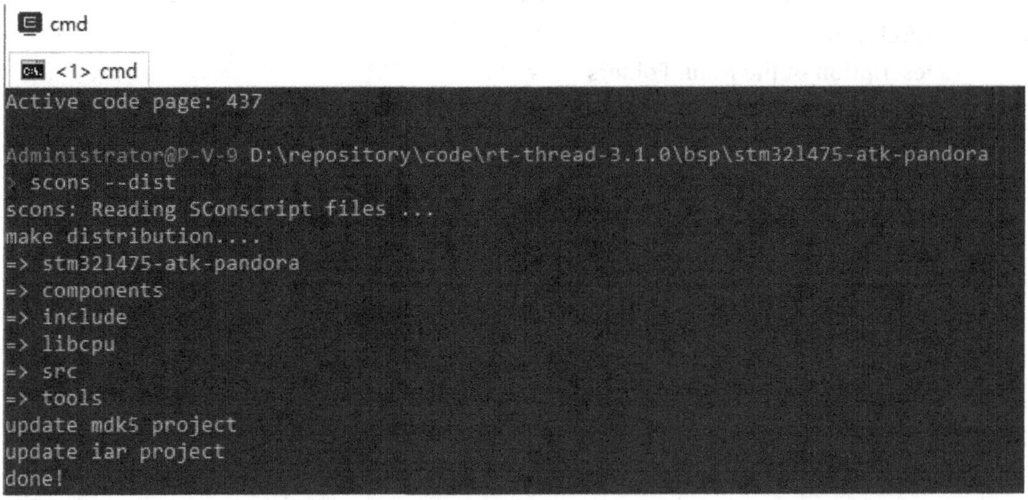

FIGURE 11.7   Use the dist command in the BSP directory.

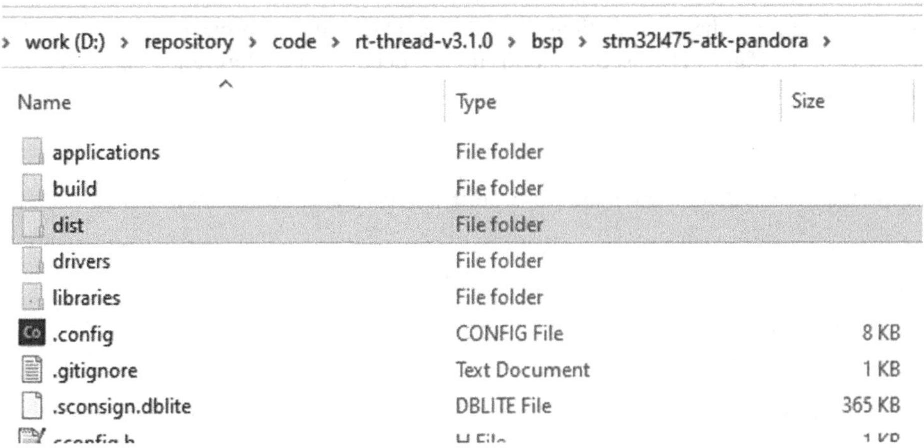

FIGURE 11.8   A new dist folder.

which can be copied to any directory, and the subsequent configuration development will be based on this.

The description of the main folders and files is shown in Table 11.1.

Open the MDK project by default, as shown in Figure 11.11. The project contains the RT-Thread kernel and other groups, which is similar to the projects provided by the other BSPs. For the detailed description of the source code grouping, please refer to Table 2.2 in Chapter 2.

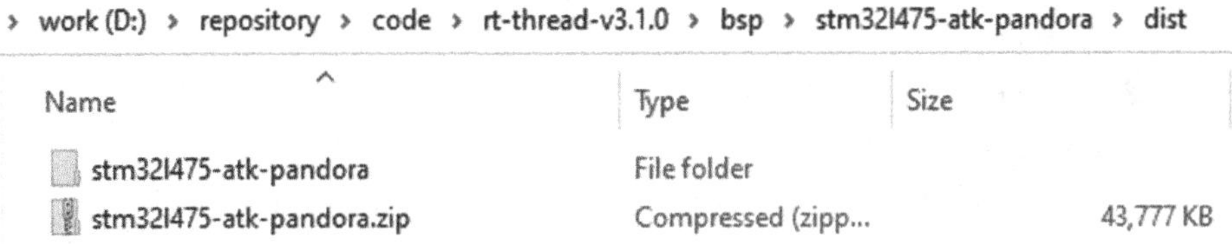

FIGURE 11.9   New project in the dist folder.

TABLE 11.1

Description of the Main Folders

Directory/File	Description
applications	Developer's application code file
drivers	Device's driver
Libraries	The firmware library downloaded from the chip's official website
rt-thread	RT-Thread source code
Kconfig	Files used for menuconfig
project.uvprojx	MDK project files used by developers
rtconfig.h	Project configuration file
SConscript	Files used for the SCons configuration tool
SConstruct	Files used for the SCons configuration tool
template. uvprojx	MDK template project file

MODIFY THE TEMPLATE PROJECT

When modifications to the MCU model or debugging options are needed, it is recommended to modify the template project directly, so that new projects generated by using SCons-related commands will also include the modifications from the template file. The project template file of MDK is template.uvprojx. Figure 11.12 shows modifying the debugging configuration option for modifying the template project file.

> bsp > stm32l475-atk-pandora > dist > stm32l475-atk-pandora

Name	Size
applications	
build	
drivers	
libraries	
rt-thread	
.config	
.gitignore	
.sconsign.dblite	
cconfig.h	
Kconfig	
project.ewd	
project.ewp	
project.eww	
project.uvoptx	
project.uvprojx	

FIGURE 11.10   Framework of the new project.

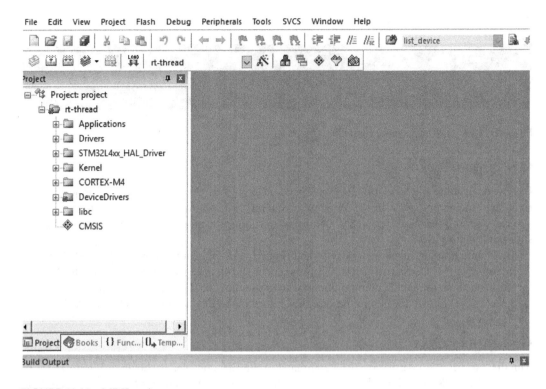

FIGURE 11.11 MDK project.

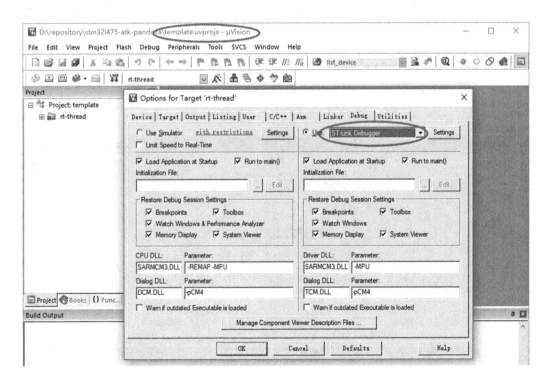

FIGURE 11.12 Modify template.uvprojx's configuration.

FIGURE 11.13   The menuconfig command.

CONFIGURE THE PROJECT

In Env, navigate to the new stm32l475-atk-pandora project directory (or go to the project folder, right-click the empty space in the Windows Explorer, and select "ConEmu Here" to open Env), you can then run the "`menuconfig`" command to run menuconfig for configuring the project, as shown in Figure 11.13.

The configuration menu is opened, and FinSH can be configured in the "RT-Thread Components → Command shell" sub-menu, as shown in Figure 11.14. Select the first option "[*] finsh shell" in the FinSH configuration menu to use the FinSH component in the project.

Select "save" to save the configuration. Then return to the Env console by selecting "exit."

GENERATE THE PROJECT

Run the "`scons --target=mdk5`" command in the Env console, as shown in Figure 11.15. In this way, the MDK project will be regenerated. The source code related to the configuration will be added to the project, and the newly generated MDK5 project will be called project.uvprojx.

Note: The newly generated project is based on the project template, which will overwrite the previous manual modifications to the project file project.uvprojx (i.e., the debugger setting mentioned in the previous section).

Open the project.uvprojx, and now you can see that the newly generated project already contains FinSH components, as shown in Figure 11.16.

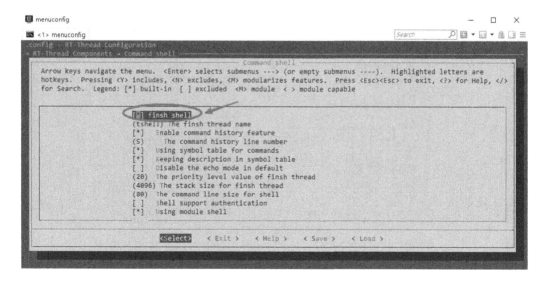

FIGURE 11.14   Configure FinSH.

```
cmd
<1> cmd                                                              Search

Administrator@P-V-9 D:\repository\code\rt-thread-3.1.0\bsp\stm321475-atk-pandora\dist\stm321475-atk-pandora
scons --target=mdk5
scons: Reading SConscript files ...
scons: done reading SConscript files.
scons: Building targets ...
scons: building associated VariantDir targets: build
CC build\applications\application.o
CC build\applications\startup.o
CC build\drivers\board.o
CC build\drivers\usart.o
CC build\kernel\components\drivers\misc\pin.o
CC build\kernel\components\drivers\serial\serial.o
CC build\kernel\components\drivers\src\completion.o
CC build\kernel\components\drivers\src\dataqueue.o
CC build\kernel\components\drivers\src\pipe.o
CC build\kernel\components\drivers\src\ringblk_buf.o
```

FIGURE 11.15 Generate new project.

RUN AND VERIFY

Connect the IoT board's uart1 to a terminal via a serial port. Compile and download the project to the development board, then press Reset. The RT-Thread logo and the version information will be printed on the terminal. FinSH is also running; press the Tab key to view all the commands, as follows:

```
 \ | /
- RT -     Thread Operating System
 / | \     3.1.0 build Sep 13 2018
 2006 - 2018 Copyright by rt-thread team

msh >
```

BUILD MORE MDK PROJECTS

Examples and projects that match this book are available on the Code page, but other MDK5 projects that will be used in the subsequent sections of this book can be built similarly using the approach in the section "Example of Building a Project with ENV". This section is based on that

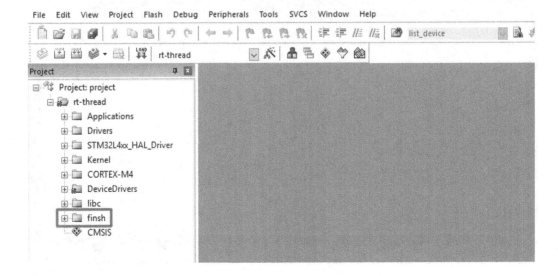

FIGURE 11.16 Project add FinSH component.

section, since it has done the preparation work based on stm32l475-atk-pandora, built the project framework, and modified the project template, so the work that follows can start directly from the steps of the configuration project.

CREATE A PERIPHERAL EXAMPLE PROJECT

Copy the project generated in the section "Example of Building a Project with ENV" to build the MDK project which is mentioned in the demo projects section of Chapter 14.

Configure Device Drivers

Run the "menuconfig" command to configure the project in the Env console, according to the options in the red box in Figure 11.17: Enable UART1 (Enable FinSH function), UART2 (Enable Serial Port Device), IIC (Enable I2C Device), QSPI1 (Enable QSPI Device), and GPIO (Enable PIN Device) via the "Hardware Drivers Config - - -> On-chip Peripheral Drivers- - ->" menu.

Configure the Peripheral Example Software Package

To configure the peripheral example package, first enable the peripheral example option under "RT-Thread online packages - - -> miscellaneous packages - - -> samples: kernel and components samples."

```
[*] a peripheral_samples package for rt-thread (NEW) --->
```

Then press Enter to select the menu option and turn on the serial device (UART2 serial port device routine), i2c device (I2C device routine), spi device (QSPI device routine), and pin device (PIN device routine), as shown in Figure 11.18.

After the project is configured and saved, return to the Env console.

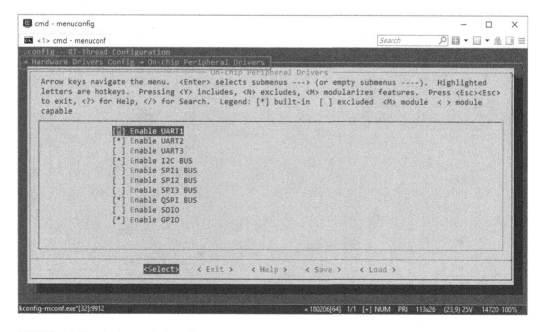

FIGURE 11.17 Configure device drivers.

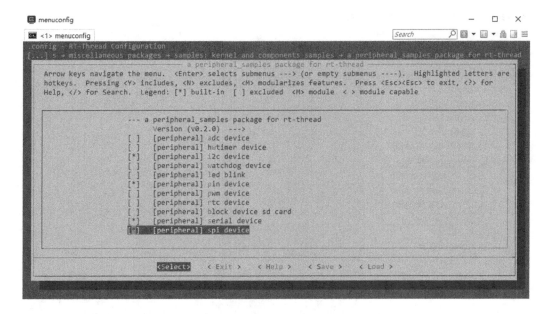

FIGURE 11.18   Configure the peripheral example software package.

Download the Software Package

Because the package's configurations are updated, the "`pkgs --update`" command needs to be run to download/update the package, as shown in Figure 11.19.

Generate the Project

Run the "`scons --target=mdk5`" command to regenerate the project in the Env console. Once it is done, open the newly generated project; you will find that the project grouping already contains the peripheral routine folder "peripheral-samples." Figure 11.20 shows the folder is the peripheral usage routine.

A project that has been generated following these steps has been placed in the chapter14 directory of the accompanying materials, Chapter 14 demo projects section would give an example.

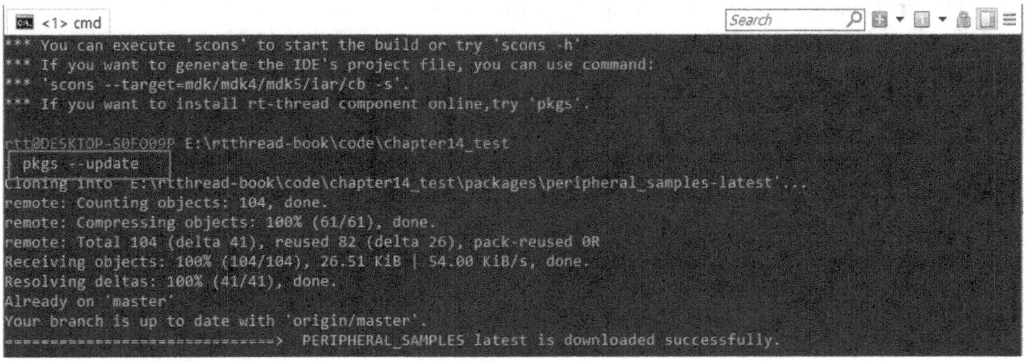

FIGURE 11.19   Download the software package.

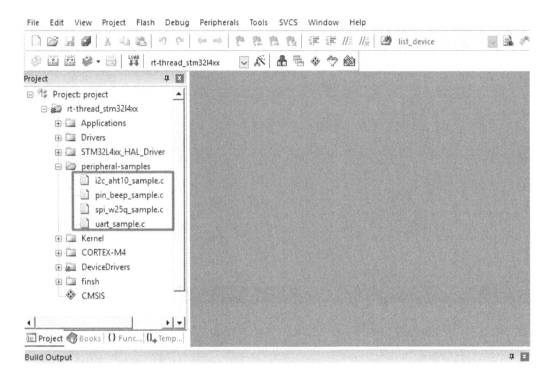

FIGURE 11.20   RT-Thread device example.

CREATE A FILE SYSTEM EXAMPLE PROJECT

Copy the project generated in the section "Example of Building a Project with ENV", to build the project that would be used in the DFS application example section of Chapter 15.

Configure to Enable DFS Components

Run the "menuconfig" command to configure the project in the Env console.

1. According to the options in the red box in Figure 11.21: Enable UART1 (Enable FinSH function), QSPI1 (Enable QSPI Device), GPIO (Enable PIN device) via the "Hardware Drivers Config - - -> On-chip Peripheral Drivers - - ->" menu.
2. Enable FLASH according to the menu directory prompt in Figure 11.22.
3. Enable the file system according to the menu directory prompt in Figure 11.23.
4. Then enter the "elm-chan's FatFs, Generic FAT Filesystem Module - - ->" menu to configure, as shown in Figure 11.24.

Configure the File System Example Package

To configure the file system example package, first, enable the file system example option under the "RT-Thread online packages - - -> miscellaneous packages - - -> samples: kernel and components samples - - ->" menu:

```
[*] a filesystem_samples package for rt-thread --->
```

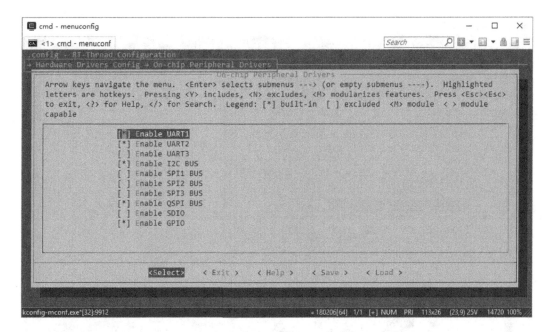

FIGURE 11.21 Hardware configuration 1.

Then press Enter to select the menu option and enable all the examples, as shown in Figure 11.25.
After the project is configured and saved, return to the Env console.

Download the Software Package

Because the relevant packages are configured, the "pkgs --update" command needs to be run in the Env console to download/update the package, as shown in Figure 11.26.

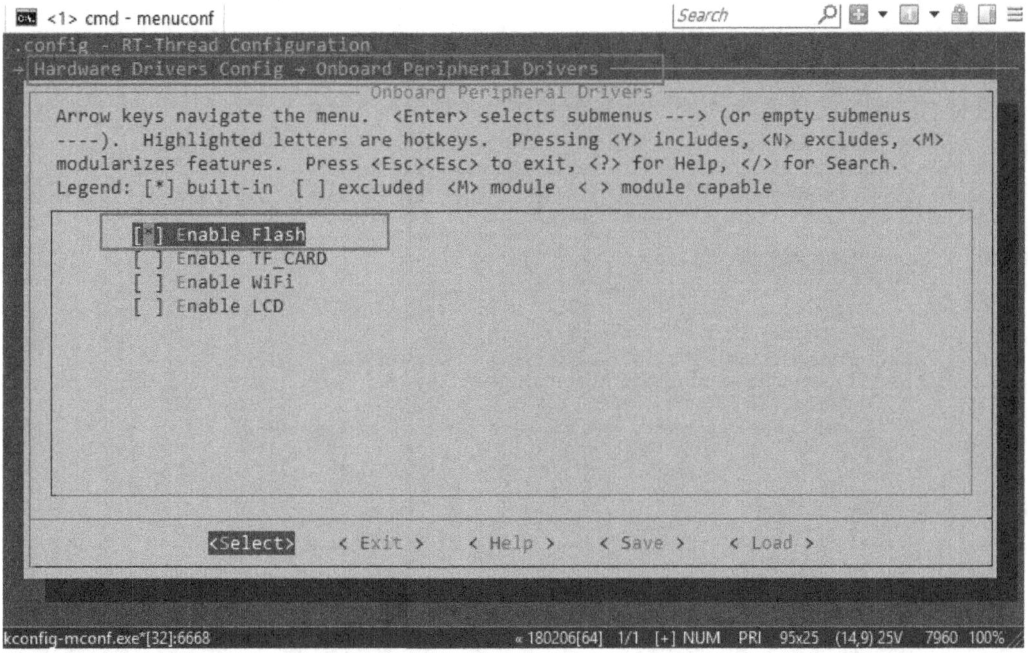

FIGURE 11.22 Hardware configuration 2.

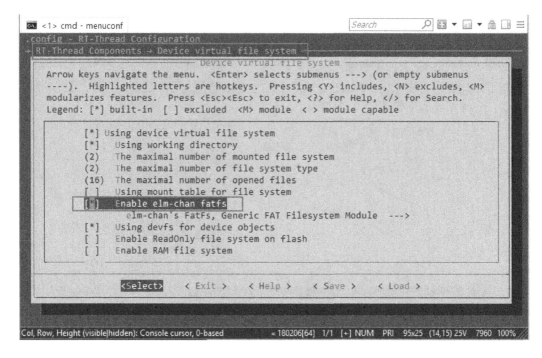

FIGURE 11.23   Enable file system.

Generate the Project

Running the "`scons --target=mdk5`" command to regenerate the project in the Env console and opening the newly generated project, you will find that the project grouping already contains the peripheral routine folder "filesystem-samples," and under the folder is the file system usage routine, as shown in Figure 11.27.

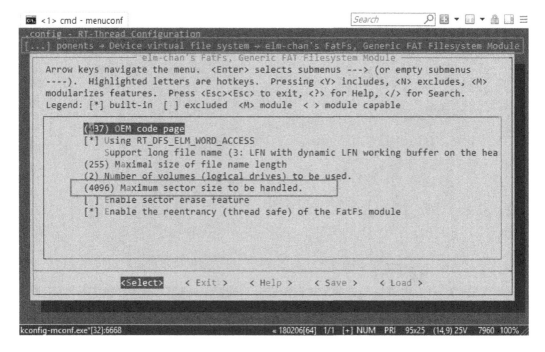

FIGURE 11.24   config the file system example.

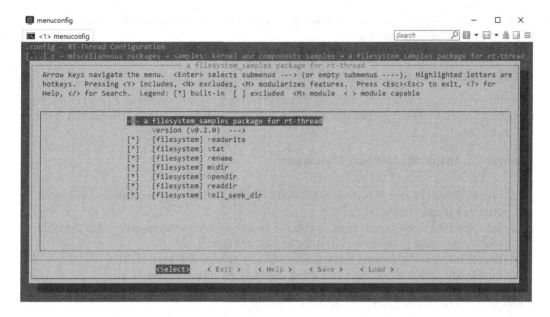

FIGURE 11.25 File system example.

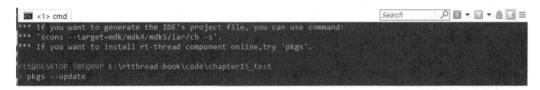

FIGURE 11.26 Download software package.

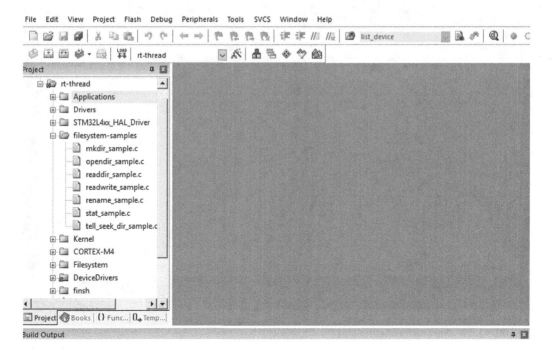

FIGURE 11.27 File system example project.

A project that has been generated following these steps has been placed in the chapter15 directory of the accompanying materials.

CREATE A NETWORK EXAMPLE PROJECT

Copy the project generated in the section "Example of Building a Project with ENV," rename it to chapter16_test, to build the project that would be used in the network application example section of Chapter 16.

Configure to Enable the Network Component

1. Enable GPIO (Enable PIN device) and SPI2 (Enable SPI device) according to the menu directory prompt in Figure 11.28.
2. Set the network abstraction layer, enable the socket abstraction layer, and enable the BSD socket, according to the menu directory prompt in Figure 11.29.
3. Enable lwip stack according to the menu directory prompt in Figure 11.30.
4. Enable ENC28J60 according to the menu directory prompt in Figure 11.31.

Configure the Network Example Software Package

1. Enter the "RT-Thread online packages --->IoT - internet of things --->" menu for configuration.

```
[*] netutils: Networking utilities for RT-Thread --->
```

Then press Enter to select this option and turn on the ping function, as shown in Figure 11.32.

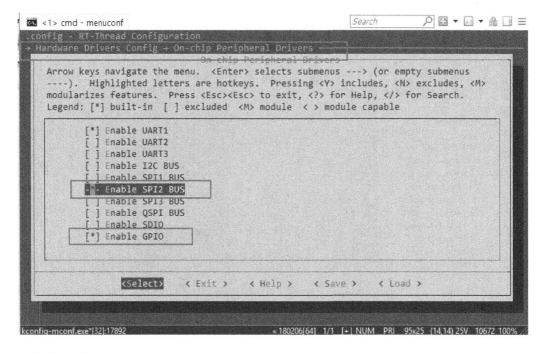

FIGURE 11.28 Network configuration 1.

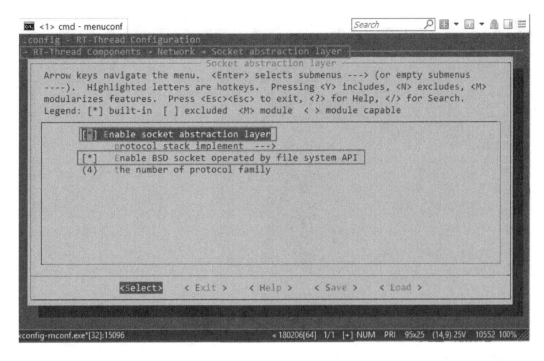

FIGURE 11.29 Network configuration 2.

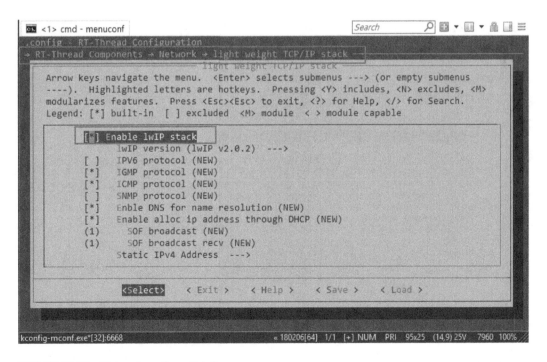

FIGURE 11.30 Network configuration 3.

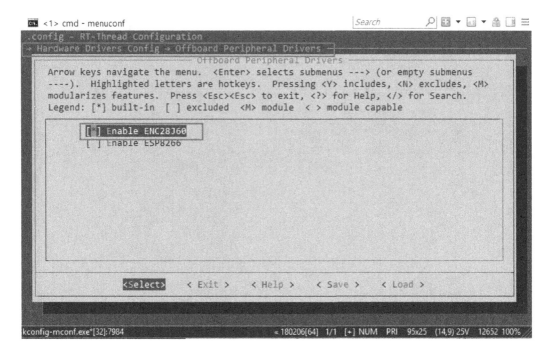

FIGURE 11.31 Network configuration 4.

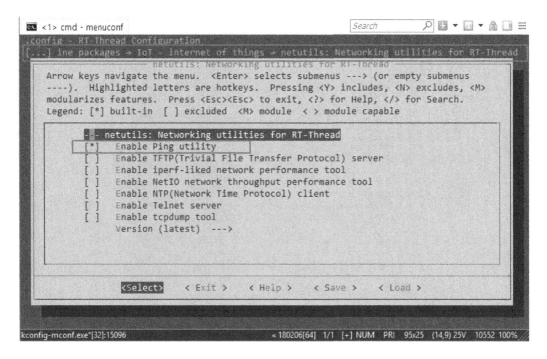

FIGURE 11.32 Configuration of network tool package.

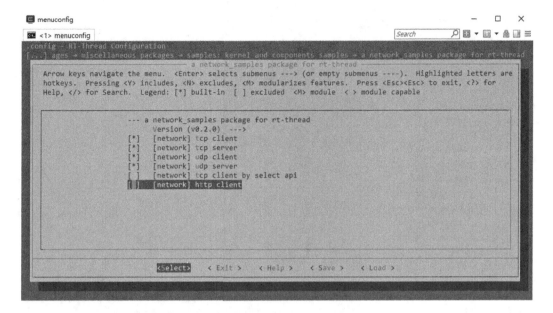

FIGURE 11.33   Configuration of network package.

2. Enable the file system example option under the "RT-Thread online packages - - -> miscellaneous packages - - -> samples: kernel and components samples - - ->"menu:

    [*] a network_samples package for rt-thread - - ->

Then press Enter to select this option and enable the example, as shown in Figure 11.33.
    After the project is configured and saved, return to the Env console.

Download the Software Package

Because the packages are configured, the "pkgs --update" command needs to be run in the Env console to download/update the package, as shown in Figure 11.34.

Generate the Project

Running the "scons --target=mdk5" command to regenerate the project in the Env console and opening the newly generated project, you will find that the project grouping already contains the peripheral routine folder "network-samples," and under the folder is the network routine, as shown in Figure 11.35.

    A project that has been generated following these steps has been placed in the chapter16 directory of the accompanying materials.

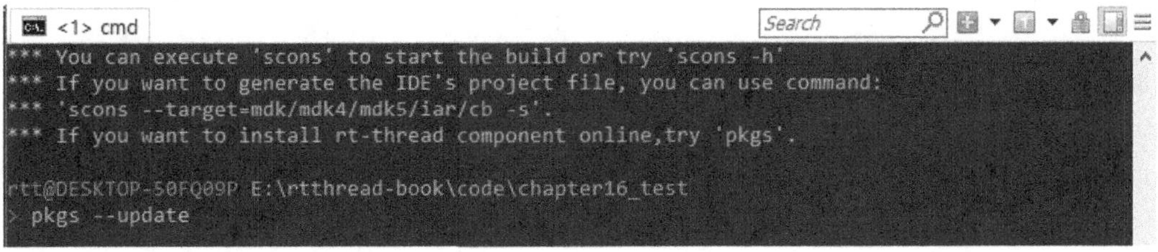

FIGURE 11.34   Download the software package.

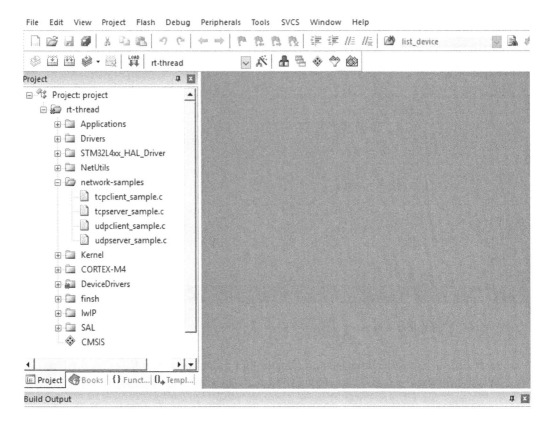

FIGURE 11.35   Network sample.

CHAPTER SUMMARY

This chapter described the functional features and the usage of Env and also introduced the Env environment, which consists of SCons compiling construction tools, menuconfig graphical system configuration tools, and software package management tools, QEMU simulators, and others.

12 FinSH Console

In the early days, before the advent of graphics systems, there was no such thing called mouse or even a keyboard. How did people interact with a computer then? The answer is using a punch card. Later, the great innovation of the computer, monitor, and keyboard became the standard configuration. However, the operating system at that time did not support graphical user interface yet. So computer pioneers developed a program that accepts commands from the user, interprets those commands, then hands over the output to operating system, and finally returns the result provided by operating system to the user. This program wraps around the operating system like a layer of shell. So it got the name "shell."

For embedded systems, a shell-like program is a very useful tool for debugging and testing.

The "shell" of RT-Thread is FinSH, pronounced ['fɪnʃ].

INTRODUCTION

FinSH is the CLI (Command Line Interface) component of RT-Thread, which provides a set of utilities to facilitate debugging, checking system status, etc. A user may access the CLI of a target device from the terminal of the host PC through a serial port, USB, or Ethernet, as shown in Figure 12.1.

The following process happened when a user typed some text (ending with ↵) in the terminal:

- The corresponding device driver hands over the input to FinSH.
- FinSH parses the text for any valid commands.
- If there are any, FinSH executes the command and hands over the result to the corresponding device driver.
- The corresponding device driver then sends the data to the host PC.

Figure 12.2 shows the process with a serial port as an example.

To improve security, FinSH provides a simple and effective authentication function. When enabled, the developer has to enter predefined password before FinSH can be started.

Last but not least, the most awesome feature of FinSH is that it supports auto-completion and history. Table 12.1 shows how to access the feature.

For the syntax of input, FinSH supports traditional command line mode and C language interpreter mode.

TRADITIONAL COMMAND LINE MODE

This mode is also known as MSH (Module SHell). In MSH mode, FinSH is more in line with the traditional shell (e.g., DOS or Bash) usage habits. For example, the developer may issue the command "cd /" to switch the working directory to root directory.

In MSH mode, the command (build-in command or loadable module name) and each argument are separated by a space as the following format shows:

```
command [arg1] [arg2] [...]
```

C LANGUAGE INTERPRETER MODE

This mode is also known as C-style mode. In C-style mode, FinSH follows the C language syntax and supports (predefined) variables. In addition, it can create variables through the command line.

TABLE 12.1
FinSH Key Mapping

Key	Description
⇥ (Tab)	When input is empty, FinSH prints out all supported commands; otherwise, it performs auto-completion
↑ and ↓ (Arrow up and down)	Selects previous inputs from history
⌫ (Backspace)	Deletes one character before the cursor
← and → (Arrow left and right)	Moves the cursor left or right

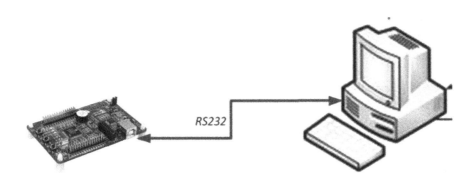

FIGURE 12.1   Accessing FinSH CLI from the host.

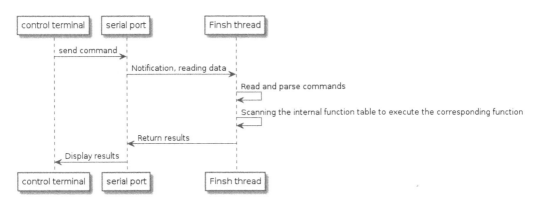

FIGURE 12.2   FinSH command execution flowchart.

In C-style mode, commands and arguments are arranged by calling a C function. For example, to list the information of all threads in the current system, the developer may issue the command "list _ thread()." After executing the given command, FinSH also prints out its return value in decimal and hexadecimal formats. For those functions without a return value (return "void"), a random number is printed out.

Initially, FinSH supported C-style mode only. However, RT-Thread developers preferred the traditional shell usage habits more. Furthermore, this mode requires quite a lot of memory space. Therefore, MSH mode comes to the rescue.

In a target device, if both the modes are enabled, MSH mode is selected by default. A developer may issue the command "exit" to switch to C-style mode and later issue the command "msh()" to switch back to MSH mode.

BUILT-IN COMMANDS

To print out the list of all built-in commands available in the current system:

- In C-style mode, issue the command "help()."
- In MSH mode, type "⇆."

The number of commands printed out is not fixed—it depends on the system configuration. For example, when the DFS (Device File System) component is enabled, the commands such as "ls," "cp," and "cd" will then appear in the list.

Due to the fact that most of the command names are shared between these two modes, MSH mode is demonstrated in the following sections if not indicated.

Following is an example of a list of built-in commands:

```
RT-Thread shell commands:
version        - show RT-Thread version information
list_thread    - list thread
list_sem       - list semaphore in system
list_event     - list event in system
list_mutex     - list mutex in system
list_mailbox   - list mail box in system
list_msgqueue  - list message queue in system
list_timer     - list timer in system
list_device    - list device in system
exit           - return to RT-Thread shell mode
help           - RT-Thread shell help
ps             - List threads in the system
time           - Execute command with time
free           - Show the memory usage in the system
```

Let's check the most useful commands in detail.

DISPLAY THREAD STATUS

The "ps" or "list _ thread" command is used to list the information of all the threads in current system, including thread priority, state, maximum stack usage, and more. Table 12.2 describes its return field.

```
msh />list_thread
thread    pri  status   sp         stack size max used left tick   error
--------  ---  ------   ---------- ---------- ------   ---------- ---
tshell    20   ready    0x00000118 0x00001000    29%   0x00000009 000
tidle     31   ready    0x0000005c 0x00000200    28%   0x00000005 000
timer      4   suspend  0x00000078 0x00000400    11%   0x00000009 000
```

TABLE 12.2
"list_thread" Output Description

Field	Description
thread	Thread name
pri	The current priority
status	The current status
sp	The current stack pointer
stack size	Stack size
max used	The maximum used stack in percentage
left tick	The number of remaining ticks
error	Error code

DISPLAY SEMAPHORE STATUS

The "list _ sem" command is used to list the information of all the semaphores in the current system, including the name of the semaphore, the value of the semaphore, and the number of threads waiting for this semaphore. Table 12.3 describes its return field.

```
msh />list_sem
semaphore v   suspend thread
-------- --- --------------
shrx      000 0
e0        000 0
```

TABLE 12.3
"list_sem" Output Description

Field	Description
semaphore	Semaphore name
v	The current value
suspend thread	The number of blocked threads

DISPLAY EVENT STATUS

The "list _ event" command is used to list the information of all the events in the current system, including the event name, the value of the event, and the number of threads waiting for this event. Table 12.4 describes its return field.

```
msh />list_event
event      set    suspend thread
-----    ---------- --------------
```

TABLE 12.4
"list_event" Output Description

Field	Description
event	Event name
set	The current value
suspend thread	The number of blocked threads

DISPLAY MUTEX STATUS

The "list_mutex" command is used to list the information of all the mutexes in the current system, including the mutex name, the owner of the mutex, and the number of nestings the owner holds on the mutex. Table 12.5 describes its return field.

```
msh />list_mutex
mutex     owner      hold suspend thread
--------  --------  ---- --------------
fat0      (NULL)    0000 0
sal_lock  (NULL)    0000 0
```

TABLE 12.5
"list_mutex" Output Description

Field	Description
mutex	Mutex name
owner	The thread currently holding the mutex
hold	The nest count on the mutex held by a thread
suspend thread	The number of blocked threads

DISPLAY MAILBOX STATUS

The "list_mailbox" command is used to list the information of all the mailboxes in the current system, including the mailbox name, the number of messages in the mailbox, and the maximum number of messages the mailbox can hold. Table 12.6 describes its return field.

```
msh />list_mailbox
mailbox   entry size suspend thread
--------  ----  ---- --------------
etxmb     0000  0008 1:etx
erxmb     0000  0008 1:erx
```

TABLE 12.6
"list_mailbox" Output Description

Field	Description
mailbox	Mailbox name
entry	The current number of messages
size	The maximum number of messages
suspend thread	The number of blocked threads

DISPLAY MESSAGE QUEUE STATUS

The "list_msgqueue" command is used to list the information of all the message queues in the current system, including the name of the message queue, the number of messages it contains, and the number of threads waiting for this message queue. Table 12.7 describes its return field.

```
msh />list_msgqueue
msgqueue entry suspend thread
--------  ----  --------------
```

TABLE 12.7
list_msgqueue Return Field Description

Field	Description
msgqueue	Message queue name
entry	The current number of messages
suspend thread	The number of blocked threads

DISPLAY MEMORY POOL STATUS

The "list _ mempool" command is used to list the information of all the memory pools in the current system, including the name of the memory pool, the size of the memory pool, and the maximum memory size used. Table 12.8 describes its return field.

```
msh />list_mempool
mempool block total free suspend thread
------- ----  ----  ---- --------------
signal  0012  0032  0032 0
```

TABLE 12.8
list_mempool Return Field Description

Field	Description
mempool	Memory pool name
block	Block size
total	The total number of memory blocks
free	The number of free memory blocks
suspend thread	The number of blocked threads

DISPLAY TIMER STATUS

The "list _ timer" command is used to list the information of all the timers in the current system, including the name of the timer, whether it is the periodic timer, and the number of beats of the timer timeout. Table 12.9 describes its return field.

```
msh />list_timer
timer     periodic    timeout     flag
--------  ----------  ----------  -----------
tshell    0x00000000  0x00000000  deactivated
tidle     0x00000000  0x00000000  deactivated
timer     0x00000000  0x00000000  deactivated
```

TABLE 12.9
list_timer Return Field Description

Field	Description
timer	Timer name
periodic	Period time in ticks
timeout	Timeout time in ticks
flag	Status: activated or deactivated

DISPLAY DEVICE STATUS

The "list _ device" command is used to list the information of all the devices in the current system, including the device name, device type, and the number of times the device was opened. Table 12.10 describes its return field.

```
msh />list_device
device          type       ref count
------ ------------------ ----------
e0     Network Interface 0
uart0  Character Device  2
```

TABLE 12.10
list_device Return Field Description

Field	Description
device	Device name
type	Device type
ref count	The number of times the device been currently referenced (opened)

DISPLAY DYNAMIC MEMORY STATUS

The "free" command is used to list the memory information of the current system. Table 12.11 describes its return field.

```
msh />free
total memory: 7669836
used memory : 15240
maximum allocated memory: 18520
```

TABLE 12.11
free Return Field Description

Field	Description
total memory	Total memory in bytes
used memory	Used memory in bytes
maximum allocated memory	Maximum allocated memory in bytes

USER-DEFINED FinSH COMMAND

Beside those built-in commands, developers may add their own commands to FinSH.

USER-DEFINED MSH COMMAND

The custom MSH command can be run in MSH mode. To export a command to MSH mode, you can use the following macro interface. Table 12.12 describes its parameters. Developers may use the following macro to export a function to FinSH as a command in MSH mode:

```
MSH_CMD_EXPORT(name, desc);
```

TABLE 12.12
MSH_CMD_EXPORT

Parameter	Description
name	Command (function) name
desc	A short description

An example to export a function without a parameter in MSH mode is as follows:

```
void hello(void)
{
    rt_kprintf("hello RT-Thread!\n");
}

MSH_CMD_EXPORT(hello , say hello to RT-Thread);
```

And an example to export a function with parameters in MSH mode is as follows:

```
static void atcmd(int argc, char**argv)
{
    ......
}
MSH_CMD_EXPORT(atcmd, atcmd sample: atcmd <server|client>);
```

USER-DEFINED C-STYLE COMMAND AND VARIABLE

Developers may use the following macro to export a function to FinSH as a command in C-style mode.

Export custom commands to C-style mode can use the following interface. Table 12.13 describes its parameters.

```
FINSH_FUNCTION_EXPORT(name, desc);
```

TABLE 12.13
FINSH_FUNCTION_EXPORT

Parameter	Description
name	Exporting function name
desc	A short description

An example to export a function without a parameter in C-style mode is as follows:

```
void hello(void)
{
    rt_kprintf("hello RT-Thread!\n");
}

FINSH_FUNCTION_EXPORT(hello , say hello to RT-Thread);
```

In a similar way, developers may use the following macro to FinSH in C-style. Table 12.14 describes its parameters.

```
FINSH_VAR_EXPORT(name, type, desc);
```

TABLE 12.14
FINSH_VAR_EXPORT

Parameter	Description
name	Variable name
type	Variable type
desc	A short description

An example to export a variable in C-style mode is as follows:

```
static int dummy = 0;
FINSH_VAR_EXPORT(dummy, finsh_type_int, dummy variable for finsh)
```

USER-DEFINED COMMAND WITH ALIAS

The maximum length of the FinSH command name is defined by the macro "FINSH _ NAME _ MAX" (finsh.h). Its default value is 16 bytes. This may cause problems when exporting a function whose name is longer than the limit. So the following macro is introduced. Table 12.15 describes its parameters.

```
FINSH_FUNCTION_EXPORT_ALIAS(name, alias, desc);
```

TABLE 12.15
FINSH_FUNCTION_EXPORT_ALIAS

Parameter	Description
name	Exporting function name
alias	Command name
desc	A short description

Please note that "FINSH_FUNCTION_EXPORT_ALIAS" is shared with MSH and C-style modes. If the command name (alias) starts with "__cmd_," the function will be exported to MSH mode; otherwise, it will be exported to C-style mode.

Following is an example to export "hello" function C-style mode with the command name "ho."

```
void hello(void)
{
    rt_kprintf("hello RT-Thread!\n");
}

FINSH_FUNCTION_EXPORT_ALIAS(hello , ho, say hello to RT-Thread);
```

CONFIGURATION

Table 12.16 shows all the FinSH-related configuration options (rtconfig.h).

TABLE 12.16
FinSH Configuration

Macro Definition	Value Type	Description	Default
RT_USING_FINSH	None	Enable FinSH	on
FINSH_THREAD_NAME	String	FinSH thread name	"tshell"
FINSH_USING_HISTORY	None	Enable historical tracing back	on
FINSH_HISTORY_LINES	Integer type	The maximum number of historical inputs can be traced back	5
FINSH_USING_SYMTAB	None	Enable symbol table in FinSH	on
FINSH_USING_DESCRIPTION	None	Enable showing command description	on
FINSH_USING_MSH	None	Enable MSH mode	on
FINSH_USING_MSH_ONLY	None	Use only MSH mode	on
FINSH_ARG_MAX	Integer type	The maximum number of command parameters	10
FINSH_USING_AUTH	None	Enable simple authentication	off
FINSH_DEFAULT_PASSWORD	String	The password used for authentication	off

Following is an example configuration of these options.

```
/* Open FinSH */
#define RT_USING_FINSH

/* Define the thread name as tshell */
#define FINSH_THREAD_NAME "tshell"

/* Open history command */
#define FINSH_USING_HISTORY
/* Record 5 lines of history commands */
#define FINSH_HISTORY_LINES 5

/* Enable the use of the Tab key */
#define FINSH_USING_SYMTAB
/* Turn on description */
#define FINSH_USING_DESCRIPTION

/* Define FinSH thread priority to 20 */
#define FINSH_THREAD_PRIORITY 20
/* Define the stack size of the FinSH thread to be 4KB */
#define FINSH_THREAD_STACK_SIZE 4096
/* Define the command character length to 80 bytes */
#define FINSH_CMD_SIZE 80

/* Open msh function */
#define FINSH_USING_MSH
/* Use msh function by default */
#define FINSH_USING_MSH_DEFAULT
/* The maximum number of input parameters is 10 */
#define FINSH_ARG_MAX 10
```

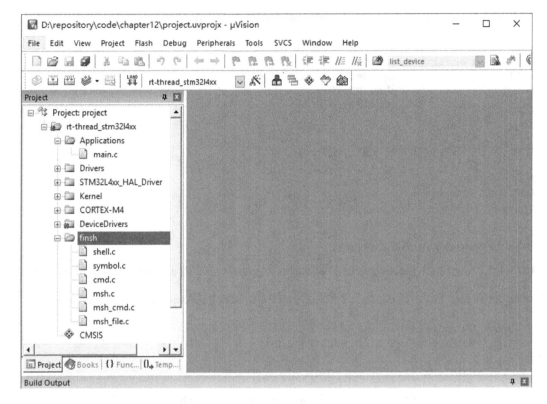

FIGURE 12.3 FinSH example MDK5 project.

APPLICATION EXAMPLES

The example code of this section can be found in the `chapter12` directory of the complementary code package. Figure 12.3 shows the example MDK5 project.

After compiling and downloading the project to the target board, the following messages should be observed in the terminal:

```
 \ | /
- RT -     Thread Operating System
 / | \     3.1.0 build Aug 27 2018
 2006 - 2018 Copyright by rt-thread team
```

MSH Command without Argument

In the following example, a function named "hello" is exported to FinSH as a command in MSH mode:

```
#include <rtthread.h>

void hello(void)
{
    rt_kprintf("hello RT-Thread!\n");
}

MSH_CMD_EXPORT(hello , say hello to RT-Thread);
```

After the system starts, press "⇆" (tab) in the terminal to show the "hello" command in the list:

```
msh />
RT-Thread shell commands:
hello              - say hello to RT-Thread
version            - show RT-Thread version information
list_thread        - list thread
......
```

Issue the "hello" command and observe the result:

```
msh />hello
hello RT_Thread!
msh />
```

MSH COMMAND WITH ARGUMENTS

In Code Listing 12.1, a function named "atcmd" is exported to FinSH as a command in MSH mode.

Code listing 12.1 Example of User Defined MSH Command with Parameters

```
#include <rtthread.h>
static void atcmd(int argc, char**argv)
{
    if (argc < 2)
    {
        rt_kprintf("Please input'atcmd <server|client>'\n");
        return;
    }

    if (!rt_strcmp(argv[1], "server"))
    {
        rt_kprintf("AT server!\n");
    }
    else if (!rt_strcmp(argv[1], "client"))
    {
        rt_kprintf("AT client!\n");
    }
    else
    {
        rt_kprintf("Please input'atcmd <server|client>'\n");
    }
}

MSH_CMD_EXPORT(atcmd, atcmd sample: atcmd <server|client>);
```

After the system starts, press "⇆" (tab) in the terminal to show the "atcmd" command in the list:

```
msh />
RT-Thread shell commands:
hello              - say hello to RT-Thread
atcmd              - atcmd sample: atcmd <server|client>
version            - show RT-Thread version information
list_thread        - list thread
......
```

Issue the "atcmd" command and observe the result:

```
msh />atcmd
Please input 'atcmd <server|client>'
msh />
```

Issue the "atcmd server" command and observe the result:

```
msh />atcmd server
AT server!
msh />
```

Issue the "atcmd client" command and observe the result:

```
msh />atcmd client
AT client!
msh />
```

CHAPTER SUMMARY

This chapter introduced FinSH features, working mechanisms, commands, configuration options, and application examples.

The purpose is to let more developers become aware and make use of RT-Thread FinSH. It will be your right-hand man when doing debugging and testing.

13 I/O Device Framework

Most embedded systems include some I/O (input/output) devices, such as data displays on instruments, serial communication on industrial devices, Flash or SD cards for saving data on data acquisition devices, and Ethernet interfaces for network devices.

This chapter describes how RT-Thread manages different I/O devices. After this chapter, you will understand the I/O device model of RT-Thread and be familiar with the different functions of the I/O device management interface.

I/O DEVICE INTRODUCTION

I/O DEVICE FRAMEWORK

RT-Thread provides an I/O device framework, as shown in Figure 13.1. It is located between the hardware and the application. It is divided into three layers. From top to bottom are the I/O device management layer, device driver framework layer, and device driver layer.

An application obtains the correct device driver through the I/O device management interface and then uses this device driver to perform data (or control) interaction with the bottom I/O hardware device.

The I/O device management layer implements the encapsulation of device drivers. The application accesses the bottom devices through a standard interface provided by the I/O device layer. The upgrade and replacement of the device driver will not affect the upper-layer application. In this way, the hardware-related code in the device can exist independently from the application. So both parties only need to focus on the respective function implementation, thereby reducing the coupling and complexity of the code and improving the reliability of the system.

The device driver framework layer is an abstraction of the same kind of hardware device drivers. The same characteristic of its kind is extracted, and the unique characteristics are left out of interface, which should be implemented by the driver.

The device driver layer is a set of programs that drive the hardware devices to work, enabling access to them. It is responsible for creating and registering I/O devices. For devices with simple operation logic, you can register devices directly into the I/O Device Manager without going through the device driver framework layer. The sequence diagram is shown in Figure 13.2. There are two main points:

- The device driver creates a device instance with hardware access capabilities based on the device model definition and registers the device using the rt _ device _ register() interface into the I/O Device Manager.
- The application finds the device through the rt _ device _ find() interface and then uses the I/O device management interface to access the hardware.

For other devices, such as watchdogs, the created device instance will be registered to the corresponding device driver framework, and then the device driver framework will register into the I/O Device Manager. Figure 13.3 shows the use sequence of the watchdog device. The main points are as follows:

- The watchdog device driver creates a watchdog device instance with hardware access capability based on the watchdog device model definition. Then it registers the watchdog device through the rt _ hw _ watchdog _ register() interface into the watchdog device driver framework.

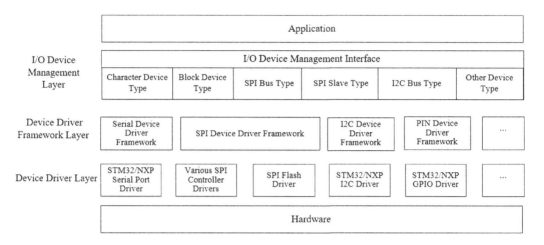

FIGURE 13.1   I/O device framework.

- The watchdog device driver framework registers the watchdog device into the I/O Device Manager via the rt _ device _ register() interface.
- The application accesses the watchdog device hardware through the I/O device management interface.

I/O Device Model

The device model in RT-Thread is based on the kernel object model, which is considered a kind of object and is included in the scope of the object manager. Each device object is derived from the base object. Each concrete device can inherit the properties of its parent class object and derive its own properties. Figure 13.4 is a schematic diagram of the inheritance and derivation relationship of the device object.

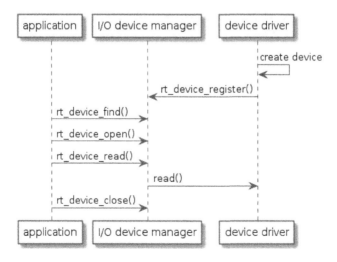

FIGURE 13.2   Simple I/O device using a sequence diagram.

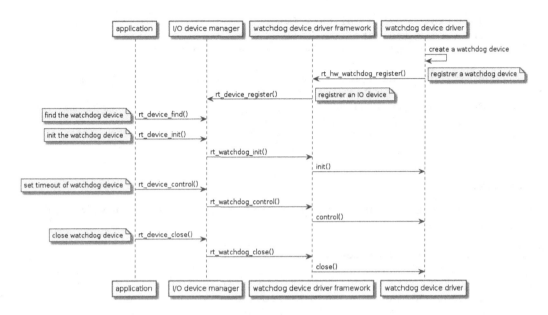

FIGURE 13.3   Watchdog device use sequence diagram.

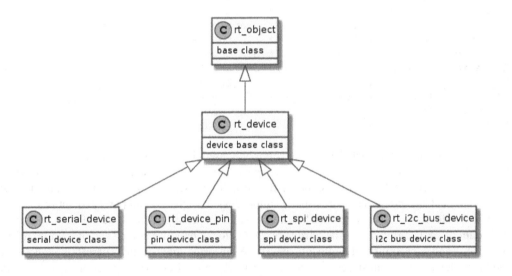

FIGURE 13.4   Device inheritance diagram.

The specific definitions of device objects are as follows:

```
struct rt_device
{
    struct rt_object          parent;      /* kernel object base class */
    enum rt_device_class_type type;        /* device type */
    rt_uint16_t               flag;        /* device parameter */
    rt_uint16_t               open_flag;   /* device open flag */
    rt_uint8_t                ref_count;   /* number of times the device
was cited */
    rt_uint8_t                device_id;   /* device ID,0 - 255 */
```

```
    /* data transceiving callback function */
    rt_err_t (*rx_indicate)(rt_device_t dev, rt_size_t size);
    rt_err_t (*tx_complete)(rt_device_t dev, void *buffer);

    const struct rt_device_ops *ops;      /* device operate methods */

    /* device's private data */
    void *user_data;
};
typedef struct rt_device *rt_device_t;
```

I/O DEVICE TYPE

RT-Thread supports multiple I/O device types. The main device types are as shown in Code Listing 13.1.

Code listing 13.1 I/O Device Types

```
RT_Device_Class_Char             /* character device       */
RT_Device_Class_Block            /* block device          */
RT_Device_Class_NetIf            /* network interface device    */
RT_Device_Class_MTD              /* memory device        */
RT_Device_Class_RTC              /* RTC device        */
RT_Device_Class_Sound            /* sound device         */
RT_Device_Class_Graphic          /* graphic device          */
RT_Device_Class_I2CBUS           /* I2C bus device       */
RT_Device_Class_USBDevice        /* USB device */
RT_Device_Class_USBHost          /* USB host device    */
RT_Device_Class_SPIBUS           /* SPI bus device       */
RT_Device_Class_SPIDevice        /* SPI device         */
RT_Device_Class_SDIO             /* SDIO device        */
RT_Device_Class_Miscellaneous    /* miscellaneous devices        */
```

Character devices and block devices are commonly used device types, and their classification is based on the transmission processing between device data and the system. Character mode devices allow for unstructured data transfers, that is, data usually transfers in the form of serial mode, one byte at a time. Character devices are usually simple devices such as serial ports and buttons.

A block device transfers one data block at a time, for example, 512 bytes of data at a time. This data block is enforced by the hardware. Data blocks may use some type of data interface or some mandatory transport protocol; otherwise, an error may occur. Therefore, sometimes the block device driver must perform additional work on read or write operations, as shown in Figure 13.5.

When the system serves a write operation with a large amount of data, the device driver must first divide the data into multiple packets, each with the block size specified by the device. In the actual process, the last part of the data size may be smaller than the normal device block size. Each block in the above figure is written to the device using a separate write request, while the first three blocks are directly written. However, the last data block size is smaller than the device block size, so that the device driver must process the last data block differently than the first three blocks. First, the device driver needs to perform a read operation of the corresponding device block. Then, it needs to overwrite the write data onto the read data block. Finally it writes the "composited" data block back to the device as a whole block. For example, for block 4 in Figure 13.5, the driver needs to read out the device block corresponding to block 4 and then overwrite the data to be written to the data read from the device block, and then merge them into a new block. Finally, it needs to write back to the block device.

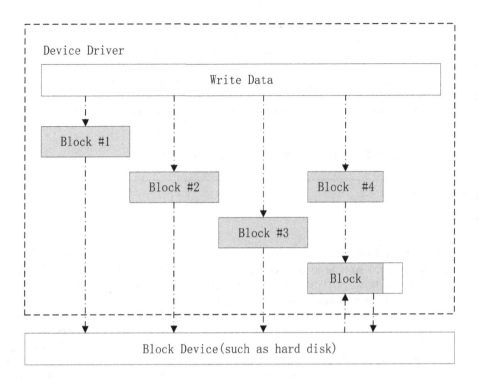

FIGURE 13.5 Writing data in a block device.

CREATE AND REGISTER THE I/O DEVICE

The driver layer is responsible for creating device instances and registering them into the I/O Device Manager. You can create device instances in a statically declared manner or dynamically create them with the following interfaces. Table 13.1 shows its parameters in detail.

```
rt_device_t rt_device_create(int type, int attach_size);
```

TABLE 13.1
rt_device_create

Parameters	Description
type	Device type; the device type values listed in "I/O Device Type" section can be used here
attach_size	User data size
Return	--
Device Handle	Create successfully
RT_NULL	Creation failed, dynamic memory allocation failed

When this interface is called, the system allocates a device control block from the dynamic heap memory, the size of which is the sum of `struct rt_device` and `attach_size`, and the type of the device is set by the parameter type. After the device is created, implementing its access to the hardware is needed.

```
struct rt_device_ops
{
    /* common device interface */
    rt_err_t   (*init)   (rt_device_t dev);
```

```
    rt_err_t   (*open)   (rt_device_t dev, rt_uint16_t oflag);
    rt_err_t   (*close)  (rt_device_t dev);
    rt_size_t  (*read)   (rt_device_t dev, rt_off_t pos, void *buffer,
rt_size_t size);
    rt_size_t  (*write)  (rt_device_t dev, rt_off_t pos, const void *buffer,
rt_size_t size);
    rt_err_t   (*control)(rt_device_t dev, int cmd, void *args);
};
```

A description of each method of operation is shown in Table 13.2.

TABLE 13.2
Methods

Method Name	Method Description
init	Initialize the device. After the device is initialized, the flag of the device control block is set to the active state (RT_DEVICE_FLAG_ACTIVATED). If the flag in the device control block has been set to the active state, then the initialization interface will be returned immediately when running again and will not be reinitialized.
open	Open the device. Some devices are not started when the system is started, or the device needs to send and receive data. However, if the upper application is not ready, the device should not be enabled by default and start receiving data. Therefore, it is recommended to enable the device in the bottom driver when the open interface is called.
close	Close the device. When the device is open, the device control block maintains an open count. The count will add 1 when the device is opened, and the count will subtract 1 when the device is closed. A real shutdown operation is operated when the counter turns to 0.
read	Read data from the device. The parameter pos is the offset of the read data, but some devices do not necessarily need to specify the offset, such as serial devices. In this case, the device driver should ignore this parameter. But for block devices, pos and size are measured in the block size of the block device. For example, the block size of the block device is 512 byte, and in the parameter pos = 10, size = 2, then the driver should return the 10th block in the device (starting from the 0th block) for a total of 2 blocks of data. The type returned by this interface is rt_size_t, which is the number of bytes read or the number of blocks dependent on the device type. Normally, the value of the size in the parameter should be returned. If it returns zero, set the corresponding errno value.
write	Write data to the device. The parameter pos is the offset of the write data. Similar to read operations, for block devices, pos and size are measured in the block size of the block device. The type returned by this interface is rt_size_t, which is the number of bytes or blocks of data actually written. Normally, the value of the size in the parameter should be returned. If it returns zero, set the corresponding errno value.
control	Control the device according to the cmd command. Commands are often implemented by the bottom device drivers. For example, the parameter RT_DEVICE_CTRL_BLK_GETGEOME means to get the size information of the block device.

When a dynamically created device is no longer needed, it can be destroyed using the following interface. Table 13.3 shows its parameters.

```
void rt_device_destroy(rt_device_t device);
```

TABLE 13.3
rt_device_destroy

Parameters	Description
device	Device handle

After the device is created, it needs to be registered into the I/O Device Manager to be able to accessed by the application. The interfaces for registering the device are as follows. Table 13.4 shows its parameters.

```
rt_err_t rt_device_register(rt_device_t dev, const char* name, rt_uint16_t
flags);
```

TABLE 13.4
rt_device_register

Parameters	Description
dev	Device handle
name	Device name; the maximum length of the device name is specified by the macro RT_NAME_MAX defined in rtconfig.h, and the extra characters will be automatically truncated
flags	Device mode flag
Return	--
RT_EOK	Registration success
-RT_ERROR	Registration failed; dev is empty or the name already exists

Repeatedly registering registered devices and registering devices with the same name should be avoided.

Flags parameters support the following parameters (multiple parameters can be combined by bit OR logic, as shown in Code Listing 13.2).

Code listing 13.2 Value of Parameter Flag

```
#define RT_DEVICE_FLAG_RDONLY       0x001 /* read only */
#define RT_DEVICE_FLAG_WRONLY       0x002 /* write only  */
#define RT_DEVICE_FLAG_RDWR         0x003 /* read and write  */
#define RT_DEVICE_FLAG_REMOVABLE    0x004 /* can be removed  */
#define RT_DEVICE_FLAG_STANDALONE   0x008 /* stand alone   */
#define RT_DEVICE_FLAG_SUSPENDED    0x020 /* suspended  */
#define RT_DEVICE_FLAG_STREAM       0x040 /* stream mode  */
#define RT_DEVICE_FLAG_INT_RX       0x100 /* interrupt reception */
#define RT_DEVICE_FLAG_DMA_RX       0x200 /* DMA reception */
#define RT_DEVICE_FLAG_INT_TX       0x400 /* interrupt sending */
#define RT_DEVICE_FLAG_DMA_TX       0x800 /* DMA sending */
```

The RT_DEVICE_FLAG_STREAM parameter indicates the device stream mode, which is used to output a character string to the serial terminal. In this mode, when the output character is "\n", it automatically fills in a "\r" before "\n" to make a branch.

Using the "list _ device" command on the FinSH command line can list the successfully registered devices. The information available includes device name, device type, and number of times the device is opened:

```
msh />list_device
device          type               ref count
--------    -------------------    ----------
e0          Network Interface      0
sd0         Block Device           1
rtc         RTC                    0
uart1       Character Device       0
uart0       Character Device       2
msh />
```

When the device is unregistered, the device will be removed from the Device Manager and the device will no longer be found through the device. Unregistering a device does not release the memory occupied by the device control block. The function to unregister the device is as follows. Table 13.5 shows its parameters.

```
rt_err_t rt_device_unregister(rt_device_t dev);
```

TABLE 13.5
rt_device_unregister

Parameters	Description
dev	Device handle
Return	--
RT_EOK	Successful

Code Listing 13.3 is an example of registering a watchdog device. After calling the rt _ hw _ watchdog _ register() interface, the device is registered to the I/O Device Manager via the rt _ device _ register() interface.

Code listing 13.3 An Example of Registering a Watchdog

```
const static struct rt_device_ops wdt_ops =
{
    rt_watchdog_init,
    rt_watchdog_open,
    rt_watchdog_close,
    RT_NULL,
    RT_NULL,
    rt_watchdog_control,
};

rt_err_t rt_hw_watchdog_register(struct rt_watchdog_device *wtd,
                                 const char                *name,
                                 rt_uint32_t               flag,
                                 void                      *data)
{
    struct rt_device *device;
    RT_ASSERT(wtd != RT_NULL);

    device = &(wtd->parent);

    device->type       = RT_Device_Class_Miscellaneous;
    device->rx_indicate = RT_NULL;
    device->tx_complete = RT_NULL;

    device->ops        = &wdt_ops;
    device->user_data  = data;

    /* register a character device */
    return rt_device_register(device, name, flag);
}
```

ACCESS I/O DEVICES

Applications can access the hardware device through the I/O device management interface when the device driver is implemented. The mapping relationship between the I/O device management interface and the operations on the I/O device is shown in Figure 13.6.

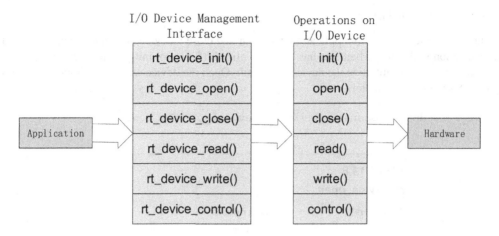

FIGURE 13.6 Mapping between the I/O device management interface and the operations on the I/O device.

FIND DEVICE

An application obtains the device handle based on the device name, which in turn allows the device to operate. To find a device, use the following function. Table 13.6 shows its parameters.

```
rt_device_t rt_device_find(const char* name);
```

TABLE 13.6
rt_device_find

Parameters	Description
name	Device name
Return	--
Device Handle	Finding the corresponding device will return the corresponding device handle
RT_NULL	No corresponding device object found

INITIALIZE DEVICE

Once the device handle is obtained, the application can initialize the device using the following function. Table 13.7 shows its parameters.

```
rt_err_t rt_device_init(rt_device_t dev);
```

TABLE 13.7
rt_device_init

Parameters	Description
dev	Device handle
Return	--
RT_EOK	Device initialization succeeded
Error Code	Device initialization failed

When a device has been successfully initialized, calling this interface later will not perform a repeat initialization of the device.

OPEN AND CLOSE THE DEVICE

Through the device handle, an application can open and close a device. When a device is opened, the application will detect whether it has been initialized. If it is not initialized, it will call the initialization interface to initialize the device by default. Open the device with the following function. Table 13.8 shows its parameters.

```
rt_err_t rt_device_open(rt_device_t dev, rt_uint16_t oflags);
```

TABLE 13.8
rt_device_open

Parameters	Description
dev	Device handle
oflags	Open device in oflag mode
Return	--
RT_EOK	Device successfully turned on
-RT_EBUSY	Device will not allow being repeatedly opened if the RT_DEVICE_FLAG_STANDALONE parameter is included in the parameters specified when the device is registered
Other Error Code	Device failed to be turned on

oflags supports the following parameters:

```
#define RT_DEVICE_OFLAG_CLOSE 0x000   /* device was already closed(internal
use)*/
#define RT_DEVICE_OFLAG_RDONLY 0x001 /* open the device in read-only mode
*/
#define RT_DEVICE_OFLAG_WRONLY 0x002 /* open the device in write-only mode
*/
#define RT_DEVICE_OFLAG_RDWR 0x003   /* open the device in read-and_write
mode */
#define RT_DEVICE_OFLAG_OPEN 0x008   /* device was already closed(internal
use) */
#define RT_DEVICE_FLAG_STREAM 0x040  /* open the device in stream mode */
#define RT_DEVICE_FLAG_INT_RX 0x100  /* open the device in interrupt
reception mode */
#define RT_DEVICE_FLAG_DMA_RX 0x200  /* open the device in DMA mode */
#define RT_DEVICE_FLAG_INT_TX 0x400  /* open the device in interrupt
sending mode */
#define RT_DEVICE_FLAG_DMA_TX 0x800  /* open the device in DMA mode */
```

Notes: If the upper application needs to set the device's receive callback function, it must open the device as RT_DEVICE_FLAG_INT_RX or RT_DEVICE_FLAG_DMA_RX; otherwise, the callback function will not be called.

After the application opens the device and completes reading and/or writing, if no further operations are needed, you can close the device using the following functions. Table 13.9 shows its parameters.

```
rt_err_t rt_device_close(rt_device_t dev);
```

TABLE 13.9
rt_device_close

Parameters	Description
dev	Device handle
Return	--
RT_EOK	Device successfully closed
-RT_ERROR	Device has been completely closed and cannot be closed repeatedly
Other Error Code	Failed to close device

Notes: Device interfaces `rt_device_open()` **and** `rt_device_close()` **need to be used in pairs. Opening a device requires closing the device, so that the device will be completely closed; otherwise, the device will remain on.**

CONTROL DEVICE

By sending the control cmd, the application can also control the device with the following function. Table 13.10 shows its parameters.

```
rt_err_t rt_device_control(rt_device_t dev, rt_uint8_t cmd, void* arg);
```

TABLE 13.10
rt_device_control

Parameters	Description
dev	Device handle
cmd	Command control word; this parameter is usually related to the device driver
arg	Controlled parameter
Return	--
RT_EOK	Function executed successfully
-RT_ENOSYS	Execution failed, dev is empty
Other Error Code	Execution failed

A generic device command for the parameter cmd can be defined as follows:

```
#define RT_DEVICE_CTRL_RESUME      0x01    /* resume device */
#define RT_DEVICE_CTRL_SUSPEND     0x02    /* suspend device */
#define RT_DEVICE_CTRL_CONFIG      0x03    /* configure device */
#define RT_DEVICE_CTRL_SET_INT     0x10    /* set interrupt */
#define RT_DEVICE_CTRL_CLR_INT     0x11    /* clear interrupt */
#define RT_DEVICE_CTRL_GET_INT     0x12    /* obtain interrupt status */
```

READ AND WRITE DEVICE

An application can read data from the device by using the following function. Table 13.11 shows its parameters.

```
rt_size_t rt_device_read(rt_device_t dev, rt_off_t pos,void* buffer,
rt_size_t size);
```

TABLE 13.11
rt_device_read

Parameters	Description
dev	Device handle
pos	Read data offset
buffer	Memory buffer pointer; the data read will be saved in the buffer
size	Size of the data read
Return	--
Actual Size of the Data Read	If it is a character device, the return size is in bytes; if it is a block device, the returned size is in block units
0	Needs to read the current thread's errno to determine the error status

Calling this function will read the data from the dev device to the buffer. The maximum length of this buffer is defined as *size. pos* has different meanings depending on the device class.

Writing data to the device can be done by the following function. Table 13.12 shows its parameters.

```
rt_size_t rt_device_write(rt_device_t dev, rt_off_t pos,const void*
buffer, rt_size_t size);
```

TABLE 13.12
rt_device_write

Parameters	Description
dev	Device handle
pos	Write data offset
buffer	Memory buffer pointer, placing the data to be written in
size	Size of the written data
Return	--
Actual Size of the Data Written	If it is a character device, the return size is in bytes; if it is a block device, the returned size is in block units
0	Needs to read the current thread's errno to determine the error status

Calling this function will write the data from the buffer to the *dev* device. The maximum length of the written data is defined as *size. pos* has different meanings depending on the device class.

DATA TRANSCEIVING AND CALLBACK

The following function can be used to set a callback function to indicate the data is being received from the hardware. It notifies the upper application thread that the data has arrived. Table 13.13 shows its parameters.

```
rt_err_t rt_device_set_rx_indicate(rt_device_t dev, rt_err_t
(*rx_ind)(rt_device_t dev,rt_size_t size));
```

TABLE 13.13

rt_device_set_rx_indicate

Parameters	Description
dev	Device handle
rx_ind	Callback function pointer
Return	--
RT_EOK	Set successfully

The callback function is provided by the developer. When the hardware device receives the data, it will call the callback function and pass the length of received data to the upper-layer application in the *size* parameter. The upper application thread should read the data from the device as soon as it receives the indication.

When the application calls rt _ device _ write() to write data, if the bottom hardware can support automatic sending, the upper application can set a callback function. This callback function is called after the bottom hardware data has been sent (e.g., when the DMA transfer is completed or the FIFO has been written to complete, triggered interrupt). Use the following function to set the device with a send completion indication. The function parameters and return values are as follows. Table 13.14 shows its parameters.

```
rt_err_t rt_device_set_tx_complete(rt_device_t dev, rt_err_t
(*tx_done)(rt_device_t dev,void *buffer));
```

TABLE 13.14

rt_device_set_tx_complete

Parameters	Description
dev	Device handle
tx_done	Callback function pointer
Return	--
RT_EOK	Set successfully

The callback function is provided by the developer. When the hardware device sends the data, the driver calls the function and passes the sent data block address buffer as a parameter to the upper application. When the upper-layer application (thread) receives the indication, this means the transaction is complete and the buffer is no longer used by the driver.

ACCESS DEVICE SAMPLE

Code Listing 13.4 is an example of accessing a device. First, find the watchdog device through the rt _ device _ find() interface, obtain the device handle, then initialize the device through the rt _ device _ init() interface, and set the watchdog device timeout through the rt _ device _ control() interface.

Code listing 13.4 Device Sample

```
#include <rtthread.h>
#include <rtdevice.h>

#define IWDG_DEVICE_NAME    "iwg"

static rt_device_t wdg_dev;
```

```
static void idle_hook(void)
{
    /* feed the dog in the callback function of the idle thread */
    rt_device_control(wdg_dev, RT_DEVICE_CTRL_WDT_KEEPALIVE, NULL);
    rt_kprintf("feed the dog!\n ");
}

int main(void)
{
    rt_err_t res = RT_EOK;
    rt_uint32_t timeout = 1000;    /* timeout */

    /* find the watchdog device based on the device name, and obtain the
device handle */
    wdg_dev = rt_device_find(IWDG_DEVICE_NAME);
    if (!wdg_dev)
    {
        rt_kprintf("find %s failed!\n", IWDG_DEVICE_NAME);
        return RT_ERROR;
    }
    /* initialize device */
    res = rt_device_init(wdg_dev);
    if (res != RT_EOK)
    {
        rt_kprintf("initialize %s failed!\n", IWDG_DEVICE_NAME);
        return res;
    }
    /* set watchdog timeout */
    res = rt_device_control(wdg_dev, RT_DEVICE_CTRL_WDT_SET_TIMEOUT,
&timeout);
    if (res != RT_EOK)
    {
        rt_kprintf("set %s timeout failed!\n", IWDG_DEVICE_NAME);
        return res;
    }
    /* set idle thread callback function */
    rt_thread_idle_sethook(idle_hook);

    return res;
}
```

CHAPTER SUMMARY

This chapter describes the working mechanism of I/O device management and the functionalities of the I/O device management interface. This chapter could be summarized as follows:

1. If a device has no driver, the driver needs to be implemented according to the driver framework. The focus is on the implementation of the device operation method.
2. Avoid duplicating registration of registered devices, as well as registering devices with the same name.
3. The open and close interfaces of the device should be paired.
4. Device objects that are created dynamically will need to be freed by the destruction interface to release the memory.

14 General Peripheral Interface

With the fast growth of large-scale integrated circuits, some peripheral modules, such as UART, SPI, I2C, GPIO, etc., have become standard on-chip peripherals of the microcontroller.

In RT-Thread, those standard peripherals are abstracted as a set of common device driver interfaces for various microcontrollers. The following sections discuss UART, PIN, SPI, and I2C devices.

UART DEVICE DRIVER

UART (Universal Asynchronous Receiver/Transmitter) is used for serial communication between two hardware devices.

To achieve two-way communication, normally there are two wires for transmitting and receiving. And the same set of parameters, for example, baud rate and data format (Figure 14.1), shall be used by both peers.

- Start bit: Indicating the start of the data frame (logic level: "0").
- Data frame length: The possible values are 5-bit, 6-bit, 7-bit, 8-bit, and 9-bit, depending on the underlying hardware support. However, 8-bit is the most common choice, as ASCII characters are represented in 8-bit.
- Parity check bit: Used for integrity checking. For the bits in the data frame, the occurrences of bits whose value is "1" are counted. If that count is odd, the parity check bit value is set to "1" for even parity or set to "0" for odd parity (to make the total count of occurrences of "1"s in the whole frame an even number for even parity or odd number for odd parity), and vice versa.
- Stop bit: Indicating the end of the data frame (logic level: "1").
- Baud rate: The data transfer rate in bits per second (bps).

UART DEVICE MANAGEMENT

RT-Thread supports the polling, interrupt, and DMA methods of UART data transfer.

The UART device driver is derived from the device object ("struct rt_device"):

```
struct rt_serial_device
{
    struct rt_device         parent;        /* Device class */
    const struct rt_uart_ops *ops;          /* PIN device operation
method, provided by PIN device driver */
    struct serial_configure  config;        /* Serial port device
configuration parameters  */
    void *serial_rx;
    void *serial_tx;
};
typedef struct rt_serial_device rt_serial_t;
```

Figure 14.2 shows the UART device invoke diagram.

1. UART driver creates an instance of the UART device and registers it in the UART device driver framework (rt_hw_serial_register()).

Start Bit	LSB	1	2	3	4	5	6	MSB	Odd/Even /Non-polar	Stop Bit

Data

FIGURE 14.1   UART data format.

2. UART device driver framework registers the device in I/O Device Manager (rt_device_register()).
3. The application program then is able to access the UART hardware through the I/O device management interfaces.

CREATE AND REGISTER THE UART DEVICE

To create a UART device, the developer shall implement the UART device operations defined in the "rt_uart_ops" structure:

```
struct rt_uart_ops
{
    rt_err_t (*configure)(struct rt_serial_device *serial, struct serial_
configure *cfg);
    rt_err_t (*control)(struct rt_serial_device *serial, int cmd, void
*arg);
    int (*putc)(struct rt_serial_device *serial, char c);
    int (*getc)(struct rt_serial_device *serial);
    rt_size_t (*dma_transmit)(struct rt_serial_device *serial, rt_uint8_t
*buf, rt_size_t size, int direction);
};
```

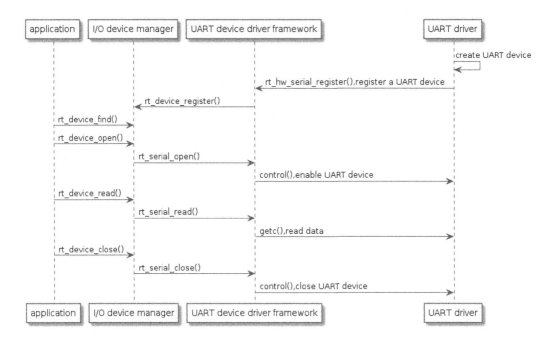

FIGURE 14.2   UART device invoke diagram.

TABLE 14.1
UART Device Operations

Method	Description
configure	Configure the UART hardware's baud rate, data format, etc.
control	Control the UART hardware state. For example, to enable or disable interrupts, the developer could issue the command (parameter "cmd") RT_DEVICE_CTRL_SET_INT or RT_DEVICE_CTRL_CLR_INT
putc	Transmit a data frame
getc	Receive a data frame
dma_transmit (optional)	To transmit or receive the data in a buffer through DMA

The operating method of the UART device is described in Table 14.1.

To register the just-created UART device, the developer shall call the following function. The parameters and return values are shown in Table 14.2.

```
rt_err_t rt_hw_serial_register(struct rt_serial_device *serial,
                                const char    *name,
                                rt_uint32_t   flag,
                                void          *data);
```

TABLE 14.2
rt_hw_serial_register

Parameters	Description
serial	Pointer to a UART device instance
name	Pointer to the device name
flag	UART device mode flag (please refer to Code Listing 14.1)
data	Pointer to user data
Return	— —
RT_EOK	Registration is successful
-RT_ERROR	Registration has failed (the selected name has already been taken)

Supported flag parameter values (some of the choices may be combined with "|", as shown in Code Listing 14.1).

Code listing 14.1 Supported Flag Parameter Values

```
#define RT_DEVICE_FLAG_RDONLY       0x001    /* Read-only Device  */
#define RT_DEVICE_FLAG_WRONLY       0x002    /* Write-only Device */
#define RT_DEVICE_FLAG_RDWR         0x003    /* Read-write Device */
#define RT_DEVICE_FLAG_REMOVABLE    0x004    /* Removable device  */
#define RT_DEVICE_FLAG_STANDALONE   0x008    /* Standalone Device */
#define RT_DEVICE_FLAG_SUSPENDED    0x020    /* Suspended Device  */
#define RT_DEVICE_FLAG_STREAM       0x040    /* Device in Stream mode */
#define RT_DEVICE_FLAG_INT_RX       0x100    /* Device in interrupt-
receiving mode */
#define RT_DEVICE_FLAG_DMA_RX       0x200    /* Device in DMA receiving
mode */
#define RT_DEVICE_FLAG_INT_TX       0x400    /* Device in interrupt-
sending mode */
#define RT_DEVICE_FLAG_DMA_TX       0x800    /* Device in DMA sending
mode */
```

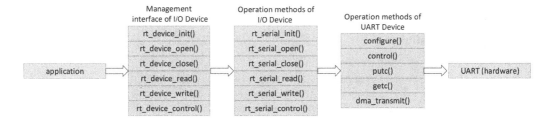

FIGURE 14.3 The relationship between I/O device management interfaces and UART hardware.

The most common flag parameter value is "RT_DEVICE_FLAG_RDWR I RT_DEVICE_FLAG_INT_RX" for a UART device working in polling transmitting and interrupt receiving mode.

ACCESS UART DEVICE

Figure 14.3 shows the relationship between I/O device management interfaces and UART hardware.

For example, when the developer is involving the "rt_device_read()" interface to read the data received by UART hardware, the serial port device driver is triggered for framework operation, "rt_serial_read()," which in turn triggers the UART device operation, "getc()."

EXAMPLE OF USING THE UART DEVICE

The specific use of the UART device can be seen in Code Listing 14.2.

Code listing 14.2 Examples of Using the UART Device

```
/*
 * Program List: This is a UART device use routine
 * Routine exports uart_sample command to control terminal
 * Command call format: uart_sample uart2
 * Command interpretation: The second parameter of the command is the
name of the UART device to be used, and the default UART device is used
for null
 * Program functions: output strings via serial port , "hello RT-Thread!,"
and then misplace the input characters
*/
#include <rtthread.h>

#define SAMPLE_UART_NAME        "uart2"

/*Semaphore used to receive messages */
static struct rt_semaphore rx_sem;
static rt_device_t serial;

/*Receiving data callback function */
static rt_err_t uart_input(rt_device_t dev, rt_size_t size)
{
    /* The serial port receives the data and causes an interrupt, calls
this callback function, and then sends the receive semaphore */
    rt_sem_release(&rx_sem);

    return RT_EOK;
}
```

```
static void serial_thread_entry(void *parameter)
{
    char ch;

    while (1)
    {
        /* Read one byte of data from the serial port, and wait to receive
the semaphore */
        while (rt_device_read(serial, -1, &ch, 1) != 1)
        {
            /* Receiving semaphore in a permanent waiting way, after
receiving the semaphore and read the data again */
            rt_sem_take(&rx_sem, RT_WAITING_FOREVER);
        }
        /* The read data should be misplace output by serial port */
        ch = ch + 1;
        rt_device_write(serial, 0, &ch, 1);
    }
}

static int uart_sample(int argc, char *argv[])
{
    rt_err_t ret = RT_EOK;
    char uart_name[RT_NAME_MAX];
    char str[] = "hello RT-Thread!\r\n";

    if (argc == 2)
    {
        rt_strncpy(uart_name, argv[1], RT_NAME_MAX);
    }
    else
    {
        rt_strncpy(uart_name, SAMPLE_UART_NAME, RT_NAME_MAX);
    }

    /* Find the UART device in the system */
    serial = rt_device_find(uart_name);
    if (!serial)
    {
        rt_kprintf("find %s failed!\n", uart_name);
        return RT_ERROR;
    }

    /* Initialize the semaphore */
    rt_sem_init(&rx_sem, "rx_sem", 0, RT_IPC_FLAG_FIFO);
    /* Open the UART device in the way of read-write and
interrupt-receiving*/
    rt_device_open(serial, RT_DEVICE_OFLAG_RDWR | RT_DEVICE_FLAG_INT_RX);
    /* Set to receive callback function */
    rt_device_set_rx_indicate(serial, uart_input);
    /* Send string */
    rt_device_write(serial, 0, str, (sizeof(str) - 1));

    /* Create serial thread */
    rt_thread_t thread = rt_thread_create("serial", serial_thread_entry,
RT_NULL, 1024, 25, 10);
```

```
    /* start the thread after successfully created */
    if (thread != RT_NULL)
    {
        rt_thread_startup(thread);
    }
    else
    {
        ret = RT_ERROR;
    }

    return ret;
}
/* Export to the msh command list */
MSH_CMD_EXPORT(uart_sample, uart device sample);
```

PIN DEVICE DRIVER

The PIN device in RT-Thread means General-Purpose Input Output (GPIO).

PIN DEVICE MANAGEMENT

RT-Thread supports the following PIN device features:

- Interrupt triggering mode (as shown in Figure 14.4) configuration.
- Input/output switching.
- Output mode (push-pull, open-drain, pull-up, and pull-down) configuration.
- Input mode (floating, pull-up, pull-down, and Analog) configuration.

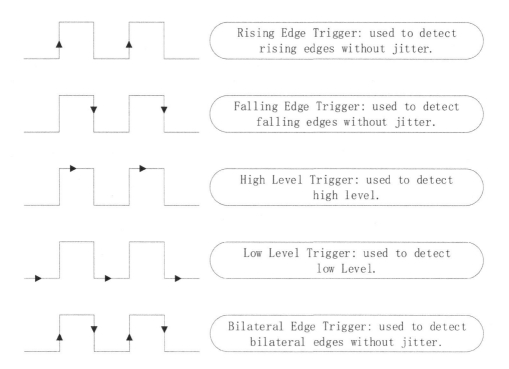

FIGURE 14.4   PIN device interrupt trigger modes.

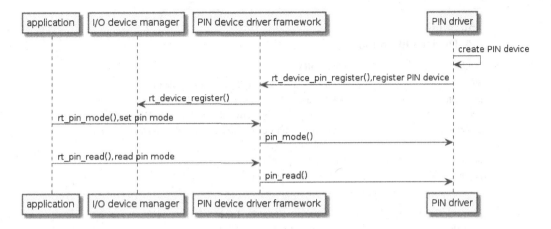

FIGURE 14.5   Pin device invoke diagram.

The PIN device is derived from the device object ("struct rt_device"):

```
struct rt_device_pin
{
    struct rt_device parent;              /* Device class */
    const struct rt_pin_ops *ops;         /* PIN device operation method,
provided by PIN device driver */
};
```

Figure 14.5 shows the PIN device invoke diagram.

1. The PIN driver creates an instance of a PIN device and registers it in the PIN device driver framework (rt_device_pin_register()).
2. The PIN device driver framework registers the PIN device in I/O Device Manager (rt_device_register()).
3. The application program then is able to access the GPIO hardware through the PIN device interfaces.

CREATE AND REGISTER THE PIN DEVICE

To create a PIN device, the developer shall implement the PIN device operations defined in the "rt_pin_ops" structure:

```
struct rt_pin_ops
{
    void (*pin_mode)(struct rt_device *device, rt_base_t pin, rt_base_t
mode);
    void (*pin_write)(struct rt_device *device, rt_base_t pin, rt_base_t
value);
    int (*pin_read)(struct rt_device *device, rt_base_t pin);
    rt_err_t (*pin_attach_irq)(struct rt_device *device, rt_int32_t pin,
                rt_uint32_t mode, void (*hdr)(void *args), void *args);
    rt_err_t (*pin_detach_irq)(struct rt_device *device, rt_int32_t pin);
    rt_err_t (*pin_irq_enable)(struct rt_device *device, rt_base_t pin,
rt_uint32_t enabled);
};
```

TABLE 14.3
PIN Device Operations

Method	Description
pin_mode	Set input/output mode (please refer to Code Listing 14.3)
pin_write	Set output level: PIN_LOW (low level) or PIN_HIGH (high level)
pin_read	Get input level: PIN_LOW (low level) or PIN_HIGH (high level)
pin_attach_irq	Set interrupt callback function and interrupt mode (please refer to Code Listing 14.4)
pin_detach_irq	Unset interrupt callback function
pin_irq_enable	Enable/disable interrupt

The operation method of the PIN device is described in Table 14.3.

To register the just-created PIN device, the developer shall call the following function. Table 14.4 shows its parameters.

```
int rt_device_pin_register(const char *name, const struct rt_pin_ops *ops,
void *user_data);
```

TABLE 14.4
rt_device_pin_register

Parameters	Description
name	Pointer to device name
ops	Pointer to PIN device operation structure
user_data	Pointer to user data
Return	— —
RT_EOK	Registration is successful
-RT_ERROR	Registration has failed (the selected name has already been taken)

ACCESS THE PIN DEVICE

In the application side, a developer could access the GPIO hardware through PIN device management interfaces. Figure 14.6 shows the relationship between the PIN device management interfaces and GPIO hardware.

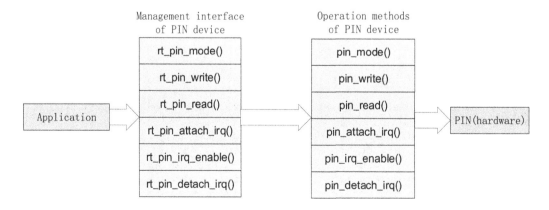

FIGURE 14.6   The relationship between PIN device management interfaces and GPIO hardware.

Set Input/Output Mode

Setting the input/output mode is the first thing to call before actually using a PIN device. Table 14.5 shows its parameters.

```
void rt_pin_mode(rt_base_t pin, rt_base_t mode);
```

TABLE 14.5
rt_pin_mode

Parameter	Description
pin	PIN number
mode	PIN operation mode

Currently RT-Thread supports the following five modes listed in Code Listing 14.3. (The low-level implementation is the user's PIN device driver.)

Code listing 14.3 Available PIN Modes

```
#define PIN_MODE_OUTPUT 0x00          /* Output */
#define PIN_MODE_INPUT 0x01          /* Input */
#define PIN_MODE_INPUT_PULLUP 0x02    /* input Pull up  */
#define PIN_MODE_INPUT_PULLDOWN 0x03  /* input Pull down  */
#define PIN_MODE_OUTPUT_OD 0x04       /* output Open drain  */
```

Set Output Level

The function to set the PIN output level is as follows. Table 14.6 shows its parameters.

```
void rt_pin_write(rt_base_t pin, rt_base_t value);
```

TABLE 14.6
rt_pin_write

Parameter	Description
pin	PIN number
value	The logic level of GPIO output
	PIN_LOW, low level
	PIN_HIGH, high level

Get Input Level

The functions to read the PIN level are as follows. Table 14.7 shows its parameters.

```
int rt_pin_read(rt_base_t pin);
```

TABLE 14.7
rt_pin_read

Parameter	Description
pin	PIN number
Return	— —
PIN_LOW	Low level
PIN_HIGH	High level

Set Interrupt Callback Function

The callback function is executed when the interrupt occurs with the selected PIN. Table 14.8 shows its parameters.

```
rt_err_t rt_pin_attach_irq(rt_int32_t pin, rt_uint32_t mode,
                           void (*hdr)(void *args), void *args);
```

TABLE 14.8
rt_pin_attach_irq

Parameter	Description
pin	PIN number
mode	Interrupt triggering mode (please refer to Code Listing 14.4)
hdr	Pointer to interrupt callback function
args	Pointer to interrupt callback function parameter
Return	— —
RT_EOK	Setting is successful
Error Code	Setting has failed

Code listing 14.4 Interrupt Trigger Mode of PIN

```
#define PIN_IRQ_MODE_RISING 0x00          /* Rising edge trigger */
#define PIN_IRQ_MODE_FALLING 0x01          /* Falling edge trigger */
#define PIN_IRQ_MODE_RISING_FALLING 0x02 /* Edge trigger (triggered on
both rising and falling edges)*/
#define PIN_IRQ_MODE_HIGH_LEVEL 0x03       /* High level trigger */
#define PIN_IRQ_MODE_LOW_LEVEL 0x04      /* Low level trigger */
```

Enable/Disable Interrupt

After binding the PIN interrupt callback function, use the following function to enable the PIN interrupt. Table 14.9 shows its parameters.

```
rt_err_t rt_pin_irq_enable(rt_base_t pin, rt_uint32_t enabled);
```

TABLE 14.9
rt_pin_irq_enable

Parameter	Description
pin	PIN number
enabled	Status: PIN_IRQ_ENABLE or PIN_IRQ_DISABLE
Return	— —
RT_EOK	Setting is successful
Error Code	Setting has failed

Unset Interrupt Callback Function

You can use the following function to detach the PIN interrupt callback function. Table 14.10 shows its parameters.

```
rt_err_t rt_pin_detach_irq(rt_int32_t pin);
```

TABLE 14.10
rt_pin_detach_irq

Parameter	Description
pin	PIN number
Return	— —
RT_EOK	Setting is successful
Error Code	Setting has failed

EXAMPLE OF USING A PIN DEVICE

Code Listing 14.5 provides an example of using the PIN device, implementing the following functions:

1. Set the beep pin as output with default low logic level.
2. Set the key0 and key1 pins as input with corresponding interrupts enabled and interrupt callback functions.
3. When the key0 is pressed, the beep should start to make a sound. When the key1 is pressed, the beep should stop.

Code listing 14.5 Example of Using a PIN Device

```
/*
 * Program listing: This is a PIN device usage routine
 * The routine exports the pin_beep_sample command to the control terminal
 * Command call format:pin_beep_sample
 * Program function: control the buzzer by controlling the level state of
the corresponding pin of the buzzer by pressing the button
*/

#include <rtthread.h>
#include <rtdevice.h>

/* Pin number, determined by looking at the device driver file drv_gpio.c
*/
#ifndef BEEP_PIN_NUM
    #define BEEP_PIN_NUM              37  /* PB2 */
#endif
#ifndef KEY0_PIN_NUM
    #define KEY0_PIN_NUM              57  /* PD10 */
#endif
#ifndef KEY1_PIN_NUM
    #define KEY1_PIN_NUM              56  /* PD9 */
#endif

void beep_on(void *args)
{
    rt_kprintf("turn on beep!\n");

    rt_pin_write(BEEP_PIN_NUM, PIN_HIGH);
}

void beep_off(void *args)
{
```

```
    rt_kprintf("turn off beep!\n");

    rt_pin_write(BEEP_PIN_NUM, PIN_LOW);
}

static void pin_beep_sample(void)
{
    /* Beep pin is in output mode */
    rt_pin_mode(BEEP_PIN_NUM, PIN_MODE_OUTPUT);
    /* Default low level */
    rt_pin_write(BEEP_PIN_NUM, PIN_LOW);

    /* KEY 0 pin is the input mode */
    rt_pin_mode(KEY0_PIN_NUM, PIN_MODE_INPUT_PULLUP);
    /* Bind interrupt, falling edge mode, callback function named beep_on
*/
    rt_pin_attach_irq(KEY0_PIN_NUM, PIN_IRQ_MODE_FALLING, beep_on,
RT_NULL);
    /* Enable interrupt */
    rt_pin_irq_enable(KEY0_PIN_NUM, PIN_IRQ_ENABLE);

    /* KEY 1 pin is input mode */
    rt_pin_mode(KEY1_PIN_NUM, PIN_MODE_INPUT_PULLUP);
    /* Binding interrupt, falling edge mode, callback function named
beep_off */
    rt_pin_attach_irq(KEY1_PIN_NUM, PIN_IRQ_MODE_FALLING, beep_off,
RT_NULL);
    /* Enable interrupt */
    rt_pin_irq_enable(KEY1_PIN_NUM, PIN_IRQ_ENABLE);
}
/* Export to the msh command list */
MSH_CMD_EXPORT(pin_beep_sample, pin beep sample);
```

SPI DEVICE DRIVER

SPI (Serial Peripheral Interface) is a high-speed, synchronous communication method commonly used for microcontrollers to access, for example, EEPROM, FLASH, real-time clock, AD converter, etc. The classic four-wire configuration is shown in Figure 14.7.

- MOSI: Master Output/Slave Input
- MISO: Master Input/Slave Output
- SCLK: Clock
- CS: Chip select, also called SS, CSB, CSN, EN, etc.; used for the master device to select/ activate the slave device

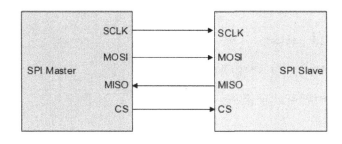

FIGURE 14.7   Four-wire SPI configuration.

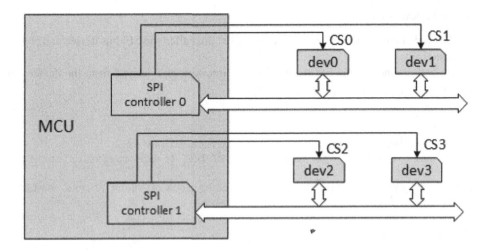

FIGURE 14.8   Example of SPI bus.

SPI works in master-slave mode.

A single microcontroller may support multiple SPI controllers/buses, as shown in Figure 14.8. The SPI controller is able to act as either a "master" or "slave." In the example, both SPI controller 0 and 1 act as "master." On each SPI bus, the "SCK," "MISO," and "MOSI" signals/wires are shared. However, every "slave" has its own "CS" signal/wire.

On an SPI bus, there may have multiple "slaves"; however, only one "master." The communication is always initiated by the "master." A "slave" can transfer data only when it is selected/activated by the "master" through the "CS" signal.

The purpose of the "SCLK" signal provided by the "master" is to keep the "master" and "slave" synchronous. There are two SPI settings related to the "SCLK" signal: CPOL (Clock Polarity) and CPHA (Clock Phase). CPOL represents the state of the initial level of the "SCLK" signal. Value 0 indicates the initial state is low, and value 1 indicates the initial level is high. CPHA represents at which clock edge the data is sampled. Value 0 indicates the data is sampled at the first clock change edge, and value 1 indicates the data is sampled at the second clock change edge. So there are four options according to the combinations of CPOL and CPHA: ①CPOL=0, CPHA=0; ②CPOL=0, CPHA=1; ③CPOL=1, CPHA=0; and ④CPOL=1, CPHA=1, as shown in Figure 14.9.

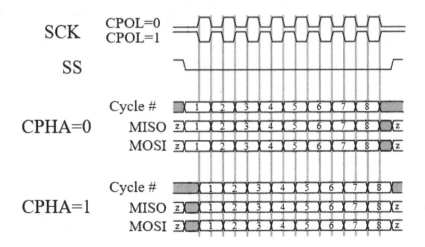

FIGURE 14.9   SPI timing diagram.

SPI DEVICE MANAGEMENT

In the RT-Thread context, the SPI "master" and "slave" are called the SPI bus device and SPI slave device, respectively.

They are both defined in the SPI device driver framework and derived from the device object ("struct rt_device").

```
struct rt_spi_bus
{
    struct rt_device parent;        /* Device class */
    const struct rt_spi_ops *ops;   /* SPI bus device operation method */

    struct rt_mutex lock;           /* Mutex lock,prevent access conflicts
*/
    struct rt_spi_device *owner;    /* Owner of SPI bus is SPI slave device
pointer*/
};
struct rt_spi_device
{
    struct rt_device parent;                /* Device class */
    struct rt_spi_bus *bus;                 /* Point to mounted SPI bus
device */
    struct rt_spi_configuration config;     /* SPI slave device transfer
mode configuration structure */
}
```

Figure 14.10 shows the SPI device invoke diagram.

1. The SPI driver creates an instance of the SPI bus device and registers it in the SPI device driver framework (rt_spi_bus_register()).
2. The SPI device driver framework registers the SPI bus device in I/O Device Manager (rt_device_register()).
3. The SPI driver also creates an instance of the SPI slave device, mounts it to the SPI bus device (rt_spi_bus_attach_device()), and registers it in the SPI device driver framework.
4. The application program (as SPI master) then is able to access the SPI slave hardware through the SPI device interfaces.

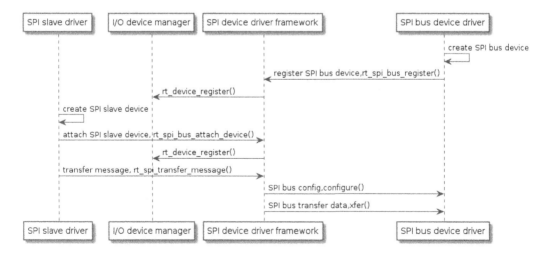

FIGURE 14.10   SPI device invoke diagram.

CREATE AND REGISTER THE SPI BUS DEVICE

To create an SPI bus device, the developer shall implement the SPI device operations defined in the "rt_spi_ops" structure. The parameters are shown in Table 14.11.

```
struct rt_spi_ops
{
rt_err_t (*configure)(struct rt_spi_device *device,
struct rt_spi_configuration *configuration);
rt_uint32_t (*xfer)(struct rt_spi_device *device,
 struct rt_spi_message *message);
};
```

TABLE 14.11
SPI Device Operations

Method	Description
configure	Configure the SPI device settings, e.g., data width, clock polarity, clock phase, clock speed, etc.
xfer	Transfer/message (including data and bus signal setting). The "message" parameter may point to a single message or a chain of messages.

To register the just-created SPI bus device, the developer shall call the following function. The parameters are shown in Table 14.12.

```
rt_err_t rt_spi_bus_register(struct rt_spi_bus      *bus,
                       const char             *name,
                       const struct rt_spi_ops *ops);
```

TABLE 14.12
rt_spi_bus_register

Parameters	Description
bus	Pointer to SPI bus device instance
name	Pointer to SPI bus device name, e.g., spi0
ops	Pointer to SPI device operation structure
Return	— —
RT_EOK	Registration has succeeded
-RT_ERROR	Registration has failed (the selected name has already been taken)

CREATE AND MOUNT THE SPI SLAVE DEVICE

To create and register an SPI slave device, the developer shall call the following function.

The creation of the SPI slave device is mainly to realize the data structure definition of the SPI slave device struct rt_spi_device and then use the rt_spi_bus_attach_device() interface to mount the SPI slave device to the SPI bus device. The interface is as follows, and the function parameters and return values are shown in Table 14.13.

```
rt_err_t rt_spi_bus_attach_device(struct rt_spi_device *device,
                                  const char           *name,
                                  const char           *bus_name,
                                  void                 *user_data)
```

TABLE 14.13

rt_spi_bus_attach_device

Parameter	Description
device	Pointer to SPI slave device instance
name	Pointer to SPI slave device name
bus_name	Pointer to SPI bus device name
user_data	Pointer to user data
Return	— —
RT_EOK	Operation has succeeded
Other Errors	Operation has failed

The naming convention for the SPI bus device is "spi[m]" and for the SPI slave device is "spi[m][m]." For example, spi10 represents device 0 mounted on SPI bus 1.

"user_data" is a pointer to user-provided data, for example, the "CS" signal–related data.

Configure the SPI Slave Device

The slave devices on an SPI bus may use different settings. So before transferring data to a slave device, the corresponding setting has to be applied (rt _ spi _ configure()). See Table 14.14 for the parameters.

```
rt_err_t rt_spi_configure(struct rt_spi_device *device,
                          struct rt_spi_configuration *cfg)
```

TABLE 14.14

rt_spi_configure

Parameter	Description
device	Pointer to SPI slave device instance
cfg	Pointer to SPI configuration structure
Return	— —
RT_EOK	Operation is successful

This function needs to be called only once for each slave device, and it saves the setting to the "config" member of the device instance.

The definition of "struct rt _ spi _ configuration" is as follows:

```
struct rt_spi_configuration
{
    rt_uint8_t mode;        /* mode */
    rt_uint8_t data_width;  /* data width, 8 bits, 16 bits, 32 bits */
    rt_uint16_t reserved;   /* reserved */
    rt_uint32_t max_hz;     /* maximum frequency */
};
```

Mode options:

```
/* Set the data transmission order whether the MSB bit is first or the
LSB bit is before */
#define RT_SPI_LSB      (0<<2)                        /* bit[2]: 0-LSB */
#define RT_SPI_MSB      (1<<2)                        /* bit[2]: 1-MSB */


/* Set the master-slave mode of the SPI */
#define RT_SPI_MASTER   (0<<3)                        /* SPI master device
*/
#define RT_SPI_SLAVE    (1<<3)                        /* SPI slave device */


/* Set clock polarity and clock phase */
#define RT_SPI_MODE_0   (0 | 0)                       /* CPOL = 0, CPHA = 0
*/
#define RT_SPI_MODE_1   (0 | RT_SPI_CPHA)             /* CPOL = 0, CPHA = 1
*/
#define RT_SPI_MODE_2   (RT_SPI_CPOL | 0)             /* CPOL = 1, CPHA = 0
*/
#define RT_SPI_MODE_3   (RT_SPI_CPOL | RT_SPI_CPHA)   /* CPOL = 1, CPHA = 1
*/


#define RT_SPI_CS_HIGH  (1<<4)                        /* Chipselect active
high */
#define RT_SPI_NO_CS    (1<<5)                        /* No chipselect */
#define RT_SPI_3WIRE    (1<<6)                        /* SI/SO pin shared */
#define RT_SPI_READY    (1<<7)                        /* Slave pulls low to
pause */
```

Data width options: 8-bit, 16-bit, or 32-bit.
Maximum frequency: The upper limit of SPI clock frequency.
Following is an example:

```
struct rt_spi_configuration cfg;
cfg.data_width = 8;
cfg.mode = RT_SPI_MASTER | RT_SPI_MODE_0 | RT_SPI_MSB;
cfg.max_hz = 20 * 1000 *1000;                              /* 20M */

rt_spi_configure(spi_dev, &cfg);
```

ACCESS THE SPI SLAVE DEVICE

On the application side, the developer could access the slave device hardware through SPI device management interfaces. The main interfaces are shown in Figure 14.11.

Transfer Data Chain

A developer could use the following interface to transmit and/or receive multiple data segments. Table 14.15 shows its parameters.

```
struct rt_spi_message *rt_spi_transfer_message(struct rt_spi_device
*device,struct rt_spi_message *message);
```

TABLE 14.15
rt_spi_transfer_message

Parameter	Description
device	Pointer to SPI slave device instance
message	Pointer to SPI message chain
Return	— —
RT_NULL	Operation has succeeded
Non-null Pointer	Operation has failed; the value is a pointer to the first unsent message in the chain

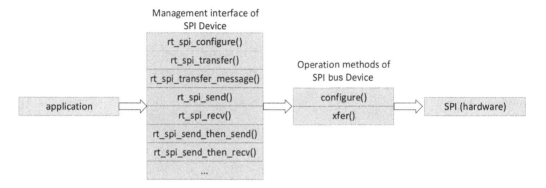

FIGURE 14.11   The relationship between SPI device interfaces and SPI slave hardware.

Except for transferring multiple data segments, this interface also provides the precise control of the "CS" signal state before ("cs_take") and after ("cs_release") each data segment. Following is the definition of the "rt _ spi _ message" structure:

```
struct rt_spi_message
{
    const void *send_buf;              /* Send buffer pointer */
    void *recv_buf;                    /* Receive buffer pointer */
    rt_size_t length;                  /* Send/receive data bytes */
    struct rt_spi_message *next;       /* Pointer to the next message to
continue sending */
    unsigned cs_take    : 1;           /* Take chip selection*/
    unsigned cs_release : 1;           /* Release chip selection */
};
```

send_buf: Pointer to the transmit buffer. It should be set to RT_NULL for the receiving-only segment.

recv_buf: Pointer to the receive buffer. It should be set to RT_NULL for the transmitting-only segment.

length: The transmit or receive buffer (which one is larger) size in bytes.

next: pointer to the next message structure in the chain. The value of the last message structure in the chain shall be RT_NULL.

cs_take: If the value is "1" (normally for the first data segment), set the "CS" signal before transferring data.

cs_release: If the value is "1" (normally for the last data segment), clear the "CS" signal after transferring data.

Transfer Data

If there is only one data segment to transfer, the developer could use the following interface. Table 14.16 shows its parameters.

```
rt_size_t rt_spi_transfer(struct rt_spi_device *device,
                          const void           *send_buf,
                          void                 *recv_buf,
                          rt_size_t            length);
```

TABLE 14.16
rt_spi_transfer

Parameter	Description
device	Pointer to SPI slave device
send_buf	Pointer to transmit buffer
recv_buf	Pointer to receive buffer
length	The transmit or receive buffer (which one is larger) size in bytes
Return	— —
0	Operation has failed
Non-0 Value	The successfully transferred data size in bytes

This interface sets and clears the "CS" signal automatically. It is equivalent to calling "rt_spi_transfer()" with the following message structure:

```
struct rt_spi_message msg;

msg.send_buf    = send_buf;
msg.recv_buf    = recv_buf;
msg.length      = length;
msg.cs_take     = 1;
msg.cs_release  = 1;
msg.next        = RT_NULL;
```

Transmit Data

If there is only one data segment to transmit, the developer could use the following interface. Any data received is ignored. Table 14.17 shows its parameters.

```
rt_size_t rt_spi_send(struct rt_spi_device *device,
                      const void           *send_buf,
                      rt_size_t            length)
```

TABLE 14.17
rt_spi_send

Parameter	Description
device	Pointer to SPI slave device
send_buf	Pointer to transmit buffer
Length	The transmit buffer size in bytes
Return	— —
0	Operation has failed
Non-0 Value	The successfully transmitted data size in bytes

It is equivalent to calling "rt_spi_transfer()" with the following message structure:

```
struct rt_spi_message msg;

msg.send_buf    = send_buf;
msg.recv_buf    = RT_NULL;
msg.length      = length;
msg.cs_take     = 1;
msg.cs_release  = 1;
msg.next        = RT_NULL;
```

Receive Data

If there is only one data segment to receive, the developer could use the following interface. The master device transmits "0xFF" for a four-wire SPI configuration. Table 14.18 shows its parameters.

```
rt_size_t rt_spi_recv(struct rt_spi_device *device,
                      void                 *recv_buf,
                      rt_size_t             length);
```

TABLE 14.18
rt_spi_recv

Parameter	Description
device	Pointer to SPI slave device
recv_buf	Pointer to receive buffer
length	The receive buffer size in bytes
Return	— —
0	Operation has failed
Non-0 Value	The successfully received data size in bytes

It is equivalent to calling "rt_spi_transfer()" with the following message structure:

```
struct rt_spi_message msg;

msg.send_buf    = RT_NULL;
msg.recv_buf    = recv_buf;
msg.length      = length;
msg.cs_take     = 1;
msg.cs_release  = 1;
msg.next        = RT_NULL;
```

Transmit Two Data Segments

The developer could use the following interface to transmit two data segments continuously. Table 14.19 shows its parameters.

```
rt_err_t rt_spi_send_then_send(struct rt_spi_device *device,
                               const void           *send_buf1,
                               rt_size_t             send_length1,
                               const void           *send_buf2,
                               rt_size_t             send_length2);
```

TABLE 14.19
rt_spi_send_then_send

Parameter	Description
device	Pointer to SPI slave device
send_buf1	Pointer to the 1st transmit buffer
send_length1	The 1st transmit buffer size in bytes
send_buf2	Pointer to the 2nd transmit buffer
send_length2	The 2nd transmit buffer size in bytes
Return	— —
RT_EOK	Operation has succeeded
-RT_EIO	Operation has failed

It is equivalent to calling "rt_spi_transfer()" with the following two message structures in the chain:

```
struct rt_spi_message msg1,msg2;

msg1.send_buf    = send_buf1;
msg1.recv_buf    = RT_NULL;
msg1.length      = send_length1;
msg1.cs_take     = 1;
msg1.cs_release  = 0;
msg1.next        = &msg2;

msg2.send_buf    = send_buf2;
msg2.recv_buf    = RT_NULL;
msg2.length      = send_length2;
msg2.cs_take     = 0;
msg2.cs_release  = 1;
msg2.next        = RT_NULL;
```

Transmit and Then Receive Data Segments

The developer could use the following interface to transmit a data segment and then receive another one. Table 14.20 shows its parameters.

```
rt_err_t rt_spi_send_then_recv(struct rt_spi_device *device,
                        const void           *send_buf,
                        rt_size_t             send_length,
                        void                 *recv_buf,
                        rt_size_t             recv_length);
```

TABLE 14.20
rt_spi_send_then_recv

Parameter	Description
device	Pointer to SPI slave device
send_buf	Pointer to transmit buffer
send_length	The transmit buffer size in bytes
recv_buf	Pointer to receive buffer
recv_length	The receive buffer size in bytes
Return	— —
RT_EOK	Operation has succeeded
-RT_EIO	Operation has failed

It is equivalent to calling "rt_spi_transfer()" with the following two message structures in the chain:

```
struct rt_spi_message msg1,msg2;

msg1.send_buf    = send_buf;
msg1.recv_buf    = RT_NULL;
msg1.length      = send_length;
msg1.cs_take     = 1;
msg1.cs_release  = 0;
msg1.next        = &msg2;

msg2.send_buf    = RT_NULL;
msg2.recv_buf    = recv_buf;
msg2.length      = recv_length;
msg2.cs_take     = 0;
msg2.cs_release  = 1;
msg2.next        = RT_NULL;
```

There are also "rt _ spi _ sendrecv8()" and "rt _ spi _ sendrecv16()" interfaces, which are wrappers of "rt _ spi _ send _ then _ recv()."

OTHER UTILITIES

Acquire SPI Bus

When multiple threads are sharing an SPI slave device, the developer could use the following interface to access SPI slave device exclusively. Table 14.21 shows its parameters.

```
rt_err_t rt_spi_take_bus(struct rt_spi_device *device);
```

TABLE 14.21
rt_spi_take_bus

Parameter	Description
device	Pointer to SPI slave device
Return	— —
RT_EOK	Operation has succeeded
Other Errors	Operation has failed

Set CS Signal

The developer could use the following interface to set the "CS" signal without transferring data. Table 14.22 shows its parameters.

```
rt_err_t rt_spi_take(struct rt_spi_device *device);
```

TABLE 14.22
rt_spi_take

Parameter	Description
device	Pointer to SPI slave device
Return	— —
0	Operation has succeeded
Other Errors	Operation has failed

Append Data to the Chain
The SPI data chain is actually a single-linked list. The developer could use the following interface to insert a message structure to the end of the list. Table 14.23 shows its parameters.

```
void rt_spi_message_append(struct rt_spi_message *list,
                           struct rt_spi_message *message);
```

TABLE 14.23
rt_spi_message_append

Parameter	Description
list	Pointer to data chain
message	Pointer to message structure

Clear CS Signal
The developer shall use the following interface to clear the "CS" signal after data transfer is completed, if "rt_spi_take()" was previously involved. Table 14.24 shows its parameters.

```
rt_err_t rt_spi_release(struct rt_spi_device *device);
```

TABLE 14.24
rt_spi_release

Parameter	Description
device	Pointer to SPI slave device
Return	— —
0	Operation has succeeded
Other Errors	Operation has failed

Release SPI Bus
The developer shall release the exclusive access of the SPI bus as soon as data transfer is done, if "rt_spi_take_bus()" was previously involved. Table 14.25 shows its parameters.

```
rt_err_t rt_spi_release_bus(struct rt_spi_device *device);
```

TABLE 14.25
rt_spi_release_bus

Parameter	Description
device	Pointer to SPI slave device
Return	— —
RT_EOK	Operation has succeeded

EXAMPLE OF USING THE SPI DEVICE

Code Listing 14.6 provides an example of reading the ID of an SPI slave device.

Code listing 14.6 Example of Using SPI Device

```
/*
 * Program listing: This is a SPI device usage routine
 * The routine exports the spi_w25q_sample command to the control terminal
 * Command call format: spi_w25q_sample spi10
 * Command explanation: The second parameter of the command is the name of
the SPI device to be used. If it is empty, the default SPI device is used.
 * Program function: read w25q ID data through SPI device
*/

#include <rtthread.h>
#include <rtdevice.h>

#define W25Q_SPI_DEVICE_NAME     "qspi10"

static void spi_w25q_sample(int argc, char *argv[])
{
    struct rt_spi_device *spi_dev_w25q;
    char name[RT_NAME_MAX];
    rt_uint8_t w25x_read_id = 0x90;
    rt_uint8_t id[5] = {0};

    if (argc == 2)
    {
        rt_strncpy(name, argv[1], RT_NAME_MAX);
    }
    else
    {
        rt_strncpy(name, W25Q_SPI_DEVICE_NAME, RT_NAME_MAX);
    }

    /* Find the spi device to get the device handle */
    spi_dev_w25q = (struct rt_spi_device *)rt_device_find(name);
    if (!spi_dev_w25q)
    {
        rt_kprintf("spi sample run failed! can't find %s device!\n",
name);
    }
    else
    {
        /* Method 1: Send the command to read the ID using rt_spi_send_
then_recv() */
        rt_spi_send_then_recv(spi_dev_w25q, &w25x_read_id, 1, id, 5);
        rt_kprintf("use rt_spi_send_then_recv() read w25q ID is:%x%x\n",
id[3], id[4]);

        /* Method 2: Send the command to read the ID using rt_spi_
transfer_message() */
        struct rt_spi_message msg1, msg2;

        msg1.send_buf   = &w25x_read_id;
```

```
            msg1.recv_buf    = RT_NULL;
            msg1.length      = 1;
            msg1.cs_take     = 1;
            msg1.cs_release  = 0;
            msg1.next        = &msg2;

            msg2.send_buf    = RT_NULL;
            msg2.recv_buf    = id;
            msg2.length      = 5;
            msg2.cs_take     = 0;
            msg2.cs_release  = 1;
            msg2.next        = RT_NULL;

            rt_spi_transfer_message(spi_dev_w25q, &msg1);
            rt_kprintf("use rt_spi_transfer_message() read w25q ID is:%x%x\n",
id[3], id[4]);

    }
}
/* Export to the msh command list */
MSH_CMD_EXPORT(spi_w25q_sample, spi w25q sample);
```

I2C DEVICE DRIVER

I2C (Inter-Integrated Circuit) is the half-duplex, bidirectional, two-wire, synchronous serial communication protocol invented by PHILIPS. There are two signals: SDA (serial data) and SCL (serial clock).

Figure 14.12 shows the I2C hardware configuration. I2C works in master-slave mode also. However, there is no "CS" signal, as is found on the SPI bus. The I2C bus uses "address" to select a slave device.

Figure 14.13 shows the I2C data format.

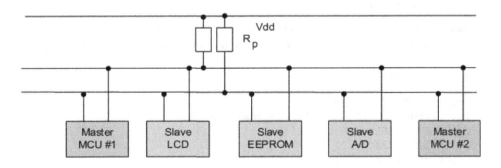

FIGURE 14.12   I2C hardware configuration.

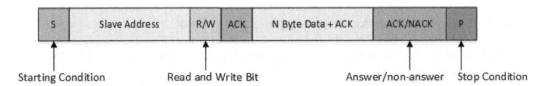

FIGURE 14.13   I2C data format.

S	A6	A5	A4	A3	A2	A1	A0	R/W	ACK

S	1	1	1	1	0	A9	A8	R/W	ACK	A7	A6	A5	A4	A3	A2	A1	A0	ACK

FIGURE 14.14 7-bit and 10-bit address format.

In the I2C idle state, both "SDA" and "SCL" signals are pulled to a high state (by pull-up resistors). When communicating, the "master" first sends a start condition, followed by the slave address and read/write control bit. After acknowledged by the "slave," the "master" then transmits/receives data (with acknowledge) to/from the "slave." After another acknowledge, the "master" finally sends the stop condition.

- **Starting Condition:** "Master" leaving "SCL" high and pulling "SDA" low, indicates the beginning of data transfer.
- **Slave Address:** The address is made up of the device address (MSB first) and read ("1")/write ("0") control bit. The device address could be 7-bit or 10-bit. For the 10-bit format, the address is sent in two segments. In the first segment, the most significant 5 bits are fixed to b'11110, followed by the most significant 2 bits of the device address and read/write control bit. The second segment includes the remaining bits of the device address, as shown in Figure 14.14.
- **Answer Signal:** After every byte sent, the counterpart answers it with ACK (acknowledge, "0") to continue or NACK (Not acknowledge, "1") to stop.
- **Data:** The data width is 8 bits, and there is no limit to the length.
- **Repeat Start Condition:** When the "master" switches to another slave read/write operation, it sends another start condition.
- **Stop Condition:** After the transition of "SCL" from "0" to "1," the "master," making a transition of "SDA" from "0" to "1," indicates the end of data transfer.

I2C DEVICE MANAGEMENT

RT-Thread supports accessing the I2C bus as master and slave devices. The I2C bus device is defined in the I2C device driver framework and derived from the device object ("struct rt_device"):

```
struct rt_i2c_bus_device
{
    struct rt_device parent;
    const struct rt_i2c_bus_device_ops *ops;
    rt_uint16_t  flags;
    rt_uint16_t  addr;
    struct rt_mutex lock;
    rt_uint32_t  timeout;
    rt_uint32_t  retries;
    void *priv;
};
```

Figure 14.15 shows the I2C device invoke diagram.

1. The I2C driver creates an instance of the I2C bus device and registers it in the I2C device driver framework (rt_i2c_bus_device_register()).
2. The I2C device driver framework registers the I2C bus device in I/O Device Manager (rt_device_register()).

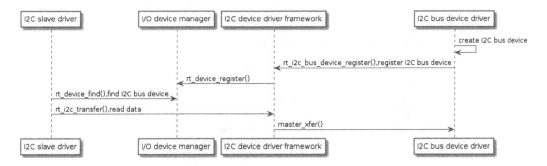

FIGURE 14.15 I2C device invoke diagram.

3. In the I2C slave driver, the user finds the I2C bus device (rt_device_find()).
4. Then it accesses the I2C bus through I2C device interfaces.

CREATE AND REGISTER THE I2C BUS DEVICE

To create an I2C bus device, the developer shall implement the I2C bus device operations defined in the "rt_i2c_bus_device_ops" structure:

```
struct rt_i2c_bus_device_ops
{
    rt_size_t (*master_xfer)(struct rt_i2c_bus_device *bus,
                             struct rt_i2c_msg msgs[],
                             rt_uint32_t num);
    rt_size_t (*slave_xfer)(struct rt_i2c_bus_device *bus,
                            struct rt_i2c_msg msgs[],
                            rt_uint32_t num);
    rt_err_t (*i2c_bus_control)(struct rt_i2c_bus_device *bus,
                                rt_uint32_t,
                                rt_uint32_t);
};
```

The operation method of the I2C bus device is described in Table 14.26.

TABLE 14.26
I2C Bus Device Operations

Operation Method	Description
master_xfer	Transfer data as "master"
slave_xfer	Transfer data as "slave"
i2c_bus_control	Control I2C bus

To register the just-created I2C bus device, the developer shall call the following function. Table 14.27 shows its parameters.

```
rt_err_t rt_i2c_bus_device_register(struct rt_i2c_bus_device *bus,
                                    const char *bus_name);
```

TABLE 14.27
rt_i2c_bus_device_register

Parameter	Description
bus	Pointer to I2C bus device
bus_name	I2C bus device name (e.g., "i2c0")
Return	— —
RT_EOK	Registration has succeeded
-RT_ERROR	Registration has failed (the selected name has already been taken)

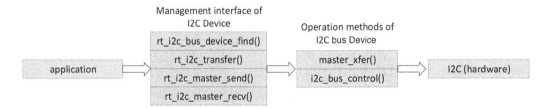

FIGURE 14.16 The relationship between I2C device interfaces and I2C hardware.

ACCESS THE I2C BUS

On the application side, the developer could access the I2C device hardware through I2C device management interfaces. The related interfaces are shown in Figure 14.16.

The developer could use the following interface to transmit and/or receive multiple data segments (as "master"). Table 14.28 shows its parameters.

```
rt_size_t rt_i2c_transfer(struct rt_i2c_bus_device *bus,
                          struct rt_i2c_msg          msgs[],
                          rt_uint32_t                num);
```

TABLE 14.28
rt_i2c_transfer

Parameter	Description
bus	Pointer to I2C bus device
msgs[]	Pointer to I2C message array
num	The message array length
Return Value	— —
The number of elements in the message array	Operation has succeeded
Error Code	Operation has failed

Except for device address and data, this interface also provides the precise control of the bus signals for each data segment. If the master needs to send a repeat start condition, it will need to send two messages.

Notes: This interface involves "rt_mutex_take()," so it cannot be used in the interrupt service routine (causing an assertion error).

Following is the definition of the "rt_i2c_msg" structure:

```
struct rt_i2c_msg
{
    rt_uint16_t addr;     /* Slave address */
    rt_uint16_t flags;    /* Reading, writing signs, etc. */
    rt_uint16_t len;      /* Read and write data bytes */
    rt_uint8_t  *buf;     /* Read and write data buffer pointer */
}
```

Slave address (addr): 7-bit or 10-bit address without read/write bit. The read/write control is set in the parameter "flags."

The following are the available values of "flags" (multiple values could be combined with "|", as shown in Code Listing 14.7).

Code listing 14.7 The Available Values of "flags"

```
#define RT_I2C_WR              0x0000        /* Write flag */
#define RT_I2C_RD              (1u << 0)     /* Read flag */
#define RT_I2C_ADDR_10BIT      (1u << 2)     /* 10-bit address mode */
#define RT_I2C_NO_START        (1u << 4)     /* No start condition */
#define RT_I2C_IGNORE_NACK     (1u << 5)     /* Ignore NACK */
#define RT_I2C_NO_READ_ACK     (1u << 6)     /* Do not send ACK when
reading */
```

EXAMPLE OF USING AN I2C DEVICE

Code Listing 14.8 provides an example of reading the sensor value through I2C, which implements the following functions:

1. Find the I2C bus device by name and then initialize the sensor, aht10.
2. Read the sensor value by calling "read_temp_humi()," which in turn involves "write_reg()" and "write_reg()" (both functions involve the I2C device interface "rt _ i2c _ transfer()").

Code listing 14.8 Example of Using the I2C Device

```
/*
 * Program listing: This is an I2C device usage routine
 * The routine exports the i2c_aht10_sample command to the control
   terminal
 * Command call format: i2c_aht10_sample i2c1
 * Command explanation: The second parameter of the command is the name of
the I2C bus device to be used. If it is empty, the default I2C bus device
is used.
 * Program function: read the temperature and humidity data of the aht10
sensor and print.
*/

#include <rtthread.h>
#include <rtdevice.h>

#define AHT10_I2C_BUS_NAME          "i2c2"  /* Sensor connected I2C bus
device name */
#define AHT10_ADDR                  0x38    /* Slave address */
```

```
#define AHT10_CALIBRATION_CMD        0xE1    /* Calibration command */
#define AHT10_NORMAL_CMD             0xA8    /* General command */
#define AHT10_GET_DATA               0xAC    /* Get data command */

static struct rt_i2c_bus_device *i2c_bus = RT_NULL;      /* I2C bus device
handle */
static rt_bool_t initialized = RT_FALSE;                 /* Sensor
initialization status */

/* Write sensor register */
static rt_err_t write_reg(struct rt_i2c_bus_device *bus, rt_uint8_t reg,
rt_uint8_t *data)
{
    rt_uint8_t buf[3];
    struct rt_i2c_msg msgs;

    buf[0] = reg; //cmd
    buf[1] = data[0];
    buf[2] = data[1];

    msgs.addr = AHT10_ADDR;
    msgs.flags = RT_I2C_WR;
    msgs.buf = buf;
    msgs.len = 3;

    /* Call the I2C device interface to transfer data */
    if (rt_i2c_transfer(bus, &msgs, 1) == 1)
    {
        return RT_EOK;
    }
    else
    {
        return -RT_ERROR;
    }
}

/* Read sensor register data */
static rt_err_t read_regs(struct rt_i2c_bus_device *bus, rt_uint8_t len,
rt_uint8_t *buf)
{
    struct rt_i2c_msg msgs;

    msgs.addr = AHT10_ADDR;
    msgs.flags = RT_I2C_RD;
    msgs.buf = buf;
    msgs.len = len;

    /* Call the I2C device interface to transfer data */
    if (rt_i2c_transfer(bus, &msgs, 1) == 1)
    {
        return RT_EOK;
    }
    else
    {
        return -RT_ERROR;
    }
}
```

```
static void read_temp_humi(float *cur_temp, float *cur_humi)
{
    rt_uint8_t temp[6];

    write_reg(i2c_bus, AHT10_GET_DATA, 0);        /* send command */
    rt_thread_mdelay(400);
    read_regs(i2c_bus, 6, temp);                  /* obtain sensor data */

    /* Humidity data conversion */
    *cur_humi = (temp[1] << 12 | temp[2] << 4 | (temp[3] & 0xf0) >> 4) *
100.0 / (1 << 20);
    /* Temperature data conversion */
    *cur_temp = ((temp[3] & 0xf) << 16 | temp[4] << 8 | temp[5]) * 200.0 /
(1 << 20) - 50;
}

static void aht10_init(const char *name)
{
    rt_uint8_t temp[2] = {0, 0};

    /* Find the I2C bus device and get the I2C bus device handle */
    i2c_bus = (struct rt_i2c_bus_device *)rt_device_find(name);

    if (i2c_bus == RT_NULL)
    {
        rt_kprintf("can't find %s device!\n", name);
    }
    else
    {
        write_reg(i2c_bus, AHT10_NORMAL_CMD, temp);
        rt_thread_mdelay(400);

        temp[0] = 0x08;
        temp[1] = 0x00;
        write_reg(i2c_bus, AHT10_CALIBRATION_CMD, temp);
        rt_thread_mdelay(400);
        initialized = RT_TRUE;
    }
}

static void i2c_aht10_sample(int argc, char *argv[])
{
    float humidity, temperature;
    char name[RT_NAME_MAX];

    humidity = 0.0;
    temperature = 0.0;

    if (argc == 2)
    {
        rt_strncpy(name, argv[1], RT_NAME_MAX);
    }
    else
    {
        rt_strncpy(name, AHT10_I2C_BUS_NAME, RT_NAME_MAX);
    }
```

```
    if (!initialized)
    {
        /* Sensor initialization */
        aht10_init(name);
    }
    if (initialized)
    {
        /* Read temperature and humidity data */
        read_temp_humi(&temperature, &humidity);

        rt_kprintf("read aht10 sensor humidity   : %d.%d %%\n",
(int)humidity, (int)(humidity * 10) % 10);
        rt_kprintf("read aht10 sensor temperature: %d.%d \n",
(int)temperature, (int)(temperature * 10) % 10);
    }
    else
    {
        rt_kprintf("initialize sensor failed!\n");
    }
}
/* Export to the msh command list */
MSH_CMD_EXPORT(i2c_aht10_sample, i2c aht10 sample);
```

DEMO PROJECTS

This section shows how to run the device driver demo applications on the RT-Thread IoT board. The code used in this section is located in the chapter14 directory of the complementary materials.

Open the MDK5 project file project.uvprojx, as shown in Figure 14.17.

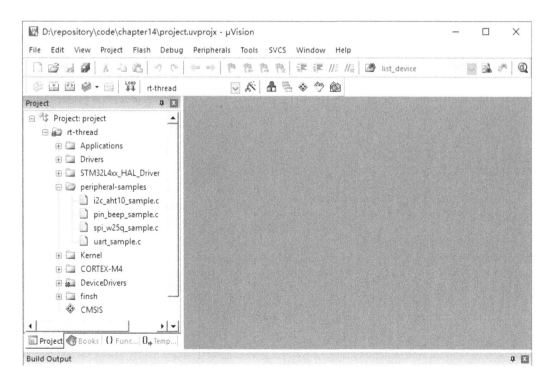

FIGURE 14.17   MDK5 project.

The device driver code is in the Device Drivers group. The device uses the sample code in the peripheral-samples group. And the sample file description is shown in Table 14.29.

TABLE 14.29
Sample Files

Sample File	Description
i2c_aht10_sample.c	Sample file using I2C device
pin_beep_sample.c	Sample file using PIN device
uart_sample.c	Sample file using UART device
spi_w25q_sample.c	Sample file using SPI device

After compiling and downloading (through ST-Link) the code to the target board, the following messages can be observed in the terminal (connected to the ST-Link virtual serial port). Pressing "Tab" will show four demo commands:

```
 \ | /
- RT -     Thread Operating System
 / | \     3.1.0 build Aug 27 2018
 2006 - 2018 Copyright by rt-thread team

msh >
RT-Thread shell commands:
i2c_aht10_sample - i2c aht10 sample
pin_beep_sample  - pin beep sample
uart_sample      - uart device sample
spi_w25q_sample  - spi w25q sample
```

The developer may issue the "list_device" command to list all devices currently registered in the system, which includes the two serial port devices (uart1 and uart2), one PIN device (pin), one SPI device (qspi10), and one I2C device (i2c1), as shown here:

```
msh >list_device
device          type             ref count
------ -------------------- ----------
qspi10 SPI Device               0
qspi1  SPI Bus                  0
i2c1   I2C Bus                  0
uart1  Character Device         2
uart2  Character Device         1
pin    Miscellaneous Device     0
```

PIN DEVICE DEMO PROJECT

The RT-Thread IoT board has a buzzer, as shown in Figure 14.18. When the "BEEP" pin outputs high, the buzzer makes sound, and when it outputs low, the buzzer stops.

Try the "pin_beep_sample" command in the terminal, which will configure the two keys on the target board to control the buzzer through PIN device interfaces (please refer to Code Listing 14.5).

```
 \ | /
- RT -     Thread Operating System
 / | \     3.1.0 build Aug 16 2018
 2006 - 2018 Copyright by rt-thread team
```

```
msh />pin_beep_sample
msh />turn on beep!
turn off beep!

msh />
```

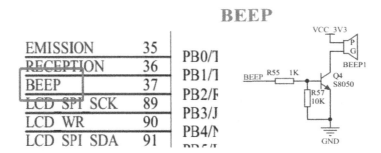

FIGURE 14.18   Schematic diagram of buzzer.

SPI DEVICE DEMO PROJECT

The RT-Thread IoT board has an SPI FLASH, W25Q128 (as shown in Figure 14.19). Its default SPI slave device name is "qspi10."

Try the "spi_w25q_sample" command in the terminal, which will find the SPI device named "qspi10" and then print out its ID through the SPI device interfaces (please refer to Code Listing 14.6).

```
 \ | /
- RT -     Thread Operating System
 / | \     3.1.0 build Aug 16 2018
 2006 - 2018 Copyright by rt-thread team

msh /> spi_w25q_sample
use rt_spi_send_then_recv() read w25q128 ID is:ef17
use rt_spi_transfer_message() read w25q128 ID is:ef17
msh />
```

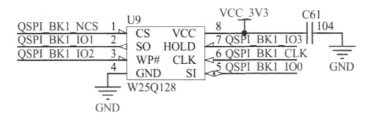

FIGURE 14.19   Schematic diagram of W25Q128.

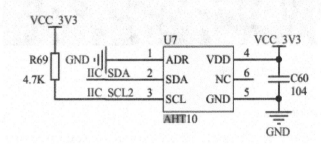

FIGURE 14.20 Schematic diagram of AHT10.

I2C Device Demo Project

The RT-Thread IoT board has a temperature and humidity sensor, which connects to the I2C port (as shown in Figure 14.20).

Try the "i2c_aht10_sample" command in the terminal, which will read the sensor and then print out the temperature and humidity values.

```
 \ | /
- RT -     Thread Operating System
 / | \     3.1.0 build Sep 16 2018
 2006 - 2018 Copyright by rt-thread team

msh />i2c_aht10_sample
read aht10 sensor humidity   : 50.0 %
read aht10 sensor temperature: 31.9
msh /> i2c_aht10_sample
read aht10 sensor humidity   : 49.5 %
read aht10 sensor temperature: 31.7
msh />
```

UART Device Demo Project

Due to "uart1" being occupied by a console shell, the UART device example uses the device "uart2" (please refer to Code Listing 14.2). Try the "uart_sample" command in the terminal, which will find the UART device, configure it, print out "hello RT-Thread!," and then echo the received data after processing (adding the value to 1).

```
 \ | /
- RT -     Thread Operating System
 / | \     3.0.4 build Jun 14 2018
 2006 - 2018 Copyright by rt-thread team

msh >uart_sample
```

To connect UART2 port to PC, a USB to TTL serial converter is required. The following figure shows the demo by using the terminal tool, PuTTY, with baud rate of 115,200 bps. The result is shown in Figure 14.21.

FIGURE 14.21   UART device driver demo.

CHAPTER SUMMARY

This chapter describes how RT-Thread device drivers work with examples. Some notes here:

1. The PIN device has to "know" the number of pins defined by the PIN device driver to work properly.
2. The SPI device driver framework provides a set of data transfer interfaces for different scenarios.
3. The slave address parameter of the I2C device interfaces does not contain a read/write bit.

15 Virtual File System

In early embedded systems, the amount of data to be stored was relatively small and the data types were relatively simple. Data was stored by directly writing it at a specified address in a storage device. However, with the evolution of embedded device capabilities, more and more data needs to be stored, and it is also increasingly complicated. It is very cumbersome to use the old method to store and manage data. So we need a new data management system to abstract the complicity of how data is stored and provide an interface in the meantime. This is the file system that we will introduce in the following paragraphs.

A file system is a set of abstract data types that implements the storage, hierarchical organization, access, and retrieval of data. It is a mechanism for providing underlying data access to users. Files and folders are two basic concepts of a file system. A file is where data is stored, and a folder helps keep an organized tree structure.

This chapter explains the RT-Thread file system and introduces you to the architecture, features, and usage of the RT-Thread virtual file system.

AN INTRODUCTION TO DFS

DFS (Device File System) is a virtual file system component provided by RT-Thread. As the name implies, it is a device virtual file system. File and folder names follow the UNIX convention. The directory structure is shown in Figure 15.1.

In the RT-Thread DFS, the file system has one root directory, which is represented by /. The *f1. bin* file in the root directory is represented by /f1.bin, and the /f1.bin directory in the 2019 directory is represented by /data/2019/f1.bin. That is, the partition symbol of the directory is /, which is exactly the same as UNIX and Linux. It is different from Windows, where \ is used as the separator of the directory.

THE ARCHITECTURE OF DFS

The main features of the RT-Thread DFS are:

- Provides a unified POSIX file system interface for applications, including read, write, poll/select, and more.
- Supports multiple types of file systems, such as FatFS, RomFS, DevFS, etc., and provides management of common files, device files, and network file descriptors.
- Supports multiple types of storage devices such as SD Card, SPI Flash, Nand Flash, etc.

The hierarchical structure of DFS is shown in Figure 15.2. It includes the POSIX interface layer, virtual file system layer, and device abstraction layer.

POSIX INTERFACE LAYER

POSIX stands for Portable Operating System Interface of UNIX. The POSIX standard defines the interface that the operating system should provide for applications. It is a general term for a series of interface standards defined by IEEE for software to run on various UNIX operating systems.

The POSIX standard is intended to achieve software portability at the source code level. In other words, a program written for a POSIX-compatible operating system should be able to compile and

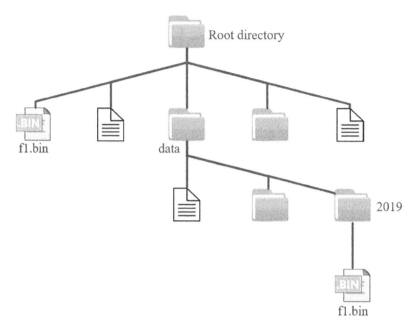

FIGURE 15.1   Figure of the directory structure.

execute on any other POSIX operating system (even from another vendor). RT-Thread supports the POSIX standard interface, so it is easy to port Linux/UNIX programs to the RT-Thread operating system.

On UNIX-like systems, normal files, device files, and network file descriptors are the same. In the RT-Thread operating system, DFS is used to achieve this uniformity. With the uniformity of such file descriptors, we can use the `poll/select` interface to uniformly poll these descriptors and bring convenience to the implementation of the program functions.

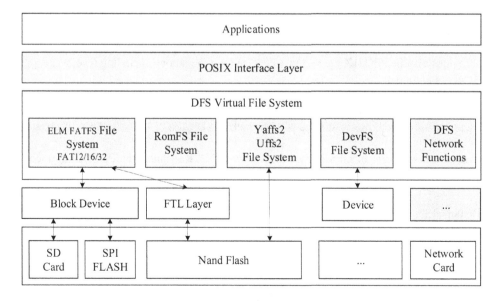

FIGURE 15.2   The hierarchical structure of DFS.

Use the `poll/select` interface to block and simultaneously detect whether a group of I/O devices that support non-blocking have events (such as readable, writable, high-priority error output, errors, etc.) until a device triggers the event or the specified wait time is exceeded. This mechanism can help callers find devices that are currently ready, reducing the complexity of programming.

VIRTUAL FILE SYSTEM LAYER

Developers can register a specific file system to DFS, such as FatFS, RomFS, DevFS, etc. Here are some common file system types:

- FatFS is a Microsoft FAT format–compatible file system developed for small embedded devices. It is written in ANSI C and has good hardware independence and portability. It is the most commonly used file system type in RT-Thread.
- The traditional RomFS file system is a simple, compact, read-only file system that does not support dynamic erasing and saving or storing data in order; thus, it supports applications to run in the XIP (execute In Place) method and saves RAM space while the system is running.
- The Jffs2 file system is a log flash file system. It is mainly used for NOR flash memory and is based on the MTD driver layer and features read and write capability, supports data compression, has a hash table–based log file system, and provides crash/power failure security protection, write balance support, etc.
- DevFS is the device file system. After the feature is enabled in the RT-Thread operating system, the devices in the system can be virtualized into files in the /dev folder, so that the device can use the interfaces such as `read` and `write` according to the operation mode of the file to operate.
- NFS (Network File System) is a technology for sharing files over a network between different machines and different operating systems. In the development and debugging phase of the operating system, this technology can be used to build an NFS-based root file system on the host and mount it on the embedded device, which can easily modify the contents of the root file system.
- UFFS is short for Ultra-low-cost Flash File System. It is an open source file system developed by Chinese developers and used for running Nand Flash in small memory environments such as embedded devices. Compared with the Yaffs file system, which is often used in embedded devices, it has the advantages of less resource consumption, faster startup speed, and it is free.

DEVICE ABSTRACTION LAYER

The device abstraction layer abstracts physical devices such as SD Card, SPI Flash, and Nand Flash into devices that are accessible to the file system. For example, the FAT file system requires that the storage device be a block device type.

Different file system types are implemented independently of the storage device driver, so the file system feature can be correctly used after the drive interface of the underlying storage device is docked with the file system.

MOUNT MANAGEMENT

The initialization process of the file system includes the following steps:

1. Initialization of the DFS component.
2. Initialization of a specific type of file system.

3. Creation of a block device on the memory.
4. Formatting the block device.
5. Mounting the block device to the DFS directory.

When the file system is no longer in use, you can uninstall it.

INITIALIZE THE DFS COMPONENT

The initialization of the DFS component is done by the dfs_init() function. The dfs_init() function initializes the relevant resources required by DFS and creates key data structures that allow DFS to find a specific file system in the system and have a way to manipulate files within a particular storage device.

REGISTERED FILE SYSTEM

After the DFS component is initialized, you also need to initialize the specific type of file system used, that is, register a specific type of file system into DFS. The registration process of the file system is as shown in Figure 15.3.

The interface to register the file system is as follows. Table 15.1 shows its parameters.

```
int dfs_register(const struct dfs_filesystem_ops *ops);
```

TABLE 15.1
dfs_register

Parameter	Description
ops	A collection of operation functions of the file system
Return	— —
0	File registered successfully
−1	File failed to register

REGISTER A STORAGE DEVICE AS A BLOCK DEVICE

Only block devices can be mounted to the file system, so you need to create the required block devices on the storage device. If the storage device is SPI Flash, you can use the "Serial Flash Universal Driver Library SFUD" component, which provides various SPI Flash drivers and abstracts the SPI Flash into a block device for mounting. The process of registering a block device is as shown in Figure 15.4.

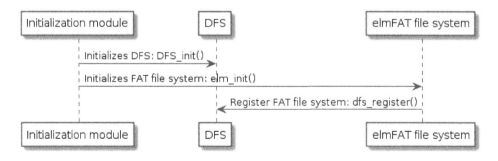

FIGURE 15.3 Register the file system.

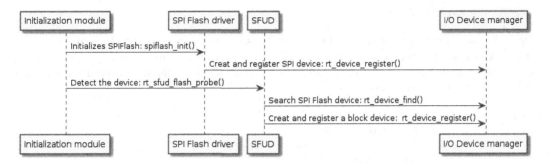

FIGURE 15.4 The timing diagram of registering a block device.

FORMAT THE FILE SYSTEM

After registering a block device, you also need to create a file system of the specified type on the block device, that is, format the file system. You can use the dfs _ mkfs() function to format the specified storage device and create a file system. The interface to format the file system is as follows. Table 15.2 shows its parameters.

```
int dfs_mkfs(const char * fs_name, const char * device_name);
```

TABLE 15.2
dfs_mkfs

Parameter	Description
fs_name	Type of file system
device_name	Name of the block device
Return	— —
0	File system formatted successfully
−1	Failure to format the file system

Take the elm-FAT file system format block device as an example. The formatting process is as shown in Figure 15.5.

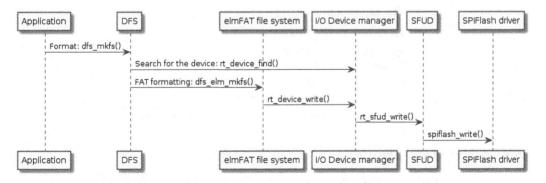

FIGURE 15.5 Formatted file system.

Mount the File System

In RT-Thread, mounting refers to attaching a storage device to an existing path. To access a file on a storage device, we must mount the partition where the file is located to an existing path and then access the storage device through this path. The interface to mount the file system is as follows. Table 15.3 shows its parameters.

```
int dfs_mount(const char    *device_name,
              const char    *path,
              const char    *filesystemtype,
              unsigned long rwflag,
              const void    *data);
```

TABLE 15.3
dfs_mount

Parameter	Description
device_name	The name of the block device that has been formatted
path	The mount path
filesystemtype	The type of the mounted file system. Possible values can refer to the dfs_mkfs() function description.
rwflag	Read and write flag bit
data	Private data for a specific file system
Return	— —
0	File system mounted successfully
−1	File system failed to be mounted

If there is only one storage device, it can be mounted directly to the root directory /.

Unmount a File System

When a file system does not need to be used anymore, it can be unmounted. The interface to unmount the file system is as follows. Table 15.4 shows its parameters.

```
int dfs_unmount(const char *specialfile);
```

TABLE 15.4
dfs_unmount

Parameter	Description
specialfile	Mount path
Return	— —
0	Unmounted the file system successfully
−1	Failed to unmount the file system

DOCUMENT MANAGEMENT

This section introduces the functions that are related to the operation of the file. The operation of the file is generally based on the file descriptor fd, as shown in Figure 15.6.

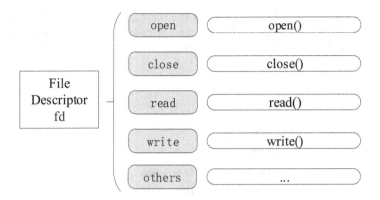

FIGURE 15.6 Common function of file management.

OPEN AND CLOSE FILES

To open or create a file, you can call the following open() function. Table 15.5 shows its parameters.

```
int open(const char *file, int flags, …);
```

TABLE 15.5
open

Parameter	Description
file	File names that are opened or created
flags	Specify the way to open the file, and values can refer to the following table
Return	— —
file descriptor	File opened successfully
−1	Failure to open the file

A file can be opened in a variety of ways, and multiple open methods can be specified at the same time. For example, if a file is opened by O_WRONLY and O_CREAT, then when the specified file that needs to be open does not exist, it will create the file first and then open it as write-only. The file opening method is as follows. Table 15.6 describes a file's open methods.

TABLE 15.6
Open Methods

Parameter	Description
O_RDONLY	Open file in read-only mode
O_WRONLY	Open file in write-only mode
O_RDWR	Open file in read-write mode
O_CREAT	If the file to be opened does not exist, you can create the file
O_APPEND	When the file is read or written, it will start from the end of the file; that is, the data written will be added to the end of the file in an additional way
O_TRUNC	Empty the contents of the file if it already exists

If you no longer need to use the file, you can use the close() function to close the file, and close() will write the data back to the disk and release the resources occupied by the file. Table 15.7 describes the parameters of the function.

```
int close(int fd);
```

TABLE 15.7

close

Parameter	Description
fd	File descriptor
Return	— —
0	File closed successfully
−1	Failure to close the file

READ AND WRITE DATA

To read the contents of a file, use the read() function. Table 15.8 describes the parameters of the function.

```
int read(int fd, void *buf, size_t len);
```

TABLE 15.8

read

Parameter	Description
fd	File descriptor
buf	Buffer pointer
len	Read number of bytes of the files
Return	— —
int	The number of bytes actually read
0	Read data has reached the end of the file or there is no readable data
−1	Read error, error code to view the current thread's errno

This function reads the len bytes of the file pointed to by the parameter fd into the memory pointed to by the buf pointer. In addition, the read/write position pointer of the file moves with the byte read.

To write data into a file, use the write() function. Table 15.9 describes the parameters of the function.

```
int write(int fd, const void *buf, size_t len);
```

This function writes len bytes in the memory pointed out by the buf pointer into the file pointed out by the parameter fd. In addition, the read and write location pointer of the file moves with the bytes written.

TABLE 15.9
`write`

Parameter	Description
fd	File descriptor
buf	Buffer pointer
len	The number of bytes written to the file
Return	— —
int	The number of bytes actually written
−1	Write error, error code to view the current thread's errno

RENAME

To rename a file, use the `rename()` function. Table 15.10 describes the parameters of the function.

```
int rename(const char *old, const char *new);
```

TABLE 15.10
`rename`

Parameter	Description
old	File's old name
new	New name
Return	— —
0	Changed the name successfully
−1	Failed to change the name

This function changes the file name specified by the parameter `old` to the file name pointed to by the parameter new. If the file specified by new already exists, the file will be overwritten.

GET STATUS

To get the file status, use the following `stat()` function. Table 15.11 describes the parameters of the function.

```
int stat(const char *file, struct stat *buf);
```

TABLE 15.11
`stat`

Parameter	Description
file	File name
buf	Structure pointer to a structure that stores file status information
Return	— —
0	Accessed status successfully
−1	Failed to access status

DELETE FILES

Delete a file in the specified directory using the unlink() function. Table 15.12 describes the parameters of the function.

```
int unlink(const char *pathname);
```

TABLE 15.12

unlink

Parameter	Description
pathname	Specify the absolute path to delete the file
Return	——
0	Deleted the file successfully
−1	Failed to deleted the file

SYNCHRONIZE FILE DATA TO STORAGE DEVICES

Synchronize all modified file data in memory to the storage device using the fsync() function. Table 15.13 describes the parameters of the function.

```
int fsync(int fildes);
```

TABLE 15.13

fsync

Parameter	Description
fildes	File descriptor
Return	——
0	Synchronized files successfully
−1	Failed to synchronize files

QUERY FILE SYSTEM–RELATED INFORMATION

Use the statfs() function to query file system–related information. Table 15.14 describes the parameters of the function.

```
int statfs(const char *path, struct statfs *buf);
```

TABLE 15.14

statfs

Parameter	Description
path	File system mount path
buf	Structure pointer for storing file system information
Return	——
0	Queried file system information successfully
−1	Failed to query file system information

MONITOR I/O DEVICE STATUS

To monitor the I/O device for events, use the `select()` function. Table 15.15 describes the parameters of the function.

```
int select( int nfds,
            fd_set *readfds,
            fd_set *writefds,
            fd_set *exceptfds,
            struct timeval *timeout);
```

TABLE 15.15
`select`

Parameter	Description
nfds	The range of all file descriptors in the collection, that is, the maximum value of all file descriptors plus 1
readfds	Collection of file descriptors that need to monitor read changes
writefds	Collection of file descriptors that need to monitor write changes
exceptfds	Collection of file descriptors that need to be monitored for exceptions
timeout	Timeout of **select**
Return	— —
positive value	A read/write event or error occurred in the monitored file collection
0	Waiting timeout, no readable or writable or erroneous files
negative value	Error

Use the `select()` interface to block and simultaneously detect whether a group of non-blocking I/O devices have events (such as readable, writable, high-priority error output, errors, etc.) until a device triggered an event or exceeded a specified wait time.

DIRECTORY MANAGEMENT

This section describes functions that directory management often uses, and operations on directories are generally based on directory addresses, as shown in Figure 15.7.

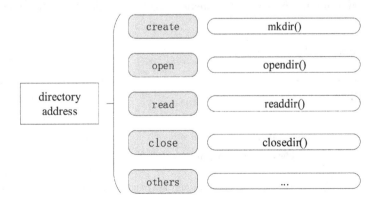

FIGURE 15.7   Functions that directory management often uses.

CREATE AND DELETE DIRECTORIES

To create a directory, you can use the mkdir() function. Table 15.16 describes the parameters of the function.

```
int mkdir(const char *path, mode_t mode);
```

TABLE 15.16
mkdir

Parameter	Description
path	The absolute address of the directory
Mode	Create a pattern
Return	— —
0	Created directory successfully
−1	Failed to create directory

This function is used to create a directory as a folder. The parameter path is the absolute path of the directory. The parameter mode is not enabled in the current version, so just fill in the default parameter 0x777.

Delete a directory using the rmdir() function. Table 15.17 describes the parameters of the function.

```
int rmdir(const char *pathname);
```

TABLE 15.17
rmdir

Parameter	Description
pathname	Absolute path to delete the directory
Return	— —
0	Deleted the directory successfully
−1	Failed to delete the directory

OPEN AND CLOSE THE DIRECTORY

Open the directory using the opendir() function. Table 15.18 describes the parameters of the function.

```
DIR* opendir(const char* name);
```

TABLE 15.18
opendir

Parameter	Description
name	Absolute address of the directory
Return	— —
DIR	Opened the directory successfully and returned a pointer to the directory stream
NULL	Failed to open

To close the directory, use the `closedir()` function. Table 15.19 describes the parameters of the function.

```
int closedir(DIR* d);
```

TABLE 15.19
`closedir`

Parameter	Description
d	Directory stream pointer
Return	— —
0	Directory closed successfully
−1	Directory closing error

This function is used to close a directory and must be used with the `opendir()` function.

READ A DIRECTORY

To read the directory, use the `readdir()` function. Table 15.20 describes the parameters of the function.

```
struct dirent* readdir(DIR *d);
```

TABLE 15.20
`readdir`

Parameter	Description
d	Directory stream pointer
Return	— —
dirent	Read successfully and returned a structure pointer to a directory entry
NULL	Read to the end of the directory

This function is used to read the directory, and the parameter d is the directory stream pointer. In addition, each time a directory is read, the pointer position of the directory stream is automatically recursed backward by one position.

GET THE READ POSITION OF THE DIRECTORY STREAM

To get the read location of the directory stream, use the `telldir()` function. Table 15.21 describes the parameters of the function.

```
long telldir(DIR *d);
```

The return value of this function records the current position of a directory stream. This return value represents the offset from the beginning of the directory file. You can use this value in the following `seekdir()` to reset the directory to the current position. In other words, the `telldir()` function can be used with the `seekdir()` function to reset the read position of the directory stream to the specified offset.

TABLE 15.21
`telldir`

Parameter	Description
d	Directory stream pointer
Return	— —
long	Read the offset of the position

SET THE LOCATION TO READ THE DIRECTORY NEXT TIME

Set the location to read the directory next time using the `seekdir()` function. Table 15.22 describes the parameters of the function.

```
void seekdir(DIR *d, off_t offset);
```

TABLE 15.22
`seekdir`

Parameter	Description
d	Directory stream pointer
offset	The offset value, displacement from this directory

This is used to set the read position of the parameter d directory stream and starts reading from this new position when readdir() is called.

RESET THE POSITION OF THE READING DIRECTORY TO THE BEGINNING

To reset the directory stream's read position to the beginning, use the `rewinddir()` function. Table 15.23 describes the parameters of the function.

```
void rewinddir(DIR *d);
```

TABLE 15.23
`rewinddir`

Parameter	Description
d	Directory stream pointer

This function can be used to set the current read position of the d directory stream to the initial position of the directory stream.

DFS CONFIGURATION OPTIONS

The specific configuration path of the file system in menuconfig is as follows:

```
RT-Thread Components  --->
    Device virtual file system  --->
```

The configuration menu description and corresponding macro definitions are shown in Table 15.24.

TABLE 15.24

The Configuration Menu of DFS

Configuration Options	Corresponding Macro Definition	Description
[*] Using device virtual file system	RT_USING_DFS	Open DFS virtual file system
[*] Using working directory	DFS_USING_WORKDIR	Open a relative path
(2) The maximal number of mounted file system	DFS_FILESYSTEMS_MAX	Maximum number of mounted file systems
(2) The maximal number of file system type	DFS_FILESYSTEM_TYPES_MAX	Maximum number of supported file systems
(4) The maximal number of opened files	DFS_FD_MAX	Maximum number of open files
[ ] Using mount table for file system	RT_USING_DFS_MNTTABLE	Open the automatic mount table
[*] Enable elm-chan fatfs	RT_USING_DFS_ELMFAT	Open the elm-FatFs file system
[*] Using devfs for device objects	RT_USING_DFS_DEVFS	Open the DevFS device file system
[ ] Enable ReadOnly file system on flash	RT_USING_DFS_ROMFS	Open the RomFS file system
[ ] Enable RAM file system	RT_USING_DFS_RAMFS	Open the RamFS file system
[ ] Enable UFFS file system: Ultra-low-cost Flash File System	RT_USING_DFS_UFFS	Open the UFFS file system
[ ] Enable JFFS2 file system	RT_USING_DFS_JFFS2	Open the JFFS2 file system
[ ] Using NFS v3 client file system	RT_USING_DFS_NFS	Open the NFS file system

---

By default, the RT-Thread operating system does not turn on the relative path feature in order to obtain a small memory footprint. When the "Support Relative Paths" option is not turned on, you have to use absolute path names when working with files and directory interfaces because there is no current working directory in the system. If you need the current working directory and the relative directory, you can enable the "Support Relative Paths" feature in the file system configuration.

DFS APPLICATION EXAMPLE

This section is based on the RT-Thread IoT board to run the file system on W25Q128, which is a 128Mb SPI Flash. This section also gives sample code for the file system and shows how the sample works to help the reader learn to use the functions provided by the file system.

Preparation

Hardware Preparation

The hardware interface used for this example and memory connection is QSPI1, as shown in Figure 15.8.

Software Preparation

The sample code for this section is located in the chapter15 directory of the code package's accompanying materials, as shown in Figure 15.9, by opening the MDK5 project file project. uvprojx.

SPI_FLASH

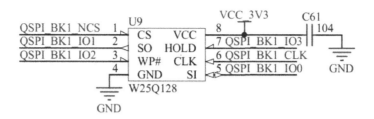

FIGURE 15.8   Schematic diagram of w25q128.

Compared to the basic MDK project, the new groupings in this project are shown in Table 15.25.

TABLE 15.25
The New Groupings in This Project

Directory Groups	Description
Filesystem	DFS component
filesystem-samples	The file system application sample code

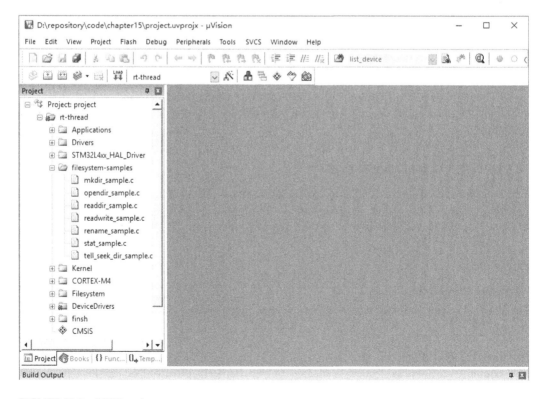

FIGURE 15.9   MDK project.

The code that mounts the block device W25Q128 to the file system is in the main() function, as follows:

```
/* Mount w25q128 device to / folder as 0 partition of the elm-FAT file
system */
if (dfs_mount("w25q128", "/", "elm", 0, 0) == 0)
{
rt_kprintf("File System initialized!\n");
}
else
{
    rt_kprintf("Failed to initialize filesystem!\n");
}
```

Compile and Run

The project is compiled and the firmware is downloaded to the board to run, and the log information shown here can be seen on the serial tool.

```
 \ | /
- RT -     Thread Operating System
 / | \       3.1.0 build Oct  7 2018
 2006 - 2018 Copyright by rt-thread team

[SFUD] Find a Winbond flash chip. Size is 16777216 bytes.
[SFUD] w25q128 flash device is initialize success.
Failed to initialize filesystem!
msh />
```

The log information indicates that w25q128 has been successfully initialized, but file system initialization failed because when an SPI Flash was first used as a storage device for the file system, it needed to be formatted to create the corresponding type of file system.

Formatting the File System

This can be done by running the following file system formatting command: mkfs, mkfs function to create a specified type of file system on the specified storage device. Use the format of mkfs (-t type) device. This example will create an elm-FAT-type file system on the w25q128 device, and the following command can be used in the FinSH console:

```
msh/>mkfs -t elm w25q128
```

Then restart the board and you can see the following log information from the output on the serial tool:

```
 \ | /
- RT -     Thread Operating System
 / | \       3.1.0 build Oct  7 2018
 2006 - 2018 Copyright by rt-thread team

[SFUD] Find a Winbond flash chip. Size is 16777216 bytes.
[SFUD] w25q128 flash device is initialize success.
Filesystem initialized!
msh />
```

The log information for "File System initialized!" indicates that the file system mounting was successful. At this point, using the "list_device" command to view the device information, you can see that the reference count of the w25q128 device changes from 0 to 1. The "list_device" command execution result is as follows:

```
msh />list_device
device  type                   ref count
------  ------------------- ----------
w25q128 Block Device            1   # The value of ref changes from 0 to 1,
indicating that the mount was successful
qspi10  SPI Device              0
qspi1   SPI Bus                 0
uart1   Character Device        2
```

FinSH Command

After the file system is successfully mounted, the files and directories can be operated. The commonly used FinSH commands for file system operations are shown in Table 15.26.

TABLE 15.26
FinSH Commands

FinSH Command	Description
ls	Display information about files and directories
cd	Enter the specified directory
cp	Copy the file
rm	Delete the file or the directory
mv	Move the file or rename it
echo	Write the specified content to the specified file, write the file when it exists, and create a new file and write when the file does not exist
cat	Display the contents of the file
pwd	Print out the current directory address
mkdir	Create a folder
mkfs	Format the file system

Use the "ls" command to view the current directory information, and the results are as follows:

```
msh />ls                          # use `ls` command to view the current
directory information
Directory /:                      # you can see that the root directory
already exists /
```

Use the "mkdir" command to create a folder, and the results are as follows:

```
msh />mkdir rt-thread             # create an rt-thread folder
msh />ls                          # view directory information as follows
Directory /:
rt-thread              <DIR>
```

Use the "echo" command to output the input string to the specified output location. The result is as follows:

```
msh />echo "hello rt-thread!!!"               # outputs the string to
standard output
```

```
hello rt-thread!!!
msh />echo "hello rt-thread!!!" hello.txt      # output the string output
to the hello.txt file
msh />ls
Directory /:
rt-thread              <DIR>
hello.txt              18
msh />
```

Use the "cat" command to view the contents of the file. The result is as follows:

```
msh />cat hello.txt                    # view the contents of the hello.txt
file and output
hello rt-thread!!!
```

Use the "rm" command to delete a folder or file. The result is as follows:

```
msh />ls                               # view the information of current
directory
Directory /:
rt-thread              <DIR>
hello.txt              18
msh />rm rt-thread                     # delete the rt-thread folder
msh />ls
Directory /:
hello.txt              18
msh />rm hello.txt                     # delete the hello.txt file
msh />ls
Directory /:
msh />
```

READ AND WRITE FILE EXAMPLES

Once the file system is working, you can run the application example. In the sample code, you first create a file text.txt using the open() function and write the string "RT-Thread Programmer!\n" in the file using the write() function, and then close the file. Use the open() function again to open the text.txt file, read the contents and print it out, and close the file in the end.

The sample code is shown in Code Listing 15.1.

Code listing 15.1 Read and Write Files

```
#include <rtthread.h>
#include <dfs_posix.h> /* this header file need to be included when you
need to operate the file */

static void readwrite_sample(void)
{
    int fd, size;
    char s[] = "RT-Thread Programmer!", buffer[80];

    rt_kprintf("Write string %s to test.txt.\n", s);

    /* open the '/text.txt' file in create and read-write mode and create
the file if it does not exist*/
```

```
    fd = open("/text.txt", O_WRONLY | O_CREAT);
    if (fd>= 0)
    {
        write(fd, s, sizeof(s));
        close(fd);
        rt_kprintf("Write done.\n");
    }

      /* open the '/text.txt' file in read-only mode */
    fd = open("/text.txt", O_RDONLY);
    if (fd>= 0)
    {
        size = read(fd, buffer, sizeof(buffer));
        close(fd);
        rt_kprintf("Read from file test.txt : %s \n", buffer);
        if (size < 0)
            return ;
    }
  }
/* export to the msh command list */
MSH_CMD_EXPORT(readwrite_sample, readwrite sample);
```

Run the example in the FinSH console, and the results are as follows:

```
msh />readwrite_sample
Write string RT-Thread Programmer! to test.txt.
Write done.
Read from file test.txt : RT-Thread Programmer!
```

During the sample run, you can see that the string "RT-Thread Programmer!" is written to the file text.txt. After that, the contents of the file are read out and printed on the terminal.

AN EXAMPLE OF CHANGING THE FILE NAME

The sample code in this section shows how to modify the file name. The program creates a function rename _ sample() that manipulates the file and exports it to the msh command list. This function calls the rename() function to rename the file named text.txt to text1.txt. The sample code is shown in Code Listing 15.2.

Code listing 15.2 Change the File Name

```
#include <rtthread.h>
#include <dfs_posix.h> /* this header file need to be included when you
need to operate the file */

static void rename_sample(void)
{
    rt_kprintf("%s => %s", "/text.txt", "/text1.txt");

    if (rename("/text.txt", "/text1.txt") < 0)
        rt_kprintf("[error!]\n");
    else
        rt_kprintf("[ok!]\n");
}
```

```
/* export to the msh command list */
MSH_CMD_EXPORT(rename_sample, rename sample);
```

Run the example in FinSH console and the results are as follows:

```
msh />echo "hello" text.txt
msh />ls
Directory /:
text.txt                  5
msh />rename_sample
/text.txt => /text1.txt [ok!]
msh />ls
Directory /:
text1.txt                 5
```

In the example demonstration, we first create a file named text.txt using the echo command, and then run the sample code to change the file name of the file text.txt to text1.txt.

GET FILE STATUS EXAMPLE

The sample code shows how to get the file status. The program creates a function stat _ sample() that manipulates the file and exports it to the msh command list. This function calls the stat() function to get the file size information of the text.txt file. The sample code is shown in Code Listing 15.3.

Code listing 15.3 Get File Status

```
#include <rtthread.h>
#include <dfs_posix.h> /* this header file need to be included when you
need to operate the file */

static void stat_sample(void)
{
    int ret;
     struct stat buf;
     ret = stat("/text.txt", &buf);
     if(ret == 0)
     rt_kprintf("text.txt file size = %d\n", buf.st_size);
     else
     rt_kprintf("text.txt file not fonud\n");
}
/* export to the msh command list */
MSH_CMD_EXPORT(stat_sample, show text.txt stat sample);
```

Run the example in FinSH console and the results are as follows:

```
msh />echo "hello" text.txt
msh />stat_sample
text.txt file size = 5
```

During the example run, the file text.txt is first created with the "echo" command, then the sample code is run, and the file size information for the file text.txt is printed.

CREATE A DIRECTORY EXAMPLE

The sample code in this section shows how to create a directory. The program creates a function mkdir_
sample() that manipulates the file and exports it to the msh command list, which calls the mkdir()
function to create a folder called dir_test. The sample code is shown in Code Listing 15.4.

Code listing 15.4 Create a Directory

```
#include <rtthread.h>
#include <dfs_posix.h> /* this header file need to be included when you
need to operate the file */

static void mkdir_sample(void)
{
    int ret;

    /* create a directory */
    ret = mkdir("/dir_test", 0x777);
    if (ret < 0)
    {
        /* fail to create a directory */
        rt_kprintf("dir error!\n");
    }
    else
    {
        /* create a directory successfully */
        rt_kprintf("mkdir ok!\n");
    }
}
/* export to the msh command list */
MSH_CMD_EXPORT(mkdir_sample, mkdir sample);
```

Run the example in FinSH console and the result is as follows:

```
msh />mkdir_sample
mkdir ok!
msh />ls
Directory /:
dir_test                    <DIR>    # <DIR> it indicates that the type of
the directory is a folder
```

This example demonstrates creating a folder named dir_test in the root directory.

READ DIRECTORY EXAMPLE

The sample code shows how to read the directory. The program creates a function readdir_
sample() that manipulates the file and exports it to the msh command list. This function calls the
readdir() function to get the contents of the dir_test folder and print it out. The sample
code is shown in Code Listing 15.5.

Code listing 15.5 Read Directory

```
#include <rtthread.h>
#include <dfs_posix.h> /* this header file need to be included when you
need to operate the file */
```

```
static void readdir_sample(void)
{
    DIR *dirp;
    struct dirent *d;

    /* open the / dir_test directory */
    dirp = opendir("/dir_test");
    if (dirp == RT_NULL)
    {
        rt_kprintf("open directory error!\n");
    }
    else
    {
        /* read the directory */
        while ((d = readdir(dirp)) != RT_NULL)
        {
            rt_kprintf("found %s\n", d->d_name);
        }

        /* close the directory */
        closedir(dirp);
    }
}
/* exports to the msh command list */
MSH_CMD_EXPORT(readdir_sample, readdir sample);
```

Run the example in FinSH console and the result is as follows:

```
msh />ls
Directory /:
dir_test                    <DIR>
msh />cd dir_test
msh /dir_test>echo "hello" hello.txt        # create a hello.txt file
msh /dir_test>cd ..                         # switch to the parent folder
msh />readdir_sample
found hello.txt
```

In this example, first create a hello.txt file under the dir_test folder and exit the dir_test folder. At this point, run the sample program to print out the contents of the dir_test folder.

AN EXAMPLE OF SETTING THE LOCATION OF THE READ DIRECTORY

The sample code in this section shows how to set the location to read the directory next time. The program creates a function `telldir _ sample()` that manipulates the file and exports it to the msh command list. This function first opens the root directory, then reads all the directory information in the root directory and prints the directory information. Meanwhile, use the `telldir()` function to record the location information of the third directory entry. Before reading the directory information in the root directory for the second time, use the `seekdir()` function to set the read location to the address of the third directory entry previously recorded. At this point, read the information in the root directory again, and the directory information is printed out. The sample code is shown in Code Listing 15.6.

Code listing 15.6 An Example of Setting the Location of the Read Directory

```
#include <rtthread.h>
```

```
#include <dfs_posix.h> /* this header file need to be included when you
need to operate the file */

/* assume that the file operation is done in one thread */
static void telldir_sample(void)
{
    DIR *dirp;
    int save3 = 0;
    int cur;
    int i = 0;
    struct dirent *dp;

    /* open the root directory */
    rt_kprintf("the directory is:\n");
    dirp = opendir("/");

    for (dp = readdir(dirp); dp != RT_NULL; dp = readdir(dirp))
    {
        /* save the directory pointer for the third directory entry */
        i++;
        if (i == 3)
            save3 = telldir(dirp);

        rt_kprintf("%s\n", dp->d_name);
    }

    /* go back to the directory pointer of the third directory entry you
just saved */
    seekdir(dirp, save3);

    /* Check if the current directory pointer is equal to the pointer to
the third directory entry that was saved. */
    cur = telldir(dirp);
    if (cur != save3)
    {
        rt_kprintf("seekdir (d, %ld); telldir (d) == %ld\n", save3, cur);
    }

    /* start printing from the third directory entry */
    rt_kprintf("the result of tell_seek_dir is:\n");
    for (dp = readdir(dirp); dp != NULL; dp = readdir(dirp))
    {
        rt_kprintf("%s\n", dp->d_name);
    }

    /* close the directory */
    closedir(dirp);
}
/* exports to the msh command list */
MSH_CMD_EXPORT(telldir_sample, telldir sample);
```

In this demo, you need to manually create the five folders from hello _ 1 to hello _ 5 in the root directory with the "mkdir" command, making sure that there is a folder directory under the root directory for running the sample.

Run the example in FinSH console and the results are as follows:

```
msh />ls
Directory /:
hello_1                <DIR>
hello_2                <DIR>
hello_3                <DIR>
hello_4                <DIR>
hello_5                <DIR>
msh />telldir_sample
the directory is:
hello_1
hello_2
hello_3
hello_4
hello_5
the result of tell_seek_dir is:
hello_3
hello_4
hello_5
```

After running the sample, you can see that the first time you read the root directory information, it starts from the first folder and prints out all the directory information in the root directory. When the directory information is printed for the second time, since the starting position of the reading is set to the position of the third folder by using the seekdir() function, the second time when reading the root directory starts from the third folder. Start reading until the last folder—only the directory information from hello_3 to hello_5 is printed.

CHAPTER SUMMARY

This chapter describes the concepts, features, workflow, configuration options, and usage of the DFS (virtual file system). It could be summarized as follows:

1. RT-Thread uses DFS to provide the functionality of the file system to developers.
2. Using a file system simplifies the storage and management of large amounts of complex data in the system and manages the data in the form of folders and files.
3. The DFS virtual file system provides the POSIX interface to the upper layer. Developers can call functions directly to complete the operation without having to worry about what type of file system the files are stored in or on which storage device they are stored.

16 Network Framework

With the popularity of the Internet, people's lives are increasingly dependent on its application. More and more products need to connect to the Internet, and device networking has become a trend. To achieve the connection between the device and the network, devices need to follow the TCP/IP protocol. Devices can either run the network protocol stack or use additional hardware IC (chips with hardware network protocol stack interfaces) to connect to the Internet.

When the device is connected to the network, it is like plugging in the wings. You can use the network to upload data in real time. The developers can see the current running status and collected data of the device in a hundred thousand miles and remotely control the device to complete specific tasks. You can also play online music, make online calls, and act as a LAN storage server through your device.

This chapter will explain the related content of the RT-Thread network framework and introduce you to the concept, function, and usage of the network framework. After reading this chapter, you will be familiar with the concept and implementation principle of the RT-Thread network framework and be familiar with network programming using Socket interfaces.

TCP/IP INTRODUCTION TO NETWORK PROTOCOLS

TCP/IP is an abbreviation for Transmission Control Protocol/Internet Protocol. It is not a single protocol, but a general term for the whole protocol family. It includes the IP protocol, ICMP protocol, TCP protocol, HTTP, FTP, POP3, HTTPS protocol, etc., which define how devices connect to the Internet and the standards by which data is transferred between them.

OSI REFERENCE MODEL

OSI (Open System Interconnect), as shown in Figure 16.1, is an open system interconnection. Generally referred to as the OSI reference model, it is a network interconnection model devised by the ISO (International Organization for Standardization) in 1985. The architecture standard defines a seven-layer framework for network interconnection (physical layer, data link layer, network layer, transport layer, session layer, presentation layer, and application layer), that is, the ISO open system interconnection reference model. The first to third layers belong to the lower three layers of the OSI reference model and are responsible for creating links for network communication connections; the fourth to seventh layers are the upper four layers of the OSI reference model and are responsible for end-to-end data communication. The capabilities of each layer are further detailed in this framework to achieve interconnectivity, interoperability, and application portability in an open system environment.

TCP/IP REFERENCE MODEL

The TCP/IP communication protocol uses a four-layer hierarchical structure. Each layer calls the network provided by its next layer to fulfill its own needs. The four layers are:

- **Application layer**: Different types of network applications have different communication rules, so the application layer protocols are various, such as Simple Mail Transfer Protocol (SMTP), File Transfer Protocol (FTP), and Telnet.
- **Transport layer**: This layer provides data transfer services between nodes, such as Transmission Control Protocol (TCP), User Datagram Protocol (UDP), etc. TCP and UDP add data to the data packet and transmit it to the next layer. This layer is responsible for transmitting data and determining that the data has been delivered and received.

- **Network layer**: Responsible for providing basic data packet transfer functions, so that each packet can reach the destination host (but not check whether it is received correctly), such as Internet Protocol (IP).
- **Network interface layer**: Management of actual network media, defining how to use actual networks (such as Ethernet, Serial Line, etc.) to transmit data.

DIFFERENCE BETWEEN TCP/IP REFERENCE MODEL AND OSI REFERENCE MODEL

Figure 16.1 shows the TCP/IP reference model and the OSI reference model in a diagram.

Both the OSI reference model and the TCP/IP reference model are hierarchical and based on the concept of a separate protocol stack. The OSI reference model has seven layers, while the TCP/IP reference model has only four layers; that is, the TCP/IP reference model has no presentation layer and session layer, and the data link layer and physical layer are merged into a network interface layer. However, there is a certain correspondence between the two layers. Due to the complexity of the OSI system and the design prior to implementation, many designs are too idealistic and not very convenient for software implementation. Therefore, there are not many systems that fully implement the OSI reference model, and the scope of application is limited. The TCP/IP reference model was first implemented in a computer system. It has a stable implementation on UNIX and Windows platforms and provides a simple and convenient programming interface on which a wide range of applications are developed. Nowadays, the TCP/IP reference model has become the international standard and industry standard for Internet connectivity.

IP ADDRESS

The IP address refers to the Internet Protocol address and is a uniform address format that assigns a logical address to each network and each host on the Internet to mask physical address differences provided by the Internet Protocol. The common Local Area Network (LAN) IP address is 192.168.X.X.

SUBNET MASK

The subnet mask (also called netmask and address mask) is used to indicate which bits of an IP address identify the subnet where the host is located and which bits are identified as the bit mask of the host. The subnet mask cannot exist alone; it must be used in conjunction with an IP address. The subnet mask has only one effect, which is to divide an IP address into two parts: network address and host address. The subnet mask is the bit of 1, the IP address is the network address, the subnet

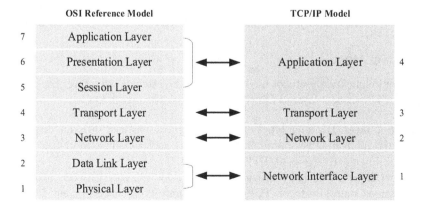

FIGURE 16.1 TCP/IP reference model and OSI reference model.

mask is the bit of 0, and the IP address is the host address. Taking the IP address 192.168.1.10 and the subnet mask 255.255.255.0 as an example, the first 24 bits of the subnet mask (converting decimal to binary) is 1, so the first 24 bits of the IP address 192.168.1 represent the network address. The remaining 0 is the host address.

MAC ADDRESS

The MAC (Media Access Control or Medium Access Control) address is translated as media access control, or the physical address—the hardware address—used to define the location of network devices. In the OSI model, the third layer, the network layer, is responsible for the IP address; the second layer, the data link layer, is responsible for the MAC address. A host will have at least one MAC address.

INTRODUCTION TO THE NETWORK FRAMEWORK OF RT-THREAD

To support various network protocol stacks, RT-Thread has developed a **Socket Abstraction Layer (SAL)** component. RT-Thread can seamlessly access various protocol stacks, including several commonly used TCP/IP protocol stacks, such as the Lightweight IP (LwIP) protocol stack, which is commonly used in embedded development, and the AT Socket protocol stack component developed by RT-Thread. These stacks can convert the data from the network layer to the transport layer.

The main features of the RT-Thread network framework are as follows:

- Support the standard network sockets BSD Socket interfaces using poll/select.
- Abstract and unify multiple network protocol stack interfaces.
- Support various physical network cards and network communication module hardware.
- The resource occupancy of the SAL component is small: ROM 2.8K and RAM 0.6K.

RT-Thread's network framework adopts a layered design with four layers, and each layer has different responsibilities. Figure 16.2 shows the RT-Thread network framework structure.

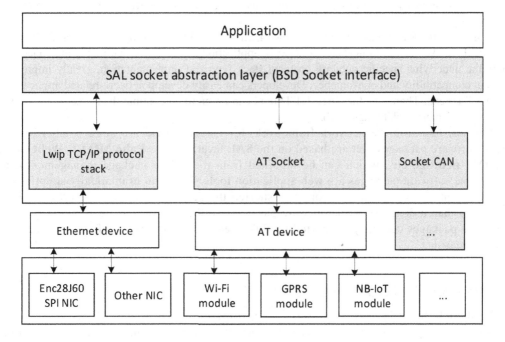

FIGURE 16.2 RT-Thread network framework structure.

The network framework provides a standard BSD Socket interface to the applications. Developers use the BSD Socket interface to operate without worrying about how the underlying network is implemented, and there is no need to care about which network protocol stack the network data passes through. The socket abstraction layer provides the upper application layer. The interfaces are `accept`, `connect`, `send`, `recv`, etc.

Below the SAL layer is the protocol stack layer. The main protocol stacks supported in the current network framework are as follows:

- **LwIP** is an open source TCP/IP protocol stack implementation that reduces RAM usage while maintaining the main functionality of the TCP/IP protocol, making the LwIP protocol stack ideal for use in embedded systems.
- **AT Socket** is a component for modules that support AT instructions. The AT command uses a standard serial port for data transmission and reception and converts complex device communication methods into simple serial port programming, which greatly simplifies the hardware design and software development costs of the product, which makes it convenient for almost all network modules such as GPRS, 3G/4G, NB-IoT, Bluetooth, WiFi, GPS, and other modules to access the RT-Thread network framework and develop network applications through the standard BSD Socket method, greatly simplifying the development of upper-layer applications.
- **Socket CAN** is a way of programming CAN. It is easy to use and easy to program. By accessing the SAL layer, developers can implement Socket CAN programming on RT-Thread.

Below the protocol stack layer is an abstract device layer that is connected to various network protocol stacks by abstracting hardware devices into Ethernet devices or AT devices.

The bottom layer is a variety of network chips or modules (e.g., Ethernet chips with a built-in protocol stack such as W5500/CH395, WiFi module with AT command, GPRS module, NB-IoT module, etc.). These hardware modules are the carrier that truly performs the network communication function and is responsible for communicating with various physical networks.

In general, the RT-Thread network framework allows developers to only focus on using the standard BSD Socket network interface for network application development, without being concerned with the underlying specific network protocol stack type and implementation, greatly improving system compatibility and convenience. Developers can easily develop network-related applications with RT-Thread. Thanks to the SAL, RT-Thread can be more compatible than others in different areas of the Internet of Things (IoT).

Besides, based on the network framework, RT-Thread provides a large number of network software packages that are based on the SAL layer, such as **Paho MQTT**, **WebClient**, **cJSON**, **netutils**, etc., which can be obtained from the online package management center. These software packages are web application tools. Then can dramatically simplify the development of network applications and shorten the development cycle. At present, there are more than a dozen network software packages. Table 16.1 lists some common network software packages currently supported by RT-Thread. The number of software packages is still increasing.

NETWORK FRAMEWORK WORKFLOW

Using the RT-Thread network framework, you first need to initialize the SAL and then register various network protocol clusters to ensure that the application can communicate using the socket network socket interface. This section mainly uses LwIP as an example.

TABLE 16.1

Network Packages

Package Name	Description
Paho MQTT	Based on Eclipse open source Paho MQTT, it has a lot of functions and provides performance optimization, such as increased automatic reconnection after disconnection, pipe model, support for non-blocking interface, support for TLS encrypted transmission, etc.
WebClient	Easy-to-use HTTP client with support for HTTP GET/POST and other common request functions, support for HTTPS, breakpoint retransmission, etc.
mongoose	Embedded web server network library, similar to Nginx in the embedded world. Licensing is not friendly enough, and the business needs to be charged,
WebTerminal	Access FinSH/MSH Shell packages in the browser or on the mobile device.
cJSON	Ultra-lightweight JSON parsing library.
ljson	Json to struct parsing, output library.
ezXML	XML file parsing library; currently does not support parsing XML data.
nanopb	Protocol buffers format the data parsing library. The protocol buffers format takes up fewer resources than JSON and XML format resources.
GAgent	Software package for accessing the Gizwits Cloud Platform.
Marvell WiFi	Marvell WiFi driver.
Wiced WiFi	WiFi driver for Wiced interface.
CoAP	Porting libcoap's CoAP communication package.
nopoll	Ported open source WebSocket communication package.
netutils	A collection of useful network debugging gadgets, including ping, TFTP, iperf, NetIO, NTP, Telnet, etc.
OneNet	Software for accessing the China Mobile OneNet Cloud.

REGISTER THE NETWORK PROTOCOL CLUSTER

First, use the `sal_init()` interface to initialize resources such as mutex locks used in the component. The interface looks like this:

```
int sal_init(void);
```

After the SAL is initialized, use the `sal_proto_family_register()` interface to register the network protocol cluster; for example, the LwIP network protocol cluster is registered to the SAL. The sample code is as follows:

```
static const struct proto_family LwIP_inet_family_ops = {
    "LwIP",
    AF_INET,
    AF_INET,
    inet_create,
    LwIP_gethostbyname,
    LwIP_gethostbyname_r,
    LwIP_freeaddrinfo,
    LwIP_getaddrinfo,
};

int LwIP_inet_init(void)
{
    sal_proto_family_register(&LwIP_inet_family_ops);

    return 0;
}
```

AF_INET stands for IPv4 address; for example, 127.0.0.1; AF stands for "Address Family" and INET stands for "Internet."

The sal_proto_family_register() interface is defined as follows. Table 16.2 describes the parameters of the function.

```
int sal_proto_family_register(const struct proto_family *pf);
```

TABLE 16.2
sal_proto_family_register

Parameters	Description
pf	Protocol cluster structure pointer
Return	— —
0	Registration successful
−1	Registration failed

NETWORK DATA RECEIVING PROCESS

After the LwIP is registered to the SAL, the application can send and receive network data through the network socket interface. In LwIP, several main threads are created, and they are tcpip thread, erx receiving thread, and etx sending thread. The network data receiving process is shown in Figure 16.3. The application receives data by calling the standard socket interface recv() in blocking mode. When the Ethernet hardware device receives the network data packet, it stores the packet in a receiving buffer and then sends mail to notify the erx thread that the data has arrived through the Ethernet interrupt program. The erx thread applies for the pbuf memory block according to the received data length and puts the data into the pbuf's payload data and then sends the pbuf memory block to the tcpip thread via mailbox, and the tcpip thread returns the data to the application that is blocking the receiving data.

NETWORK DATA SENDING PROCESS

The network data sending process is shown in Figure 16.4. When there is data to send, the application calls the standard network socket interface send() to hand the data to the tcpip thread. The tcpip thread sends a message to wake up the etx thread. The etx thread first determines if the Ethernet is sending data. If data is not being sent, it will put the data to be sent into the send buffer and then send the data through the Ethernet device. If data is being sent, the etx thread suspends itself until the Ethernet device is idle before sending the data out.

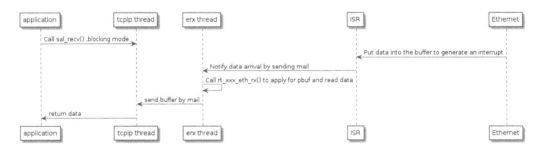

FIGURE 16.3 Call flowchart of the data-receiving function.

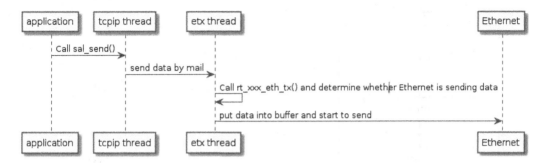

FIGURE 16.4 Call flowchart of data-sending function.

NETWORK SOCKET PROGRAMMING

The application uses Socket (network socket) interface programming to implement network communication functions. Socket is a set of application program interfaces that shields the communication details of each protocol so that the application does not need to know the protocol itself to use the interfaces provided by Socket to communicate between interconnected different hosts.

TCP Socket Communication Process

TCP (Transmission Control Protocol) is a connection-oriented protocol used to ensure reliable data transmission. Through the TCP protocol transmission, a sequential error-free data stream is obtained. The TCP-based socket programming flow diagram is shown in Figure 16.5. A connection must be established between the sender and the receiver's two sockets in order to communicate on the TCP protocol. When a socket (usually a server socket) waits for a connection to be established another socket can request a connection. Once the two sockets are connected, they can perform two-way data transmission, and both sides can send or receive data. A TCP connection is a reliable connection that guarantees that packets arrive in order, and if packet loss occurs, the packet is automatically resent.

For example, TCP is like you are calling your friend in real life. When you call your friend, you must wait for your friend to answer. Only when your friend answers your call can he or she establish a connection with you. The two parties can talk and pass information to each other. Of course, the information passed at this time is reliable, because if your friend can't hear what you said, he or she will ask you to repeat the content again. When either party wants to end the call, they will bid farewell to the other party and wait until the other party bids farewell to them before they hang up and end the call.

UDP Socket Communication Process

User Datagram Protocol (UDP) is a connectionless protocol. Each datagram is a piece of separate information, including the complete source address and destination address. It is transmitted to the destination on the network in any possible path. Therefore, whether the destination can be reached, the time to reach the destination, and the correctness of the content cannot be guaranteed. The UDP-based socket programming flow is shown in Figure 16.6.

For example, UDP is equivalent to walkie-talkie communication in real life. After you set up the channel, you can directly say the information you want to express. The data is sent out by the walkie-talkie, but you don't know whether your message has been received by others or not, unless someone else responds to you with a walkie-talkie. So this method is not reliable.

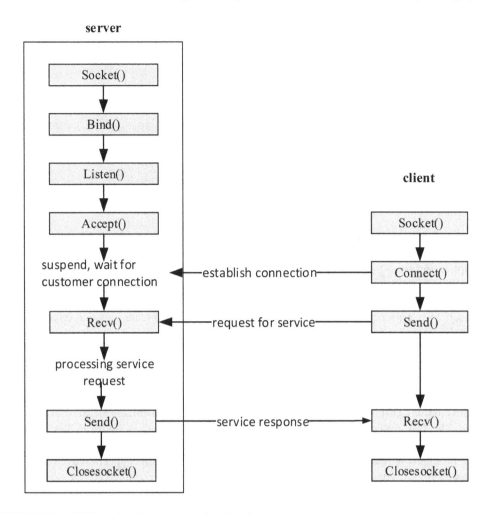

FIGURE 16.5   TCP-based socket programming flowchart.

CREATE A SOCKET

Before communicating, the communicating parties first use the socket() interface to create a socket and assign a socket descriptor and its resources based on the specified address family, data type, and protocol. The interface is as follows. Table 16.3 describes the parameters of the function.

```
int socket(int domain, int type, int protocol);
```

TABLE 16.3
socket

Parameters	Description
domain	Protocol family
type	Specify the communication type, including the values SOCK_STREAM and SOCK_DGRAM
protocol	Protocol allows specifying a protocol for the socket, which is set to 0 by default
Return	— —
>=0	Successful, returns an integer representing the socket descriptor
−1	Failure

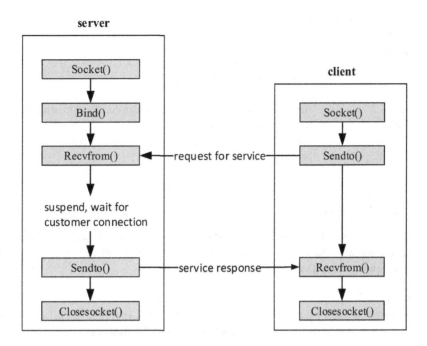

FIGURE 16.6   UDP-based socket programming flowchart.

Communication types include SOCK_STREAM and SOCK_DGRAM.

SOCK_STREAM indicates connection-oriented TCP data transfer. Data can arrive at another device without any errors. If it is damaged or lost, it can be resent, but it is relatively slow.

SOCK_DGRAM indicates the connectionless UDP data transfer method. The device only transmits data and does not perform data verification. If the data is damaged during transmission or does not reach another device, there is no way to remedy it. In other words, if the data is wrong, it cannot be retransmitted. Because SOCK_DGRAM does less validation work, it is more efficient than SOCK_STREAM.

The sample code for creating a TCP-type socket is as follows:

```
/* Create a socket, type is SOCKET_STREAM, TCP type */
   if ((sock = socket(AF_INET, SOCK_STREAM, 0)) == -1)
   {
       /* failed to create socket*/
       rt_kprintf("Socket error\n");

       return;
   }
```

Binding Socket

A binding socket is used to bind a port number and an IP address to a specified socket. When using socket() to create a socket, only the protocol family is given and no address is assigned. Before the socket receives a connection from another host, it must bind it with an address and port number using bind(). The interface is as follows. Table 16.4 describes the parameters of the function.

```
int bind(int s, const struct sockaddr *name, socklen_t namelen);
```

TABLE 16.4
bind

Parameters	Description
s	Socket descriptor
name	Pointer to the sockaddr structure representing the address to be bound
namelen	Length of sockaddr structure
Return	— —
0	Successful
−1	Failure

Establishing a TCP Connection

For server-side applications, after using bind() to bind the socket, you also need to use the listen() function to let the socket enter the passive listening state, and then call the accept() function to respond to the possible request from a client at any time.

Listening Socket

The listening socket is used by the TCP server to listen for the specified socket connection. The interface is as follows. Table 16.5 describes the parameters of the function.

```
int listen(int s, int backlog);
```

TABLE 16.5
listen

Parameters	Description
s	Socket descriptor
backlog	Indicates the maximum number of connections that can wait at a time
Return	— —
0	Successful
−1	Failure

Accept the Connection

When the application listens for connections from other clients, the connection must be initialized with the accept() function, which creates a new socket for each connection and removes the connection from the listen queue. The interface is as follows. Table 16.6 describes the parameters of the function.

```
int accept(int s, struct sockaddr *addr, socklen_t *addrlen);
```

TABLE 16.6
accept

Parameters	Description
s	Socket descriptor
addr	Client device address information
addrlen	Client device address structure length
Return	— —
>=0	Successful, return the newly created socket descriptor
−1	Failure

Establish Connection

Used by the client to establish a connection with the specified server. The interface is as follows. Table 16.7 describes the parameters of the function.

```
int connect(int s, const struct sockaddr *name, socklen_t namelen);
```

TABLE 16.7

connect

Parameters	Description
s	Socket descriptor
name	Server address information
namelen	Server address structure length
Return	— —
0	Successful
−1	Failure

When the client connects to the server, first set the server address and then use the connect() function to connect. The sample code is as follows:

```
struct sockaddr_in server_addr;
/* Initialize the pre-connected server address */
server_addr.sin_family = AF_INET;
server_addr.sin_port = htons(port);
server_addr.sin_addr = *((struct in_addr *)host->h_addr);
rt_memset(&(server_addr.sin_zero), 0, sizeof(server_addr.sin_zero));

/* Connect to the server */
if (connect(sock, (struct sockaddr *)&server_addr, sizeof(struct
sockaddr)) == -1)
{
    /* Connection failed */
    closesocket(sock);

    return;
}
```

DATA TRANSMISSION

TCP and UDP have different data transmission methods. TCP needs to establish a connection before data transmission: use the send() function for data transmission and use the recv() function for data reception. While UDP does not need to establish a connection, it uses the sendto() function, sends data, and receives data using the recvfrom() function.

TCP Data Transmission

After the TCP connection is established, data can be sent using the send() function. The interface is as follows. Table 16.8 describes the parameters of the function.

```
int send(int s, const void *dataptr, size_t size, int flags);
```

TABLE 16.8

send

Parameters	Description
s	Socket descriptor
dataptr	The data pointer to send
size	Length of data sent
flags	Flag, generally 0
Return	— —
>0	Successful, return the length of the sent data
<=0	Failed

TCP Data Reception

After the TCP connection is established, use `recv()` to receive the data. The interface is as follows. Table 16.9 describes the parameters of the function.

```
int recv(int s, void *mem, size_t len, int flags);
```

TABLE 16.9

recv

Parameters	Description
s	Socket descriptor
mem	Received data pointer
len	Received data length
flags	Flag, generally 0
Return	— —
>0	Successful, return the length of the received data
=0	The destination address has been transferred and the connection is closed
<0	Failure

UDP Data Transmission

In the case where a connection is not established, you can use the `sendto()` function to send UDP data to the specified destination address, as shown here. Table 16.10 describes the parameters of the function.

```
int sendto(int s, const void *dataptr, size_t size, int flags,
           const struct sockaddr *to, socklen_t tolen);
```

TABLE 16.10

sendto

Parameters	Description
s	Socket descriptor
dataptr	Data pointer sent
size	Length of data sent
flags	Flag, generally 0
to	Target address structure pointer
tolen	Target address structure length
Return	— —
>0	Successful, return the length of the sent data
<=0	Failure

UDP Data Reception

To receive UDP data, use the `recvfrom()` function. Table 16.11 describes the parameters of the function.

```
int recvfrom(int s, void *mem, size_t len, int flags,
          struct sockaddr *from, socklen_t *fromlen);
```

TABLE 16.11
recvfrom

Parameters	Description
s	Socket descriptor
mem	Received data pointer
len	Received data length
flags	Flag, generally 0
from	Receive address structure pointer
fromlen	Receive address structure length
Return	— —
>0	Successful, return the length of the received data
0	The receiving address has been transferred and the connection is closed
<0	Failure

CLOSE THE NETWORK CONNECTION

After the network communication is over, you need to close the network connection. There are two ways to use `closesocket()` and `shutdown()`.

The `closesocket()` interface is used to close an existing socket connection, release the socket resource, and clear the socket descriptor from memory. After the socket is closed it cannot be used again. The connection and cache associated with the socket are also lost. The TCP protocol will automatically close the connection. The interface is as follows. Table 16.12 describes the parameters of the function.

```
int closesocket(int s);
```

TABLE 16.12
closesocket

Parameters	Description
s	Socket descriptor
Return	— —
0	Successful
−1	Failure

Network connections can also be turned off using the `shutdown()` function. The TCP connection is full-duplex. You can use the `shutdown()` function to implement a half-close. It can close the read or write operation of the connection, or both ends, but it does not release the socket resource. The interface is as follows. Table 16.13 describes the parameters of the function.

```
int shutdown(int s, int how);
```

TABLE 16.13
shutdown

Parameters	Description
s	Socket descriptor
how	SHUT_RD closes the receiving end of the connection and no longer receives data
	SHUT_WR closes the connected sender and no longer sends data
	SHUT_RDWR is closed at both ends
Return	— —
0	Successful
−1	Failure

NETWORK FUNCTION CONFIGURATION

The main functional configuration options of the network framework are shown in Tables 16.14 and 16.15, which can be configured according to different functional requirements:

TABLE 16.14
SAL Component Configuration Options

Macro Definition	Value Type	Description
RT_USING_SAL	Boolean	Enable SAL
SAL_USING_LWIP	Boolean	Enable LwIP component
SAL_USING_AT	Boolean	Enable the AT component
SAL_USING_POSIX	Boolean	Enable POSIX interface
SAL_PROTO_FAMILIES_NUM	Integer	The maximum number of protocol families supported

TABLE 16.15
LwIP Configuration Options

Macro Definition	Value Type	Description
RT_USING_LWIP	Boolean	Enable LwIP protocol
RT_USING_LWIP_IPV6	Boolean	Enable IPV6 protocol
RT_LWIP_IGMP	Boolean	Enable the IGMP protocol
RT_LWIP_ICMP	Boolean	Enable the ICMP protocol
RT_LWIP_SNMP	Boolean	Enable the SNMP protocol
RT_LWIP_DNS	Boolean	Enable DHCP function
RT_LWIP_DHCP	Boolean	Enable DHCP function
IP_SOF_BROADCAST	Integer	Filtering broadcasting packets sent by IP
IP_SOF_BROADCAST_RECV	Integer	Filtering broadcasting packets received by IP
RT_LWIP_IPADDR	String	IP address
RT_LWIP_GWADDR	String	Gateway address
RT_LWIP_MSKADDR	String	Subnet mask
RT_LWIP_UDP	Boolean	Enable UDP protocol
RT_LWIP_TCP	Boolean	Enable TCP protocol
RT_LWIP_RAW	Boolean	Enable RAW interface
RT_MEMP_NUM_NETCONN	Integer	Support numbers of network links

TABLE 16.15 (*Continued*)
LwIP Configuration Options

Macro Definition	Value Type	Description
RT_LWIP_PBUF_NUM	Integer	pbuf number of memory blocks
RT_LWIP_RAW_PCB_NUM	Integer	Maximum number of connections for RAW
RT_LWIP_UDP_PCB_NUM	Integer	Maximum number of connections for UDP
RT_LWIP_TCP_PCB_NUM	Integer	Maximum number of connections for TCP
RT_LWIP_TCP_SND_BUF	Integer	TCP send buffer size
RT_LWIP_TCP_WND	Integer	TCP sliding window size
RT_LWIP_TCPTHREAD_PRIORITY	Integer	TCP thread priority
RT_LWIP_TCPTHREAD_MBOX_SIZE	Integer	TCP thread mailbox size
RT_LWIP_TCPTHREAD_STACKSIZE	Integer	TCP thread stack size
RT_LWIP_ETHTHREAD_PRIORITY	Integer	Receive/send thread's priority
RT_LWIP_ETHTHREAD_STACKSIZE	Integer	Receive/send thread's stack size
RT_LwIP_ETHTHREAD_MBOX_SIZE	Integer	Receive/send thread's mailbox size

NETWORK APPLICATION EXAMPLE

This section will be based on the RT thread IoT development board to implement the network function. The IoT board has no Ethernet interface, so it connects to the network through the external ENC28J60 Ethernet module.

PREPARATION

Hardware Preparation

The ENC28J60 is a stand-alone Ethernet controller with an SPI interface, compatible with IEEE 802.3, and integrating MAC and 10 BASE-T PHY, with a maximum speed of 10Mb/s.

The ENC28J60 is connected to the development board via the WIRELESS socket on the board and uses SPI2 to communicate with the board. The schematic diagram and physical figure are shown in Figure 16.7.

Then connect the Internet cable: one end is connected to the ENC28j60 module, and the other end should be inserted in the router or switch LAN port.

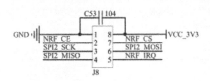

FIGURE 16.7 ENC28J60: Its schematic diagram and hardware diagram.

Software Preparation

The sample code for this section is located in the chapter16 directory of the accompanying materials, as shown in Figure 16.8, when the MDK5 project file project.uvprojx is opened.

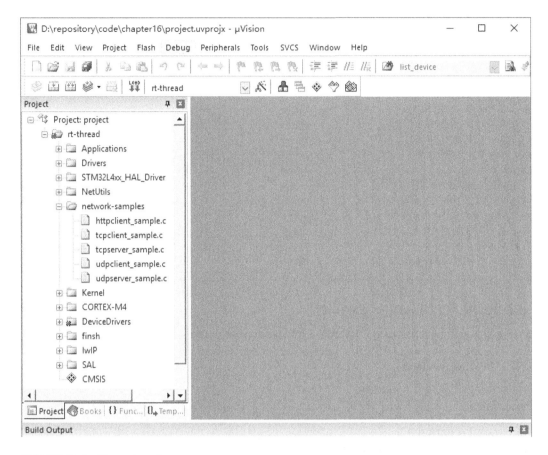

FIGURE 16.8   Network project.

Compared to the basic MDK project, the new groupings in this project are shown in Table 16.16.

TABLE 16.16
New Groupings

Directory Groups	Description
NetUtils	NetUtils software package with Ping source code inside
network-samples	Network application sample code; the sample code used in this section is under this directory, including the TCP Client sample and the UDP Client sample
LwIP	LwIP TCP/IP protocol stack components
SAL	Socket abstraction layer

The initialization code of the ENC28J60 Ethernet module is in Drivers/drv_enc28j60.c, where the ENC28J60 is mounted to the SPIbus as an SPI device via the enc28j60_attach() function, and then pin 85 (PD4) is attached as an interrupt pin for ENC28J60 to trigger the interrupt when pulled down, running the interrupt function of ENC28J60.

```
#include <drivers/pin.h>
#include <enc28j60.h>
```

```
#define ENC28J60_IRQ_PIN 85

int enc28j60_init(void)
{
    /* attach enc28j60 to spi. spi21 cs - PD6 */
    enc28j60_attach("spi21");

    /* init interrupt pin */
    rt_pin_mode(ENC28J60_IRQ_PIN, PIN_MODE_INPUT_PULLUP);
    rt_pin_attach_irq(ENC28J60_IRQ_PIN, PIN_IRQ_MODE_FALLING, enc28j60_isr,
RT_NULL);
    rt_pin_irq_enable(ENC28J60_IRQ_PIN, PIN_IRQ_ENABLE);

    return 0;
}
INIT_COMPONENT_EXPORT(enc28j60_init);
```

Compile and Run

The project is compiled and then the firmware is downloaded to the development board to run. The blue LED on the ENC28J60 Ethernet module will be lit, and the following log information can be seen on the serial port tool.

```
 \ | /
- RT -      Thread Operating System
 / | \      3.1.0 build Oct  3 2018
 2006 - 2018 Copyright by rt-thread team

LwIP-2.0.2 initialized!
[I/SAL_SOC] Socket Abstraction Layer initialize success.
msh >
```

The log information above indicates that both the LwIP component and the SAL have been successfully initialized and that the FinSH console has also been successfully run.

View the IP Address

In the console, you can use the ifconfig command to check the network status. The IP address is 192.168.12.26, and the FLAGS status is LINK_UP, indicating that the network is configured:

```
msh >ifconfig
network interface: e0 (Default)
MTU: 1500
MAC: 00 04 a3 12 34 56
FLAGS: UP LINK_UP ETHARP BROADCAST IGMP
ip address: 192.168.12.26
gw address: 192.168.10.1
net mask  : 255.255.0.0·
dns server #0: 192.168.10.1
dns server #1: 223.5.5.5
```

Ping Network Test

Use the "ping" command for network testing:

```
msh />ping rt-thread.org
60 bytes from 116.62.244.242 icmp_seq=0 ttl=49 time=11 ms
60 bytes from 116.62.244.242 icmp_seq=1 ttl=49 time=10 ms
```

```
60 bytes from 116.62.244.242 icmp_seq=2 ttl=49 time=12 ms
60 bytes from 116.62.244.242 icmp_seq=3 ttl=49 time=10 ms
msh />ping 192.168.10.12
60 bytes from 192.168.10.12 icmp_seq=0 ttl=64 time=5 ms
60 bytes from 192.168.10.12 icmp_seq=1 ttl=64 time=1 ms
60 bytes from 192.168.10.12 icmp_seq=2 ttl=64 time=2 ms
60 bytes from 192.168.10.12 icmp_seq=3 ttl=64 time=3 ms
msh />
```

Getting the above output indicates that the network connection is successful!

TCP CLIENT EXAMPLE

After the network is successfully connected, you can run the network example. First, run the TCP client example. This example will open a TCP server on the PC, then open a TCP client on the IoT board for both parties to communicate on the network.

In the example project, there is already a TCP client program tcpclient_sample.c. The function is to implement a TCP client that can receive and display the information sent from the server. If it receives the information starting with "q" or "Q," then it exits the program directly and closes the TCP client. The program exports the "tcpclient" command to the FinSH console. The command format is tcpclient URL PORT, where URL is the server address and PORT is the port number. The sample code is shown in Code Listing 16.1.

Code listing 16.1 TCP Client Example

```
/*
 * Program list: tcp client
 *
 * This is a tcp client routine
 * Export the tcpclient command to MSH
 * Command call format: tcpclient URL PORT
 * URL: server address PORT:: port number
 * Program function: Receive and display the information sent from the
server, and receive the information that starts with 'q' or 'Q' to exit
the program.
*/

#include <rtthread.h>
#include <sys/socket.h> /* To use BSD socket, you need to include the
socket.h header file. */
#include <netdb.h>
#include <string.h>
#include <finsh.h>

#define BUFSZ   1024

static const char send_data[] = "This is TCP Client from RT-Thread."; /*
Sending used data */
void tcpclient(int argc, char**argv)
{
    int ret;
    char *recv_data;
    struct hostent *host;
    int sock, bytes_received;
    struct sockaddr_in server_addr;
```

```
    const char *url;
    int port;

    /* Received less than 3 parameters */
    if (argc < 3)
    {
        rt_kprintf("Usage: tcpclient URL PORT\n");
        rt_kprintf("Like: tcpclient 192.168.12.44 5000\n");
        return ;
    }

    url = argv[1];
    port = strtoul(argv[2], 0, 10);

    /* Get the host address through the function entry parameter url (if it
is a domain name, it will do domain name resolution) */
    host = gethostbyname(url);

    /* Allocate buffers for storing received data */
    recv_data = rt_malloc(BUFSZ);
    if (recv_data == RT_NULL)
    {
        rt_kprintf("No memory\n");
        return;
    }

    /* Create a socket of type SOCKET_STREAM, TCP type */
    if ((sock = socket(AF_INET, SOCK_STREAM, 0)) == -1)
    {
        /* Failed to create socket */
        rt_kprintf("Socket error\n");

        /* Release receive buffer */
        rt_free(recv_data);
        return;
    }

    /* Initialize the pre-connected server address */
    server_addr.sin_family = AF_INET;
    server_addr.sin_port = htons(port);
    server_addr.sin_addr = *((struct in_addr *)host->h_addr);
    rt_memset(&(server_addr.sin_zero), 0, sizeof(server_addr.sin_zero));

    /* Connect to the server */
    if (connect(sock, (struct sockaddr *)&server_addr, sizeof(struct
sockaddr)) == -1)
    {
        /* Connection failed */
        rt_kprintf("Connect fail!\n");
        closesocket(sock);

        /* Release receive buffer */
        rt_free(recv_data);
        return;
    }
```

```
    while (1)
    {
        /* Receive maximum BUFSZ-1 byte data from a sock connection */
        bytes_received = recv(sock, recv_data, BUFSZ - 1, 0);
        if (bytes_received < 0)
        {
            /* Receive failed, close this connection */
            closesocket(sock);
            rt_kprintf("\nreceived error,close the socket.\r\n");

            /* Release receive buffer */
            rt_free(recv_data);
            break;
        }
        else if (bytes_received == 0)
        {
            /* Print the recv function returns a warning message with a
value of 0 */
            rt_kprintf("\nReceived warning,recv function return 0.\r\n");

            continue;
        }

        /* Received data, clear the end */
        recv_data[bytes_received] = '\0';

        if (strncmp(recv_data, "q", 1) == 0 || strncmp(recv_data, "Q", 1) ==
0)
        {
            /* If the initial letter is q or Q, close this connection */
            closesocket(sock);
            rt_kprintf("\n got a'q'or'Q',close the socket.\r\n");

            /* Release receive buffer */
            rt_free(recv_data);
            break;
        }
        else
        {
            /* Display the received data at the control terminal */
            rt_kprintf("\nReceived data = %s", recv_data);
        }

        /* Send data to sock connection */
        ret = send(sock, send_data, strlen(send_data), 0);
        if (ret < 0)
        {
            /* Send failed, close this connection */
            closesocket(sock);
            rt_kprintf("\nsend error,close the socket.\r\n");

            rt_free(recv_data);
            break;
        }
        else if (ret == 0)
        {
```

```
                /* Print the send function returns a warning message with a
value of 0 */
                rt_kprintf("\n Send warning,send function return 0.\r\n");
            }
        }
        return;
}
MSH_CMD_EXPORT(tcpclient, a tcp client sample);
```

To run the example, first open a network debugging assistant on your computer and open a TCP server. Select the protocol type as TCP Server, and fill in the local IP address and port 5000, as shown in Figure 16.9.

Then start the TCP client to connect to the TCP server by entering the following command in the FinSH console:

```
msh />tcpclient 192.168.12.45 5000  // Input according to actual situation
Connect successful
```

When the console outputs the log message "Connect successful," it indicates that the TCP connection was successfully established. Next, you can perform data communication. In the network debugging tool window, send Hello RT-Thread!, which means that data is sent from the TCP server to the TCP client, as shown in Figure 16.10.

FIGURE 16.9 Network debugging tool interface.

FIGURE 16.10   Network debugging tool interface.

After receiving the data on the FinSH console, the corresponding log information will be output, as shown:

```
msh >tcpclient 192.168.12.130 5000
Connect successful
Received data = hello world
Received data = hello world
Received data = hello world
Received data = hello world
Received data = hello world
 got a 'q' or 'Q',close the socket.
msh >
```

The above information indicates that the TCP client received five "hello world" data messages sent from the server. Finally, the exit command "q" was received from the TCP server, and the TCP client program exited the operation and returned to the FinSH console.

UDP CLIENT EXAMPLE

Code Listing 16.2 is an example of a UDP client. This example will open a UDP server on the PC and open a UDP client on the IoT board for network communication. A UDP client program has been implemented in the sample project. The function is to send data to the server. The sample code is as follows.

Code listing 16.2 UDP Client Example

```
/*
 * Program list: udp client
```

```
 *
 * This is a udp client routine
 * Export the udpclient command to the msh
 * Command call format: udpclient URL PORT [COUNT = 10]
 * URL: Server Address PORT: Port Number COUNT: Optional Parameter
Default is 10
 * Program function: send COUNT datas to the remote end of the service
 */

#include <rtthread.h>
#include <sys/socket.h> /* To use BSD socket, you need to include the
sockets.h header file. */
#include <netdb.h>
#include <string.h>
#include <finsh.h>

const char send_data[] = "This is UDP Client from RT-Thread.\n"; /* data */

void udpclient(int argc, char**argv)
{
    int sock, port, count;
    struct hostent *host;
    struct sockaddr_in server_addr;
    const char *url;

    /* Received less than 3 parameters */
    if (argc < 3)
    {
        rt_kprintf("Usage: udpclient URL PORT [COUNT = 10]\n");
        rt_kprintf("Like: tcpclient 192.168.12.44 5000\n");
        return ;
    }

    url = argv[1];
    port = strtoul(argv[2], 0, 10);

    if (argc> 3)
        count = strtoul(argv[3], 0, 10);
    else
        count = 10;

    /* Get the host address through the function entry parameter url (if it
is a domain name, it will do domain name resolution) */
    host = (struct hostent *) gethostbyname(url);

    /* Create a socket of type SOCK_DGRAM, UDP type */
    if ((sock = socket(AF_INET, SOCK_DGRAM, 0)) == -1)
    {
        rt_kprintf("Socket error\n");
        return;
    }

    /* Initialize the pre-connected server address */
    server_addr.sin_family = AF_INET;
    server_addr.sin_port = htons(port);
    server_addr.sin_addr = *((struct in_addr *)host->h_addr);
    rt_memset(&(server_addr.sin_zero), 0, sizeof(server_addr.sin_zero));
```

```
    /* Send count data in total */
    while (count)
    {
        /* Send data to the remote end of the service */
        sendto(sock, send_data, strlen(send_data), 0,
                (struct sockaddr *)&server_addr, sizeof(struct sockaddr));

        /* Thread sleep for a while */
        rt_thread_delay(50);

        /* Count value minus one */
        count --;
    }

    /* Turn off this socket */
    closesocket(sock);
}
```

When running the example, first open a network debugging assistant on your computer and open a UDP server. Select the protocol type as UDP and fill in the local IP address and port 5000, as shown in Figure 16.11.

Then you can enter the following command in the FinSH console to send data to the UDP server:

```
msh />udpclient 192.168.12.45 1001            // Need to enter according to
the real situation
```

The server will receive 10 messages reading "This is UDP Client from RT-Thread," as shown in Figure 16.12.

FIGURE 16.11   Network debugging tool interface.

FIGURE 16.12   Network debugging tool interface.

CHAPTER SUMMARY

This chapter described the network framework provided by RT-Thread, the basics of networking, and the Socket interfaces that are commonly used in network programming. This chapter was intended to make more developers understand and use the network framework of RT-Thread. The RT-Thread network framework makes it easier for developers to use various types of network protocol stacks for data communication and uses common interface programming to shorten application porting time, improves development efficiency, and shortens the product development cycle.

Appendix A: "Menuconfig" Options

This chapter provides a brief introduction to the main configuration options used by the graphical configuration tool "Menuconfig." In the "Env" tool, changing the working directory to the BSP root and then entering the "menuconfig" command will cause a configuration interface to appear. The configuration menu contains the following three categories:

```
RT-Thread Kernel   --->              [Kernel Configuration]
RT-Thread Components   --->          [Components Configuration]
RT-Thread online packages   --->     [Online Package Configuration]
```

KERNEL CONFIGURATION

The RT-Thread Kernel Configuration submenu is as shown in Code Listing A.1.

Code listing A.1 Kernel Configuration Submenu

```
(8) The maximal size of kernel object name          [The maximum length of
the kernel object name, in bytes]
(4) Alignment size for CPU architecture data access [Set the number of
bytes aligned]
The maximal level value of priority of thread (32)  [Maximum priority for
system threads]
(100) Tick frequency, Hz                            [Define the OS tick,
100 for 100 ticks per second, one tick for 10ms]
[*] Using stack overflow checkin                    [Check the stack for
overflow]
[*] Enable system hook                              [Enable the system
hook function]
(4) The max size of idle hook list                  [Number of idle thread
hook functions]
(256) The stack size of idle thread                 [Stack size of idle
thread]
[*] Enable software timer with a timer thread       [Enable software
timer]
(4) The priority level value of timer thread        [Priority for software
timer thread]
(512) The stack size of timer thread                [Software timer thread
stack size]
[] Enable debugging features --->                   [Debug feature sub-
menu]
Inter-Thread communication --->                     [Inter-threaded
synchronization and communication sub-menu]
Memory Management --->                              [Memory Management
sub-menu]
Kernel Device Object --->                           [Kernel Device Object
Management sub-menu]
```

Notes: The submenu of the options like "[] Enable debugging features --->" can be entered only when the option is selected (by pressing "space"). The submenu of the options like "Inter-Thread communication --->" can be entered directly.

The "Enable debugging features" option enables the information displayed in the console, for example, component initialization, scheduling, thread, timer, etc. The "Inter-thread synchronization and communication" option enables semaphore, mutex, events, mailbox, message queue, and signal. The "Kernel device object management" option enables the I/O device management framework.

COMPONENTS CONFIGURATION

The RT-Thread components configuration submenu is as shown in Code Listing A.2.

Code listing A.2 Components Configuration Submenu

```
[*]  Use components automatically initialization [Use the component auto-
initialization mechanism]
[*]  The main() function as user entry function  [Use the main() function
as a user entry function]
(2048)  Set main thread stack size               [Set the stack size of
the main thread]
C++ features  --->                               [c++ feature sub-menu]
Command shell  --->                              [shell command sub-menu]
Device virtual file system  --->                 [File system sub-menu]
Device Drivers  --->                             [Device-driver sub-
menu]
POSIX layer and C standard library  --->         [POSIX and C standard
library sub-menu]
[ ] Using light-weight process (NEW)             [Use lightweight process
modules]
Network stack  --->                              [Network protocol stack
sub-menu]
VBUS(Virtual Software BUS)  --->                 [Virtual software bus sub-
menu]
Utilities  --->                                  [Utility sub-menu]
```

FILE SYSTEM CONFIGURATION

The file system (device virtual file system) configuration submenu is as shown in Code Listing A.3.

Code listing A.3 File System Configuration Submenu

```
[*] Using device virtual file system            [Enable the DFS virtual
file system]
[*]  Using working directory                    [Use relative paths
based on the current working directory in Finsh]
(2)  The maximal number of mounted file system  [Maximum number of
mounted file systems]
(2)  The maximal number of file system type     [Maximum number of
supported file systems]
(4)  The maximal number of opened files         [Maximum number of open
files]
[*]  Enable elm-chan fatfs                       [Use the elm-chan fatfs
file system]
-*-  Using devfs for device objects             [Enable the devfs file
system]
[ ]  Enable ReadOnly file system on flash        [Use a read-only file
system in Flash]
```

```
[ ]    Enable RAM file system                        [Use the RAM file
system]
[ ]    Enable UFFS file system: Ultra-low-cost Flash File System    [Use
the UFFS file system]
[ ]    Enable JFFS2 file system                       [Use the JFFS2 file
system ]
```

FinSH Configuration

The FinSh (command shell) configuration submenu is as shown in Code Listing A.4.

Code listing A.4 FinSh Configuration Submenu

```
[*] finsh shell                          [Enable FinSH]
(tshell) The finsh thread name           [Set the name of the FinSH
thread]
[*]    Enable command history feature    [Supports the use of arrow
keys (up and down) backtracking history command in FinSH]
(5)    The command history line number   [Set the number of
backtracking command]
[*]    Using symbol table for commands   [Set the number of
backtracking command]
[*]    Keeping description in symbol table    [Add a string description
to each FinSH symbol]
[ ]    Disable the echo mode in default  [Disable echo mode]
(20)   The priority level value of finsh thread [Set the Priority for
FinSH thread]
(4096) The stack size for finsh thread   [Set the stack size for
FinSH thread]
(80)   The command line size for shell   [Set the FinSH command
line length, if the sum of the lengths of commands and parameters is
greater than the set value,it will not be input]
[ ]    shell support authentication      [Enable the permission
verification function ]
[*]    Using module shell                [Enable msh mode]
[*]    Using module shell in default     [Set msh mode to default
shell mode]
[*]    Only using module shell           [Only use msh mode]
(10)   The command arg num for shell     [Number of shell command
parameters]
```

Network Configuration

The network configuration submenu is as shown below:

```
Socket abstraction layer  --->           [SAL layer
configuration]
light weight TCP/IP stack  --->          [lwIP configuration]
Modbus master and slave stack  --->      [Modbus Configuration]
AT commands  --->                        [AT components configuration]
```

The SAL layer configuration (Socket abstraction layer) submenu is as shown here:

```
[*] Enable socket abstraction layer                   [Enable SAL layer]
[*]    Enable BSD socket operated by file system API  [Enable BSD Socket]
```

```
(4)  the maximum number of protocol family        [Maximum number of
supported protocol clusters]
```

SOFTWARE PACKAGE CONFIGURATION

The online packages (RT-Thread online packages) configuration submenu is as shown here.

```
IoT - internet of things  --->        [IoT-related packages]
security packages  --->               [Security-related packages]
language packages  --->               [Scripting language-related
packages]
multimedia packages  --->             [Multimedia packages]
tools packages  --->                  [Tool packages]
system packages  --->                 [System-related packages]
peripheral libraries and drivers  ---> [Peripheral libraries and drives]
miscellaneous packages  --->          [Unclassified packages]
```

Env CONFIGURATION

The latest version of the "Env" tool has one more option to automatically update software packages and generate MDK/IAR projects, which is not shown by default. Developers may enable it by using the "menuconfig -s" or "menconfig --setting" command. The Env configuration submenu is as shown here.

```
[*] Auto update pkgs config            [Package automatic updates
function]
[*] Auto create a mdk/iar project      [Automatically create MDK or IAR
project function]
Project type (MDK5)  --->              [Types of projects created
automatically]
[ ] pkgs download using mirror server  [Download the package using the
mirror server]
```

Package automatic updates function: After exiting "menuconfig," the "pkgs-update" command is executed automatically to download and update the online packages. Automatically create MDK or IAR project function: After exiting "menuconfig," the "scons --target=xyz" command is executed automatically.

Appendix B: SCons

INTRODUCTION TO SCONS

SCons is an open source build system written in the Python language. Just like Makefile to GNU make, scons uses "SConstruct" and "SConscript" files to manage the project to be built. These files are also Python scripts, so the Python standard library can be called in SConstruct and SConscript files for a variety of complex processing. It's one of the differences from Makefile, which is limited by its language capabilities.

A detailed SCons user manual can be found on the SCons website https://scons.org/. This section describes the basic usage of SCons and how to use the SCons tool in RT-Thread.

What Is a Construction Tool?

A software construction tool is a piece of software. It can call tool chain to compile source code into an executable binary program according to certain rules or instructions. This is the most basic and important feature of building tools. In fact, these are not the only functions of construction tools. Usually, these rules have a certain syntax and are organized into files. These files are used to control the behavior of the build tool, and you can do other things besides software building.

The most popular build tool today is GNU Make.

A lot of well-known open source software, such as the Linux kernel, is built using Make. Make detects the organization and dependencies of the file by reading the Makefile and completes the commands specified in the Makefile.

For historical reasons, the syntax of the Makefile is confusing, which is not conducive to beginners. Besides, it is not convenient to use Make on the Windows platform—you need to install the Cygwin environment. To overcome the shortcomings of Make, other build tools have been developed, such as CMake and SCons.

RT-Thread's Construction

RT-Thread was built using Make/Makefile in the earlier stage. Starting from 0.3.x, the RT-Thread development team gradually introduced the SCons build system. The only goal of introducing SCons is to get developers out of the complex Makefile configuration, IDE configuration, and focus on RT-Thread feature development.

Some may doubt the difference between the build tools described here and the IDE. The IDE completes the build through the operation of the graphical interface. Some IDEs can generate script files like Makefile or SConscript based on the source code added by you and call the tools like Make or SCons to build the source code.

Install SCons

SCons needs to be installed on the PC before using the SCons system. Because it is written in the Python language, you need to install the Python runtime environment before using SCons.

The Env configuration tool, provided by RT-Thread, includes SCons and Python, so using SCons on Windows platforms does not require the installation of these two software programs.

On Linux and BSD, Python should already be installed by default. At this time, you only need to install SCons. For example, in Ubuntu you can install SCons with the following command:

```
sudo apt-get install scons
```

BASIC FUNCTIONS OF SCONS

The RT-Thread build system supports multiple compilers, including ARM GCC, MDK, IAR, VisualStudio, and Visual DSP. The mainstream ARM Cortex M0, M3, M4 platforms all support ARM GCC, MDK, and IAR. Some BSPs may only support one compiler; you can read the currently supported compiler by reading the CROSS_TOOL option in rtconfig.py under the BSP directory.

If it is a chip on the ARM platform, you can use the Env tool and enter the "scons" command to compile the BSP directly. At this time, the ARM GCC compiler is used by default because the Env configuration tool also contains the ARM GCC compiler. Compile a BSP using the "scons" command, as shown in Figure B.1, and the SCons will be introduced based on this BSP.

If you want to use other compilers that the BSP is already supporting to compile the project, or if the BSP is a non-ARM platform chip, then you can't compile the project directly with the "scons" command. You need to install the corresponding compiler first and then specify the compiler path to use. Before compiling the project, you can use the following two commands in the Env command line interface to specify the installation path for the compiler as MDK and the compiler path as MDK.

```
set RTT_CC=keil
set RTT_EXEC_PATH=C:/Keilv5
```

COMMONLY USED SCONS COMMANDS

This section describes the SCons commands that are commonly used in RT-Thread. SCons not only completes basic compilation but also generates MDK/IAR/VS projects.

scons

Enter the BSP project directory to be compiled in the Env command line window, and then use this command to compile the project directly. If some source files are modified after the "scons"

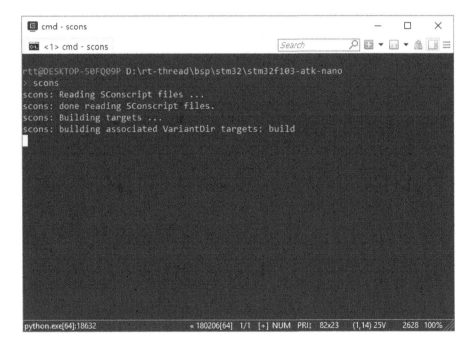

FIGURE B.1 Compile BSP using scons.

command is executed, when the "scons" command is executed again, SCons will incrementally compile.

If the following message output on Windows after the "scons" command is executed.

"scons: warning: No version of Visual Studio compiler found—C/C++ compilers most likely not set correctly."

this means the SCons can't find the Visual Studio compiler on your computer, but since we're targeting on device development and it has nothing to do with local Windows, you can ignore this warning.

"scons" can also be followed by a "-s" parameter, the command "scons -s", which differs from the "scons" command in that it does not print specific internal commands.

scons -c

Clear the compilation target. This command clears the temporary and target files generated when "scons" is executed.

scons --target=XXX

If you use MDK/IAR for project development, when you enable or disable some components, you need to use one of the following commands to regenerate the corresponding customized project, then compile and download in MDK/IAR.

```
scons --target=iar
scons --target=mdk4
scons --target=mdk5
```

In the Env command line window, enter the BSP project directory to be compiled. After using the "scons --target=mdk5" command, a new MDK project file named project.uvprojx will be generated in the BSP directory. Double-click it to open, and you can use MDK to compile and debug. Using the "scons --target=iar" command will generate a new IAR project file named project.eww. Developers who are not used to using SCons can use this approach. If project.uvproj fails to open, please delete project.uvopt and rebuild the project.

Under the "bsp/simulator" directory, you can use the following command to generate a project for vs2012 or a project for vs2005.

```
scons --target=vs2012
Scons --target=vs2005
```

If you provide template files for other IDE projects in the BSP directory, you can also use this command to generate corresponding new projects, such as ua, vs, cb, and cdk.

This command can also be followed by a "-s" parameter, such as the command "scons --target=mdk5 -s," which will not print the specific internal commands when executing this command.

Notes: To generate an MDK or IAR project file, the prerequisite is that there is a project template file in the BSP directory, and then the SCons will add the relevant source code, header file search path, compilation parameters, link parameters, etc., according to the template file. As for which chip this project is for, it is directly specified by this project template file. So in most cases, this template file is an empty project file that is used to assist SCons in generating project.uvprojx or project.eww.

scons -jN

A multi-threaded compilation target, you can use this command to speed up compilation on multi-core computers. In general, a CPU core can support two threads. Use the "scons -j4" command on a dual-core machine.

Notes: If you just want to view the compilation errors or warnings, it's best not to use the -j parameter so that the error message won't be mixed with multiple files in parallel.

scons --dist

Build a project framework. Using this command will generate the dist directory in the BSP directory. This is the directory structure of the development project, including RT-Thread source code and BSP-related projects. The irrelevant BSP folder and libcpu will be removed, and you can freely copy this work to any directory.

scons --verbose

By default, the output compiled with the "scons" command does not display compilation parameters as follows:

```
D:\repository\rt-thread\bsp\stm32f10x>scons
scons: Reading SConscript files ...
scons: done reading SConscript files.
scons: Building targets ...
scons: building associated VariantDir targets: build
CC build\applications\application.o
CC build\applications\startup.o
CC build\components\drivers\serial\serial.o
...
```

The results of using the "scons --verbose" command is as follows:

```
armcc -o build\src\mempool.o -c --device DARMSTM --apcs=interwork -ID:/
Keil/ARM/
RV31/INC -g -O0 -DUSE_STDPERIPH_DRIVER -DSTM32F10X_HD -Iapplications
-IF:\Projec
t\git\rt-thread\applications -I. -IF:\Project\git\rt-thread -Idrivers
-IF:\Proje
ct\git\rt-thread\drivers -ILibraries\STM32F10x_StdPeriph_Driver\inc -IF:\
Project
\git\rt-thread\Libraries\STM32F10x_StdPeriph_Driver\inc -
ILibraries\STM32_USB-FS
-Device_Driver\inc -IF:\Project\git\rt-thread\Libraries\STM32_USB-FS-
Device_Driv
er\inc -ILibraries\CMSIS\CM3\DeviceSupport\ST\STM32F10x -IF:\Project\git\
rt-thre
...
```

SCons ADVANCED

SCons uses SConscript and SConstruct files to organize the source structure. Usually, a project has only one SConstruct, but there will be multiple SConscripts. In general, an SConscript will be placed in each subdirectory where the source code is stored.

To make RT-Thread better support multiple compilers and to easily adjust compilation parameters, RT-Thread creates a separate file for each BSP called rtconfig.py. So the following three files exist in every RT-Thread BSP directory: rtconfig.py, SConstruct, and SConscript, and they control the compilation of the BSP. There is only one SConstruct file in a BSP, but there are multiple SConscript files. It can be said that the SConscript file is the center of the build script structure.

SConscript files are also present under most of the RT-Thread source folders. These source code folders are "found" by the SConscript file in the BSP directory to add the source code corresponding to the macro defined in rtconfig.h to the compiler. The following section will take the stm32f10x-HAL BSP as an example to explain how SCons builds the project.

SCons Built-in Functions

If you want to add some of your source code to the SCons build environment, you can usually create or modify an existing SConscript file. The SConscript file can control the addition of source files and can specify the group of files (similar to the concept of Groups in IDEs such as MDK/IAR).

SCons provides a lot of built-in functions to help us quickly add source code, and with these simple Python statements we can add or remove source code to our project as we want. The following is a brief introduction to some common functions.

GetCurrentDir()

Get the current directory.

Glob('*.c')

Get all C files in the current directory. Modify the value of the parameter to the suffix to match all files of the current directory.

GetDepend(macro)

This function is defined in the script file in the `tools` directory. It reads the configuration information from the rtconfig.h file with the macro name in rtconfig.h. This function returns true if rtconfig.h has a macro is enabled; otherwise, it returns false.

Split(str)

Split the string str into a list.

DefineGroup(name, src, depend, **parameters)

This is a function of RT-Thread based on the SCons extension. DefineGroup is used to define a component. A component can be a directory (under a file or subdirectory) and a Group or folder in some subsequent IDE project files. Table B.1 describes the parameters of the function.

TABLE B.1
Parameter Description of `DefineGroup()`

Parameter	Description
Name	Name of Group
Src	The files contained in the Group generally refer to C/C++ source files. For convenience, you can also use the Glob function to list the matching files in the directory where the SConscript file is located by using a wildcard.
depend	The options that the Group depends on when compiling. For example, the FinSH component depends on the RT_USING_FINSH macro definition. The compile option generally refers to the RT_USING_xxx macro defined in rtconfig.h. When the corresponding macro is defined in the rtconfig.h configuration file, then this group will be added to the build environment for compilation. If the dependent macro is not defined in rtconfig.h, then this Group will not be added to compile. Similarly, when using scons to generate as an IDE project file, if the dependent macros are not defined, the corresponding Group will not appear in the project file.
parameters	Configure other parameters. The values can be found in the Table B.2. You do not need to configure all parameters in actual use.

TABLE B.2

Parameters That Could Be Added

Parameter	Description
CCFLAGS	Source file compilation parameters
CPPPATH	Head file path
CPPDEFINES	Link parameter
LIBRARY	Include this parameter, the object file generated by the component will be packaged into a library file

SConscript(dirs, variant_dir, duplicate)

Read the new SConscript file. The parameter description of the SConscript() function is shown in Table B.3.

TABLE B.3

SConscript

Parameter	Description
dirs	SConscript file path
variant_dir	Specify the path to store the generated target file
duiplicate	Set whether to copy or link the source file to variant_dir

SConscript EXAMPLES

In the following, we will use a few SConscript examples to explain how to use the SCons tool.

Let's start with the SConcript file in the stm32f10x-HAL BSP directory. This file manages all the other SConscript files under the BSP, as shown here:

```
import os
cwd = str(Dir('#'))
objs = []
list = os.listdir(cwd)
for d in list:
    path = os.path.join(cwd, d)
        if os.path.isfile(os.path.join(path, 'SConscript')):
          objs = objs + SConscript(os.path.join(d, 'SConscript'))
Return('objs')
```

- `import os`: Importing the Python system programming os module, you can call the functions provided by the os module to process files and directories.
- `cwd = str(Dir('#'))`: Get the top-level directory of the project and assign it to the string variable cwd. It is the directory where the project's SConstruct is located. It has the same effect as `cwd = GetCurrentDir()`.
- `objs = []`: Defining an empty list variable objs.
- `list = os.listdir(cwd)`: Get all the subdirectories under the current directory and save them to the variable list.
- This is followed by a Python for loop that walks through all the subdirectories of the BSP and runs the SConscript files for those subdirectories. The specific operation is to take a subdirectory of the current directory. It uses `os.path.join(cwd, d)` to join directory

names into a complete path and then determine whether there is a file named SConscript in this subdirectory. If it exists, execute `objs = objs + SConscript(os.path.join(d, 'SConscript'))`. This expression uses a built-in function `SConscript()` provided by SCons, which can read in a new SConscript file and add the source code specified in the SConscript file to the source compilation list objs.

With this SConscript file, the source code required by the BSP project is added to the compilation list.

SCONSCRIPT EXAMPLE 1

So what about the other SConcript files of the stm32f10x-HAL BSP? Let's take a look at the SConcript file in the drivers directory. This directory is used to store the underlying driver code provided by RT-Thread driver framework. The contents of this SConscript file is shown in Code Listing B.1.

Code listing B.1 SConscript Example 1

```
Import('rtconfig')
from building import *

cwd = GetCurrentDir()

# add the general drivers.
src = Split("""
board.c
stm32f1xx_it.c
""")

if GetDepend(['RT_USING_PIN']):
    src += ['drv_gpio.c']
if GetDepend(['RT_USING_SERIAL']):
    src += ['drv_usart.c']
if GetDepend(['RT_USING_SPI']):
    src += ['drv_spi.c']
if GetDepend(['RT_USING_USB_DEVICE']):
    src += ['drv_usb.c']
if GetDepend(['RT_USING_SDCARD']):
    src += ['drv_sdcard.c']

if rtconfig.CROSS_TOOL == 'gcc':
    src += ['gcc_startup.s']

CPPPATH = [cwd]

group = DefineGroup('Drivers', src, depend = [''], CPPPATH = CPPPATH)

Return('group')
```

- `Import('rtconfig')`: Import the rtconfig object. The rtconfig.CROSS_TOOL used later is defined in this rtconfig module.
- `from building import *`: All the contents of the building module are imported into the current module. The DefineGroup used later is defined in this module.
- `cwd = GetCurrentDir()`: Get the current path and save it to the string variable cwd.

The next line uses the Split() function to split a file string into a list, the effect of which is equivalent to:

```
src = ['board.c', 'stm32f1xx_it.c']
```

Later, if statement and GetDepend() function are used to check whether a macro in rtconfig.h is enabled or not, and if so, src += [src _ name] is used to append the source code file to the list variable src.

- CPPPATH = [cwd]: Save the current path to a list variable CPPPATH.

The last line uses DefineGroup to create a group called Drivers, which corresponds to the grouping in the MDK or IAR. The source code file for this group is the file specified by the src variable, and the depend is empty to indicate that the group does not depend on any macros in rtconfig.h.

CPPPATH =CPPPATH indicates that the current path is to be added to the system's header file path. The CPPPATH on the left is a built-in parameter in the DefineGroup that represents the header file path. The CPPPATH on the right is defined in the previous line of this document. This allows us to reference the header file in the drivers directory in other source code.

SConscript Example 2

Let's take a look at the SConcript file in the applications directory. This file will manage the source code under the applications directory for the developer's own application code. The contents of this SConscript file are shown in Code Listing B.2.

Code listing B.2 SConscript Example 2

```
from building import *

cwd = GetCurrentDir()
src = Glob('*.c')
CPPPATH = [cwd, str(Dir('#'))]

group = DefineGroup('Applications', src, depend = [''], CPPPATH = CPPPATH)

Return('group')
```

src = Glob('*.c'): Get all the C files in the current directory.

CPPPATH = [cwd, str(Dir('#'))]: Save the current path and the path of the project's SConstruct to the list variable CPPPATH.

The last line uses DefineGroup() function to create a group called Applications. The source code file for this group is the file specified by src. If the depend is empty, this indicates that the group does not depend on any rtconfig.h macros, and the path saved by CPPPATH is added to the system header search path. Application directories and header files in the stm32f10x-HAL BSP directory can be referenced elsewhere in the source code.

To sum up, this source program will add all C programs in the current directory to the group Applications. So if you add or delete files in this directory, you can add files to the project or delete them from the project. It is suitable for adding source files in batches.

SConscript Example 3

The following is the content of the RT-Thread source code component/finsh/SConscript file, which will manage the source code under the finsh directory. The contents of this SConscript file are shown in Code Listing B.3.

Code listing B.3 SConscript Example 3

```
Import('rtconfig')
from building import *

cwd     = GetCurrentDir()
src     = Split('''
shell.c
symbol.c
cmd.c
''')

fsh_src = Split('''
finsh_compiler.c
finsh_error.c
finsh_heap.c
finsh_init.c
finsh_node.c
finsh_ops.c
finsh_parser.c
finsh_var.c
finsh_vm.c
finsh_token.c
''')

msh_src = Split('''
msh.c
msh_cmd.c
msh_file.c
''')

CPPPATH = [cwd]
if rtconfig.CROSS_TOOL == 'keil':
    LINKFLAGS = '--keep *.o(FSymTab)'

    if not GetDepend('FINSH_USING_MSH_ONLY'):
        LINKFLAGS = LINKFLAGS + '--keep *.o(VSymTab)'
else:
    LINKFLAGS = ''

if GetDepend('FINSH_USING_MSH'):
    src = src + msh_src
if not GetDepend('FINSH_USING_MSH_ONLY'):
    src = src + fsh_src

group = DefineGroup('finsh', src, depend = ['RT_USING_FINSH'], CPPPATH =
CPPPATH, LINKFLAGS = LINKFLAGS)

Return('group')
```

Let's take a look at the contents of the first Python conditional statement in the file. If the compilation tool is Keil, the variable LINKFLAGS = '--keep *.o(FSymTab)' is left blank.

DefineGroup also creates the file specified by src in the finsh directory as a finsh group. depend = ['RT_USING_FINSH'] indicates that this group depends on the macro RT_USING_FINSH in rtconfig.h. When the macro RT_USING_FINSH is expanded in rtconfig.h, the source code in the finsh group will be compiled; otherwise, SCons will not compile.

Then add the finsh directory to the system header directory so that we can reference the header files in the finsh directory in other source code.

LINKFLAGS = LINKFLAGS has the same meaning as CPPPATH = CPPPATH . The LINKFLAGS on the left represents the link parameter, and the LINKFLAGS on the right is the value defined by the previous if else statement. That is, specify the link parameters for the project.

MANAGE PROJECTS WITH SCons

So far, we've introduced the relevant SConscript of the RT-Thread source code in detail, and you've learned some common writing approaches for SConscript files. Next, we'll show you how to use SCons to manage your projects.

ADD APPLICATION CODE

As mentioned earlier, the Applications folder under BSP is used to store the developer's own application code. Currently there is only one main.c file. If your application code is not a lot, it is recommended that the relevant source files be placed under this folder. Two simple files, hello.c and hello.h, have been added under the Applications folder, as shown in Code Listing B.4.

Code listing B.4 The Content of hello.c and hello.h

```
/* file: hello.h */

#ifndef _HELLO_H_
#define _HELLO_H_

int hello_world(void);

#endif /* _HELLO_H_ */

/* file: hello.c */
#include <stdio.h>
#include <finsh.h>
#include <rtthread.h>

int hello_world(void)
{
    rt_kprintf("Hello, world!\n");

    return 0;
}

MSH_CMD_EXPORT(hello_world, Hello world!)
```

The SConcript file in the applications directory will add all source files in the current directory to the project. You need to use the "scons --target=xxx" command to add the two new files to your project. Note that the project will be regenerated each time a new file is added.

ADD A MODULE

As mentioned earlier, in the case that there are not many source code files, it is recommended that all source code files be placed in the Applications folder. If you have a lot of source code and want to create your own project module or need to use other modules that you have obtained, what is the right way?

名称 ⌃	修改日期	类型	大小
hello.c	2018/8/1 17:43	C 文件	1 KB
hello.h	2018/8/1 17:31	H 文件	1 KB
SConscript	2018/8/4 10:24	文件	1 KB

: (D:) › repository › rt-thread › bsp › stm32f10x-HAL › hello

FIGURE B.2 New added hello folder.

Also take hello.c and hello.h mentioned earlier as examples. The two files will be managed in a separate folder and have their own grouping in the MDK project file. They can be selected through menuconfig. Add a hello folder under BSP, as shown in Figure B.2.

Notice that there is an additional SConscript file in the folder. If you want to add some of your own source code to the SCons build environment, you can create a new SConscript file or modify an existing one. The contents of this new hello module SConscript file are shown in Code Listing B.5.

Code listing B.5 The Contents of This New hello module SConscript File

```
from building import *

cwd          = GetCurrentDir()
include_path = [cwd]
src          = []

if GetDepend(['RT_USING_HELLO']):
    src += ['hello.c']

group = DefineGroup('hello', src, depend = [''], CPPPATH = include_path)

Return('group')
```

With the simple lines of code above, a new group hello is created, and the source file to be added to the group can be controlled by macro definition, and the directory where the group is located is added to the system header file path. So how is the custom macro RT_USING_HELLO defined? There is a new file called Kconfig. Kconfig is used to configure the kernel. The configuration interface generated by the "menuconfig" command used when configuring the system with Env relies on the Kconfig file. The "menuconfig" command generates a configuration interface for developers to configure the kernel by reading the various Kconfig files of the project. Finally, all configuration-related macro definitions are automatically saved to the rtconfig.h file in the BSP directory. Each BSP has a rtconfig.h file. That is the configuration information of this BSP.

There is already a Kconfig file for this BSP in the stm32f10x-HAL BSP directory, and we can add the configuration options we need based on this file. The following configuration options have been added to the hello module. The # sign is followed by a comment.

```
menu "hello module"                         # create a "hello module" menu

    config RT_USING_HELLO                    # RT_USING_HELLO
configuration options
        bool "Enable hello module"           # RT_USING_HELLO is a bool
variable and display as "Enable hello module"
        default y                            # RT_USING_HELLO can take
values y and n, default y
```

```
        help                               # If use help, it would
display "this hello module only used for test"
        this hello module only used for test

    config RT_HELLO_NAME                    # RT_HELLO_NAME
configuration options
        string "hello name"                # RT_HELLO_NAME is a string
variable and the menu show as "hello name"
        default "hello"                    # default name is "hello"

    config RT_HELLO_VALUE                   # RT_HELLO_VALUE
configuration options
        int "hello value"                  # RT_HELLO_VALUE is an int
variable and the menu show as "hello value"
        default 8                          # the default value is 8

endmenu                                    # the hello menu is end
```

After entering the stm32f10x-HAL BSP directory with the Env tool, the configuration menu of the new hello module can be seen at the bottom of the main page using the "menuconfig" command, as shown in Figure B.3.

You can also modify the value of the hello value, as shown in Figure B.4.

After saving the configuration, you may exit the configuration interface and open the rtconfig.h file in the stm32f10x-HAL BSP directory. Then you can see that the configuration information of the hello module is available.

Notes: Use the "scons --target=XXX" command to generate a new project each time menuconfig is configured.

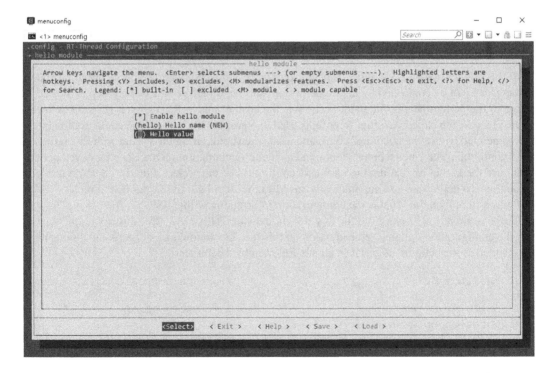

FIGURE B.3 hello module configuration menu.

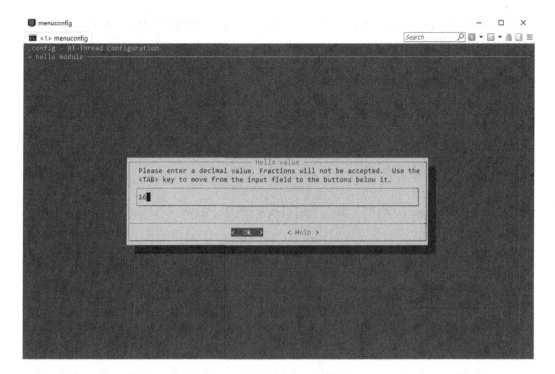

FIGURE B.4   Modify the hello value.

Because the RT_USING_HELLO macro has been defined in rtconfig.h, the source file for hello.c is added to the new project when the new project is generated.

The above simply enumerates the configuration options for adding your modules to the Kconfig file. You can also refer to the Chapter 11 Env-Assisted Development Environment, which explains how to modify and add configuration options.

ADD A LIBRARY

If you want to add an extra library to your project, you need to pay attention to the naming of the binary library by different toolchains. For example, the GCC toolchain, which identifies library names such as libabc.a, specifies abc instead of libabc when specifying a library. So you need to pay special attention to the SConscript file when linking additional libraries. Besides, it is

```
rtconfig.h

148   #define RT_USING_UART1
149   #define RT_USING_UART2
150
151   /* hello module */
152
153   #define RT_USING_HELLO
154   #define RT_HELLO_NAME "hello"
155   #define RT_HELLO_VALUE 16
156
157   #endif
158
```

FIGURE B.5   hello module-related macro definition.

also a good idea to specify the corresponding library search path when specifying additional libraries; here is an example:

```
Import('rtconfig')
from building import *

cwd = GetCurrentDir()
src = Split('''
''')

LIBPATH = [cwd + '/libs']
LIBS = ['abc']

group = DefineGroup('ABC', src, depend = [''], LIBS = LIBS,
LIBPATH=LIBPATH)
```

LIBPATH specifies the path to the library, and LIBS specifies the name of the library. If the toolchain is GCC, the name of the library should be libabc.a; if the toolchain is armcc, the name of the library should be "abc.lib". `LIBPATH = [cwd + '/libs']` indicates that the search path for the library is the 'libs' directory in the current directory.

COMPILER OPTIONS

`rtconfig.py` is an RT-Thread standard compiler configuration file that controls most of the compilation options and is a script file written in the Python language, primarily for the following tasks:

- Specify the compiler (select one from the supported multiple compilers that you are using now).
- Specify compiler parameters such as compile options, link options, and so on.

When we compile the project using the "scons" command, the project is compiled according to the compiler configuration options of `rtconfig.py`. As shown in the Code Listing B.6, part of the code is rtconfig.py in the stm32f10x-HAL BSP directory.

Code listing B.6 Part Code of rtconfig.py

```
import os

# toolchains options
ARCH='arm'
CPU='cortex-m3'
CROSS_TOOL='gcc'

if os.getenv('RTT_CC'):
    CROSS_TOOL = os.getenv('RTT_CC')

# cross_tool provides the cross compiler
# EXEC_PATH is the compiler execute path, for example, CodeSourcery, Keil
MDK, IAR

if   CROSS_TOOL == 'gcc':
    PLATFORM    = 'gcc'
    EXEC_PATH   = '/usr/local/gcc-arm-none-eabi-5_4-2016q3/bin/'
elif CROSS_TOOL == 'keil':
```

```
    PLATFORM    = 'armcc'
    EXEC_PATH   = 'C:/Keilv5'
elif CROSS_TOOL == 'iar':
    PLATFORM    = 'iar'
    EXEC_PATH   = 'C:/Program Files/IAR Systems/Embedded Workbench 6.0
Evaluation'

if os.getenv('RTT_EXEC_PATH'):
    EXEC_PATH = os.getenv('RTT_EXEC_PATH')


BUILD = 'debug'

if PLATFORM == 'gcc':
    # toolchains
    PREFIX = 'arm-none-eabi-'
    CC = PREFIX + 'gcc'
    AS = PREFIX + 'gcc'
    AR = PREFIX + 'ar'
    LINK = PREFIX + 'gcc'
    TARGET_EXT = 'elf'
    SIZE = PREFIX + 'size'
    OBJDUMP = PREFIX + 'objdump'
    OBJCPY = PREFIX + 'objcopy'

    DEVICE = '-mcpu=cortex-m3 -mthumb -ffunction-sections -fdata-sections'
    CFLAGS = DEVICE
    AFLAGS = '-c' + DEVICE + '-x assembler-with-cpp'
    LFLAGS = DEVICE + '-Wl,--gc-sections,-Map=rtthread-stm32.map,-cref,-
u,Reset_Handler -T stm32_rom.ld'
```

Where CFLAGS is the compiler option for C files, AFLAGS is the compiler option for assembly files, and LFLAGS is the linker option. The BUILD variable controls the level of code optimization. The default BUILD variable takes the value "debug," which means the source code will be compiled in debug mode, with an optimization level of 0. If you modify this variable to any other value, it will be compiled with optimization level 2. The following are all valid values (as long as it's not "debug"):

```
BUILD = ''
BUILD = 'release'
BUILD = 'hello, world'
```

It is recommended to use the debug mode to compile during the development phase without optimization and to consider the optimization when the product is stable.

The specific meaning of these options needs to refer to the compiler manual, such as armcc used above is the underlying compiler of MDK. The meaning of its compile options is detailed in MDK help.

As mentioned earlier, if you want to compile the project with another compiler, you can use the relevant commands to specify the compiler and compiler paths on the command line side of Env. However, such modification only works for the current Env process. When you enable it again, you need to reuse the command settings. And you can directly modify the rtconfig.py file to achieve the purpose of permanently configuring the compiler. In general, you only need to modify the CROSS_TOOL and EXEC_PATH options here.

- CROSS_TOOL: Specify the compiler. The optional values are keil, gcc, and iar. Browse rtconfig.py to see the compilers supported by the current BSP. If MDK is installed on your machine, you can modify CROSS_TOOL to Keil and use MDK to compile the project.

- EXEC_PATH: The installation path of the compiler. There are two points to note here:
  - When installing the compiler (such as MDK, GNU GCC, IAR, etc.), do not install it in a path with Chinese characters or with spaces. Otherwise, some errors will occur when parsing the path. Some programs are installed by default into the C:\Program Files directory which has a space in its name. It is recommended to choose other paths during installation to develop good development habits.
  - When modifying EXEC_PATH, you need to be aware of the format of the path. On the Windows platforms, the default path separator is the backslash "\, " which is used for escape characters in both C and Python. So when modifying the path, you can change "\" to "/" or add r (Python-specific syntax for raw data).

Suppose a compiler is installed under D:\Dir1\Dir2. The following are the valid values:

- EXEC_PATH = r'D:\Dir1\Dir2' Note that with the string r in front of the string, "\" can be used normally.
- EXEC_PATH = 'D:/Dir1/Dir2' Note that instead of "/", there is no r in front.
- EXEC_PATH = 'D:\\Dir1\\Dir2' Note that the escapement of "\" is used here to escape "\" itself.
- This is the wrong way to write: EXEC_PATH = 'D:\Dir1\Dir2'.

If the rtconfig.py file has the following code, comment out the following code when configuring your own compiler:

```
if os.getenv('RTT_CC'):
    CROSS_TOOL = os.getenv('RTT_CC')
... ...
if os.getenv('RTT_EXEC_PATH'):
    EXEC_PATH = os.getenv('RTT_EXEC_PATH')
```

The above two if judgments will set CROSS_TOOL and EXEC_PATH to the default value of Env.

After the compiler is configured, we can use SCons to compile the BSP of RT-Thread. Open a command line window in the BSP directory and execute the "scons" command to start the compilation process.

RT-Thread Auxiliary Compilation Script

In the tools directory of the RT-Thread source code, there are some auxiliary scripts provided by RT-Thread, such as the project files for automatically generating RT-Thread for some IDEs. The most important of these is the building.py script.

SCons Further Usage

For a complex, large-scale system, it is more than just a few files in a directory that can handle it. It is probably a combination of several folders at the first level.

In SCons, you can write SConscript script files to compile files in these relatively independent directories, and you can also use the Export and Import functions in SCons to share data between SConstruct and SConscript files (i.e., an object data in Python). For more information on how to use SCons, please refer to the SCons Official Documentation *https://scons.org/*.

Appendix C: Getting Started with QEMU (Ubuntu)

The development of embedded software is inseparable from the development board. Without physical development boards, similar virtual machines like QEMU can be used to emulate the development board. QEMU is a virtual machine that supports cross-platform virtualization. It can virtualize many development boards. To facilitate the experience of RT-Thread without a development board, RT-Thread provides a board-level support package (BSP) for the QEMU-emulated **ARM vexpress A9** development board.

PREPARATIONS

- Install Git on the PC: `sudo apt-get install git`
- Download RT-Thread source code: `git clone https://github.com/RT-Thread/rt-thread.git`
- Install QEMU: `sudo apt-get install qemu`
- Install Scons: `sudo apt-get install scons`
- Install the compiler. If the compiler version installed with the "apt-get" command is too old, it will cause compilation errors. Download and install the new version by using the following command in turn. The download link and the decompression folder name will vary according to the download version. After this, the compression packet will unzip to the /opt folder.

```
wget
https://armkeil.blob.core.windows.net/developer/Files/downloads/
gnu-rm/6-2016q4/gcc-arm-none-eabi-6_2-2016q4-20161216-linux.tar.bz2
cd/opt
sudo tar xf ~/gcc-arm-none-eabi-6_2-2016q4-20161216-linux.tar.bz2
```

- Install the ncurses library: `sudo apt-get install libncurses5-dev`
- After the compiler is installed, it is necessary to modify the `rtconfig.py` file under `rt-thread/bsp/qemu-vexpress-a9` BSP, and modify the corresponding path to the bin directory corresponding to the compiler decompressed into the opt directory. Referring to Figure C.1, the directory name varies according to the downloaded version of the compiler.

QEMU BSP CATALOGUE INTRODUCTION

The board support package (BSP) provided by RT-Thread simulates the ARM vexpress A9 development board and is located in the `qemu-vexpress-a9` folder under the `bsp` directory of RT-Thread source code. This BSP implements LCD, keyboard, mouse, SD card, Ethernet card, serial port, and other related drivers. The contents of the folder are shown in Figure C.2.

The main files and directories of `qemu-vexpress-a9` BSP are described in Table C.1.

TABLE C.1

qemu-vexpress-a9 BSP

Files/Directories	Description
applications	Developer's application code directory
Drivers	The underlying driver provided by RT-Thread
qemu.bat	Script files running on Windows platform
qemu.sh	Script files running on Linux platform
qemu-dbg.bat	Debugging script files on Windows platform
qemu-dbg.sh	Debugging script files on Linux platform
README.md	Description document of BSP
rtconfig.h	A header file of BSP

ENV TOOL INSTRUCTION

Env is RT-Thread's development assistant tool that provides compilation and build environments, graphical system configuration, and package management for the projects based on the RT-Thread operating system. For more information, please read the section "Env Introduction" and "Env Features" in Chapter 11.

First, enter the BSP directory under the Linux terminal.

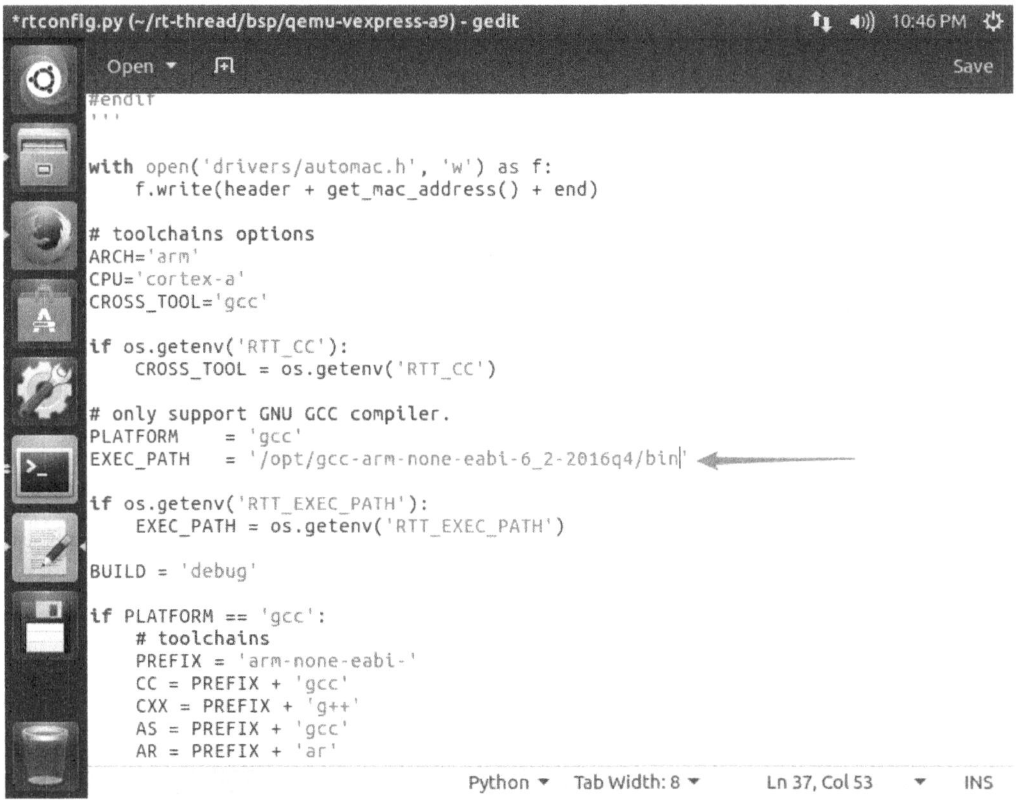

FIGURE C.1   Edit EXEC_PATH in rtconfig.py.

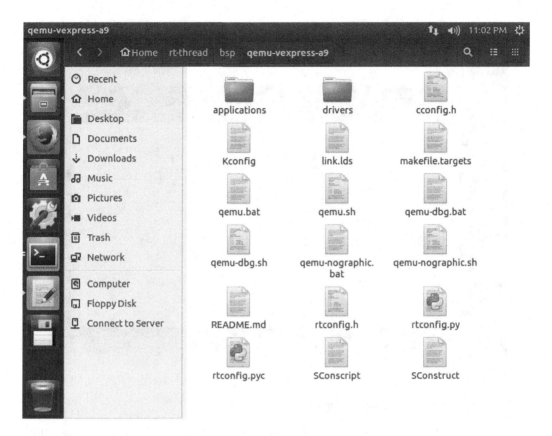

FIGURE C.2   qemu-vexpress-a9 folder.

Install Env and Configure BSP

```
scons --menuconfig
```

The Env tool will be installed and initialized after using the "scons --menuconfig" command, as shown in Figure C.3. Then it will enter the configuration interface, and you can configure BSP, as shown in Figure C.4.

The keyboard ↑ key and ↓ key are used to look up and down menu items, the "Enter" key can enter the selected directory, the "Space" key can select or cancel bool variables, and "Esc" can be used to exit the current directory.

Acquisition of Software Packages

```
source ~/.env/env.sh
scons --menuconfig
pkgs --update
```

The "env.sh" file is a file that needs to be executed. It configures the environment variables so that we can update the package with the "pkgs" command and execute it with the "source ~/.env/env.sh" command. It is recommended to put this command into a shell profile like ~/.bash_profile.

```
rtt@ubuntu:~/rt-thread/bsp/qemu-vexpress-a9$ scons --menuconfig
```

FIGURE C.3   Install Env tool.

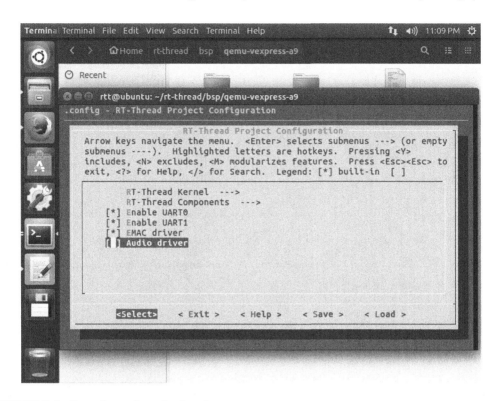

FIGURE C.4   Enter the configuration interface.

Then use the "`scons --menuconfig`" command to enter menuconfig, and select the online packages, as shown in Figure C.5 and C.6.

For example, select the kernel sample package: semaphore sample, as shown in Figure C.7.

Exit and save the configuration, as shown in Figure C.8.

An online package is selected. Download the package to the packages folder in the BSP directory using the "`pkgs --update`" command (Git needs to be installed), as shown in Figure C.9.

Use the *scons* Command to Compile the Project

Switch to the QEMU BSP directory and enter the "`scons`" command to compile the project. If the compilation is correct, the "`rtthread.elf`" file will be generated in the BSP directory, which is a target file required for QEMU to run, as shown in Figure C.10.

Use the *./qemu.sh* Command to Run QEMU

After compiling, type "./qemu.sh" to start the virtual machine, and BSP project. qemu.sh is a Linux batch file. This file is located in the BSP folder and primarily includes the execution instructions for QEMU. The first run of the project will create a blank sd.bin file under the BSP folder, which is a virtual SD card with a size of 64M. The Env command interface displays the initialization information and version number information printed during the start-up of RT-Thread system; also, the QEMU virtual machine is running, as shown in Figure C.11.

Use "Ctrl"+"C" on the keyboard to exit QEMU.

```
rtt@ubuntu:~/rt-thread/bsp/qemu-vexpress-a9$ source ~/.env/env.sh
rtt@ubuntu:~/rt-thread/bsp/qemu-vexpress-a9$
rtt@ubuntu:~/rt-thread/bsp/qemu-vexpress-a9$ scons --menuconfig
```

FIGURE C.5   Menuconfig command.

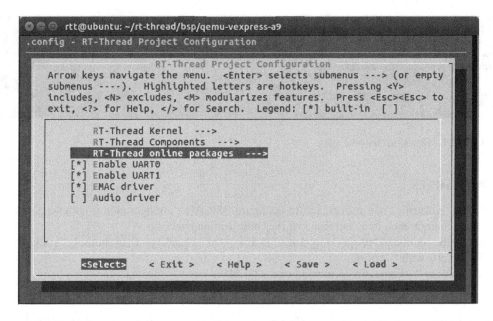

FIGURE C.6 The package menu.

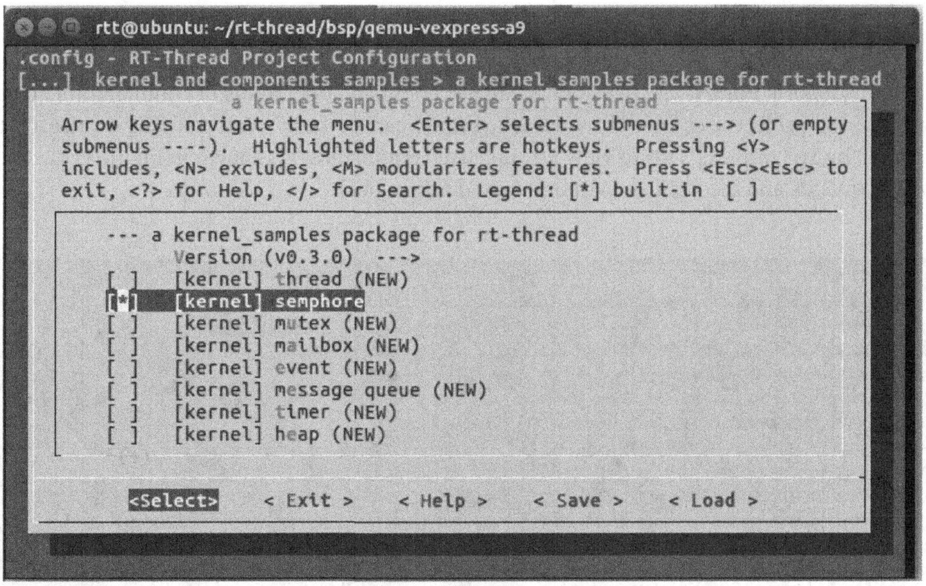

FIGURE C.7 Select a package.

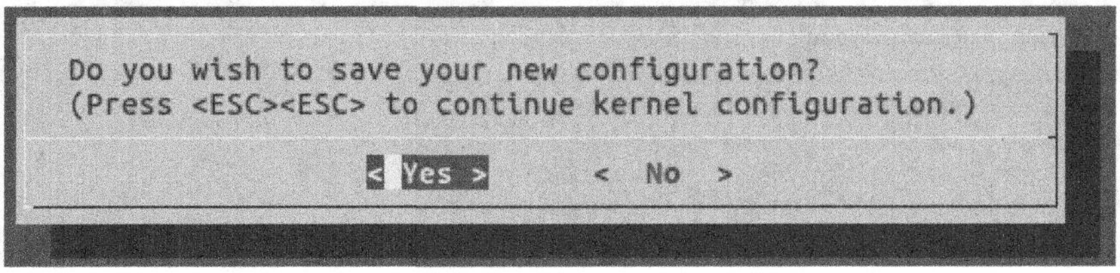

FIGURE C.8 Save the configuration.

```
rtt@ubuntu:~/rt-thread/bsp/qemu-vexpress-a9$ pkgs --update
Start to download package : kernel_samples-v0.3.0.zip
Downloaded 26 KB
Start to unpack. Please wait...
==============================> KERNEL_SAMPLES v0.3.0 is downloaded successfull
y.

Operation completed successfully.
```

FIGURE C.9   Download the package.

RUN SAMPLES

1. Use "scons --menuconfig" to configure BSP. After configuration is complete, save the configuration first, and then exit the configuration interface.
2. Use the command "pkgs --update" to download the package when choosing a package.
3. Compile with "scons,".
4. Enter the "./qemu.sh" command to run QEMU.
5. Enter the corresponding command to run the sample code.

Next, take these steps to run some sample code, such as the FinSH console, kernel sample code, file system, network sample code, etc.

RUN THE FINSH CONSOLE

RT-Thread supports FinSH, and developers can use command operations in command line mode.

Because the project uses FinSH by default, FinSH can be run directly. If BSP is already running, type "help" or press Tab to view all supported commands. As shown in Figure C.12, the left shows commands and the right shows command descriptions.

```
rtt@ubuntu: ~/rt-thread/bsp/qemu-vexpress-a9
rtt@ubuntu:~/rt-thread/bsp/qemu-vexpress-a9$ scons  ⬅
scons: Reading SConscript files ...
scons: done reading SConscript files.
scons: Building targets ...
scons: building associated VariantDir targets: build
CC build/applications/lcd_init.o
CC build/applications/main.o
CC build/applications/mnt.o
CC build/drivers/board.o
CC build/drivers/drv_keyboard.o
CC build/drivers/drv_mouse.o
CC build/drivers/drv_sdio.o
CC build/drivers/drv_smc911x.o
CC build/drivers/drv_timer.o
CC build/drivers/secondary_cpu.o
CC build/drivers/serial.o
CXX build/kernel/components/cplusplus/Mutex.o
CXX build/kernel/components/cplusplus/Semaphore.o
CXX build/kernel/components/cplusplus/Thread.o
CXX build/kernel/components/cplusplus/crt.o
CC build/kernel/components/cplusplus/crt_init.o
CC build/kernel/components/dfs/filesystems/devfs/devfs.o
CC build/kernel/components/dfs/filesystems/elmfat/dfs_elm.o
```

FIGURE C.10   Compile the project.

```
rtt@ubuntu: ~/rt-thread/bsp/qemu-vexpress-a9

rtt@ubuntu:~/rt-thread/bsp/qemu-vexpress-a9$ sudo ./qemu.sh
[sudo] password for rtt:
WARNING: Image format was not specified for 'sd.bin' and probing guessed raw.
         Automatically detecting the format is dangerous for raw images, write
 operations on block 0 will be restricted.
         Specify the 'raw' format explicitly to remove the restrictions.
pulseaudio: set_sink_input_volume() failed
pulseaudio: Reason: Invalid argument
pulseaudio: set_sink_input_mute() failed
pulseaudio: Reason: Invalid argument

 \ | /
- RT -     Thread Operating System
 / | \     4.0.2 build Jul 24 2019
 2006 - 2019 Copyright by rt-thread team
lwIP-2.0.2 initialized!
[I/sal.skt] Socket Abstraction Layer initialize success.
[I/SDIO] SD card capacity 65536 KB.
file system initialization done!
hello rt-thread
msh />
msh />
```

QEMU

FIGURE C.11 Run the project.

```
msh />help
RT-Thread shell commands:
semaphore_sample - semaphore sample
memcheck         - check memory data
memtrace         - dump memory trace information
list_fd          - list file descriptor
ifconfig         - list the information of all network interfaces
ping             - ping network host
dns              - list and set the information of dns
netstat          - list the information of TCP / IP
version          - show RT-Thread version information
list_thread      - list thread
list_sem         - list semaphore in system
list_event       - list event in system
list_mutex       - list mutex in system
list_mailbox     - list mail box in system
list_msgqueue    - list message queue in system
list_memheap     - list memory heap in system
list_mempool     - list memory pool in system
list_timer       - list timer in system
list_device      - list device in system
exit             - return to RT-Thread shell mode.
help             - RT-Thread shell help.
ls               - List information about the FILEs.
cp               - Copy SOURCE to DEST.
mv               - Rename SOURCE to DEST.
cat              - Concatenate FILE(s)
rm               - Remove(unlink) the FILE(s).
cd               - Change the shell working directory.
pwd              - Print the name of the current working directory.
```

FIGURE C.12 View all supported commands.

```
msh />list_thread
thread     cpu pri  status     sp           stack size max used left tick   error
--------   --- ---  -------    ----------   ---------- -------- ----------   ---
tshell       0  20  running 0x000002a8 0x00001000       19%  0x0000000a 000
aio        N/A  16  suspend 0x0000008c 0x00000800       07%  0x0000000a 000
sys_work   N/A  23  suspend 0x00000088 0x00000800       52%  0x00000009 000
mmcsd_de   N/A  22  suspend 0x000000b8 0x00000400       67%  0x00000012 000
tcpip      N/A  10  suspend 0x000000f0 0x00000400       87%  0x00000012 000
etx        N/A  12  suspend 0x000000b0 0x00000400       26%  0x0000000d 000
erx        N/A  12  suspend 0x000000bc 0x00000400       63%  0x00000010 000
tidle1       1  31  running 0x00000068 0x00000400       22%  0x0000000b 000
tidle0     N/A  31  ready   0x00000068 0x00000400       22%  0x00000018 000
timer      N/A   4  suspend 0x00000084 0x00000400       21%  0x0000000a 000
msh />list_timer
timer        periodic    timeout        flag
--------   ----------  ----------   ----------
tshell     0x00000000 0x00000000 deactivated
aio        0x00000000 0x00000000 deactivated
sys_work   0x00004e20 0x00004f4d deactivated
mmcsd_de   0x00000001 0x0000000a deactivated
tcpip      0x00000032 0x00010040 activated
etx        0x00000000 0x00000000 deactivated
erx        0x00000000 0x00000000 deactivated
tidle1     0x00000000 0x00000000 deactivated
tidle0     0x00000000 0x00000000 deactivated
timer      0x00000064 0x0000012c deactivated
current tick:0x00010014
```

FIGURE C.13 Threads and timers.

For example, by entering the "list _ thread" command, the currently running threads, thread status, and stack size can be seen; by entering "list _ timer", the status of the timers can be seen, as shown in Figure C.13.

RUN KERNEL SAMPLE CODE

To run kernel samples with QEMU, first, enter the qemu-vexpress-a9 directory, and then perform the following steps.

scons --menuconfig

Use the "scons --menuconfig" command to enter the configuration menu, select some kernel examples, then save and exit.

```
-> RT-Thread online packages
   -> miscellaneous packages
      -> samples: kernel and components samples
         [*] a kernel_samples package for rt-thread -->
                 Version (v0.3.0)  --->
            [*]   [kernel] thread (NEW)
            [*]   [kernel] semphore (NEW)
            [*]   [kernel] mutex (NEW)
            [*]   [kernel] mailbox (NEW)
            [*]   [kernel] event (NEW)
            [ ]   [kernel] message queue (NEW)
            [ ]   [kernel] timer (NEW)
            [ ]   [kernel] heap (NEW)
            [ ]   [kernel] memheap (NEW)
```

```
[ ]     [kernel] mempool (NEW)
[ ]     [kernel] idle hook (NEW)
[ ]     [kernel] signal (NEW)
[ ]     [kernel] interrupt (NEW)
[ ]     [kernel] priority inversion (NEW)
[ ]     [kernel] time slice (NEW)
[ ]     [kernel] sheduler hook (NEW)
[ ]     [kernel] producer consumer model (NEW)
```

pkgs --update

Use the "pkgs --update" command to download the package to the packages folder in the BSP directory, as shown in Figure C.14.

```
rtt@ubuntu:~/rt-thread/bsp/qemu-vexpress-a9$ pkgs --update
Start to download package : kernel_samples-v0.3.0.zip
Downloded 26 KB
Start to unpack. Please wait...
================================> KERNEL_SAMPLES v0.3.0 is downloaded successfull
y.

Operation completed successfully.
```

FIGURE C.14 Download the package.

scons

Use the "scons" command to compile BSP, as shown in Figure C.15.

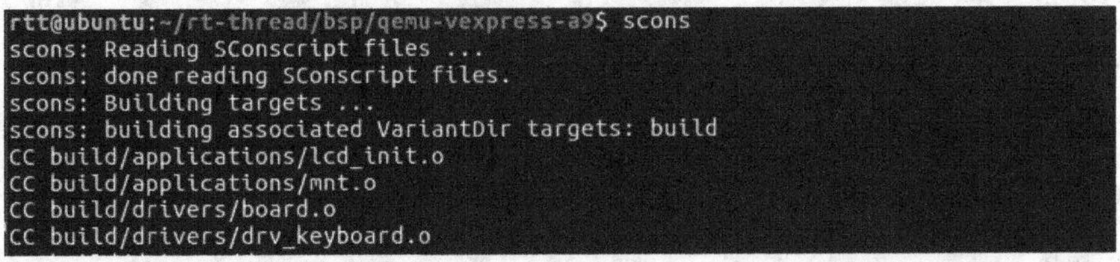

```
rtt@ubuntu:~/rt-thread/bsp/qemu-vexpress-a9$ scons
scons: Reading SConscript files ...
scons: done reading SConscript files.
scons: Building targets ...
scons: building associated VariantDir targets: build
CC build/applications/lcd_init.o
CC build/applications/mnt.o
CC build/drivers/board.o
CC build/drivers/drv_keyboard.o
```

FIGURE C.15 Compile.

./qemu.sh

Use the "./qemu.sh" command to run QEMU, as shown in Figure C.16.

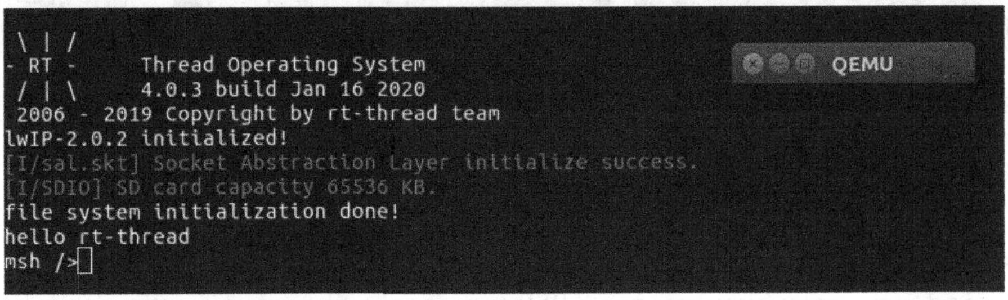

```
     \ | /
- RT -     Thread Operating System                      QEMU
 / | \     4.0.3 build Jan 16 2020
2006 - 2019 Copyright by rt-thread team
lwIP-2.0.2 initialized!
[I/sal.skt] Socket Abstraction Layer initialize success.
[I/SDIO] SD card capacity 65536 KB.
file system initialization done!
hello rt-thread
msh />
```

FIGURE C.16 Run the project.

Run Kernel Sample Code

Use the "help" command to view all system commands. It can be seen that some kernel examples have been successfully added to the system, as shown in Figure C.17.

```
msh />help
RT-Thread shell commands:
thread_sample      - thread sample
semaphore_sample - semaphore sample
mutex_sample       - mutex sample
mailbox_sample     - mailbox sample
event_sample       - event sample
memcheck           - check memory data
memtrace           - dump memory trace information
list_fd            - list file descriptor
ifconfig           - list the information of all network interfaces
ping               - ping network host
dns                - list and set the information of dns
netstat            - list the information of TCP / IP
version            - show RT-Thread version information
list_thread        - list thread
list_sem           - list semaphore in system
list event         - list event in system
```

FIGURE C.17   View all commands.

Run the thread_sample, as shown in Figure C.18.

```
msh />thread_sample
msh />thread2 count: 0
thread2 count: 1
thread2 count: 2
thread2 count: 3
thread2 count: 4
thread2 count: 5
thread2 count: 6
thread2 count: 7
thread2 count: 8
thread2 count: 9
thread2 exit
thread1 count: 0
thread1 count: 1
thread1 count: 2
thread1 count: 3
thread1 count: 4
thread1 count: 5
thread1 count: 6
thread1 count: 7
thread1 count: 8
thread1 count: 9
thread1 count: 10
```

FIGURE C.18   Thread sample.

Run the semaphore_sample, as shown in Figure C.19.

RUN THE FILE SYSTEM

rt-thread/bsp/qemu-vexpress-a9 BSP turns on the file system function by default, so RT-Thread can run directly after compiling.

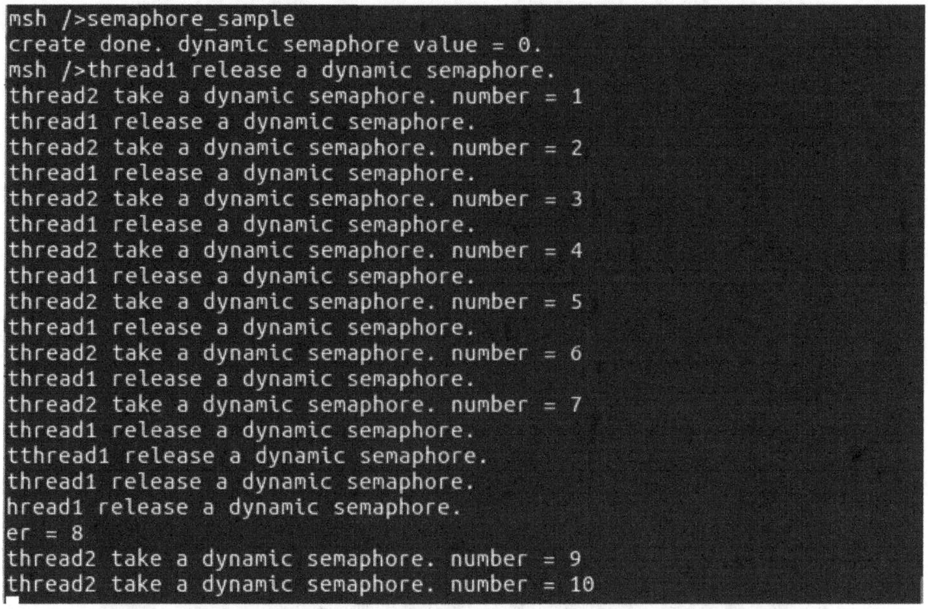

```
msh />semaphore_sample
create done. dynamic semaphore value = 0.
msh />thread1 release a dynamic semaphore.
thread2 take a dynamic semaphore. number = 1
thread1 release a dynamic semaphore.
thread2 take a dynamic semaphore. number = 2
thread1 release a dynamic semaphore.
thread2 take a dynamic semaphore. number = 3
thread1 release a dynamic semaphore.
thread2 take a dynamic semaphore. number = 4
thread1 release a dynamic semaphore.
thread2 take a dynamic semaphore. number = 5
thread1 release a dynamic semaphore.
thread2 take a dynamic semaphore. number = 6
thread1 release a dynamic semaphore.
thread2 take a dynamic semaphore. number = 7
thread1 release a dynamic semaphore.
tthread1 release a dynamic semaphore.
thread1 release a dynamic semaphore.
hread1 release a dynamic semaphore.
er = 8
thread2 take a dynamic semaphore. number = 9
thread2 take a dynamic semaphore. number = 10
```

FIGURE C.19   Semaphore sample.

1. Use the "`scons`" command to compile.
2. Use the "`./qemu.sh`" command to run QEMU.

Type "`list _ device`" to view all devices registered in the system. The virtual SD card "sd0" device can be seen, which is also shown in Figure C.20. Next, format the SD card by using the "`mkfs sd0`" command, which will format the SD card into a FatFS file system. FatFs is a Microsoft FAT–compatible file system developed for small embedded devices. It is written in ANSI C, uses the abstract hardware I/O layer, and provides continuous maintenance, so it has good portability.

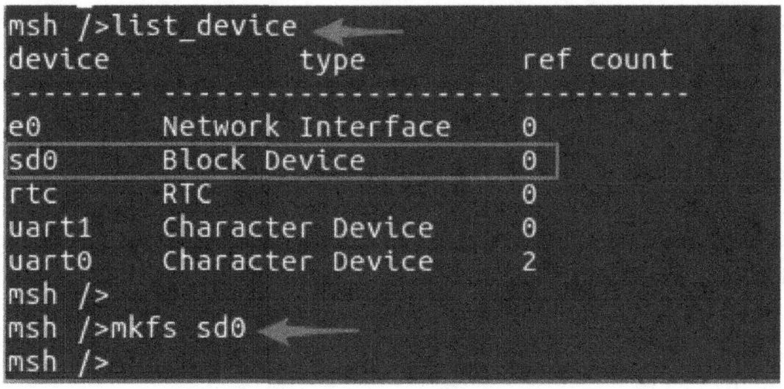

```
msh />list_device
device              type            ref count
------- ------------------- ----------
e0          Network Interface    0
sd0         Block Device         0
rtc         RTC                  0
uart1       Character Device     0
uart0       Character Device     2
msh />
msh />mkfs sd0
msh />
```

FIGURE C.20   Format SD card.

For more information on FatFS, visit: http://elm-chan.org/fsw/ff/00index_e.html.

The file system will not be loaded immediately after formatting the SD card, and it will be loaded correctly after a reboot. So exit the QEMU, and then restart the virtual machine by entering "`./qemu.sh`" on the command line interface. Entering the "`ls`" command, you can see that

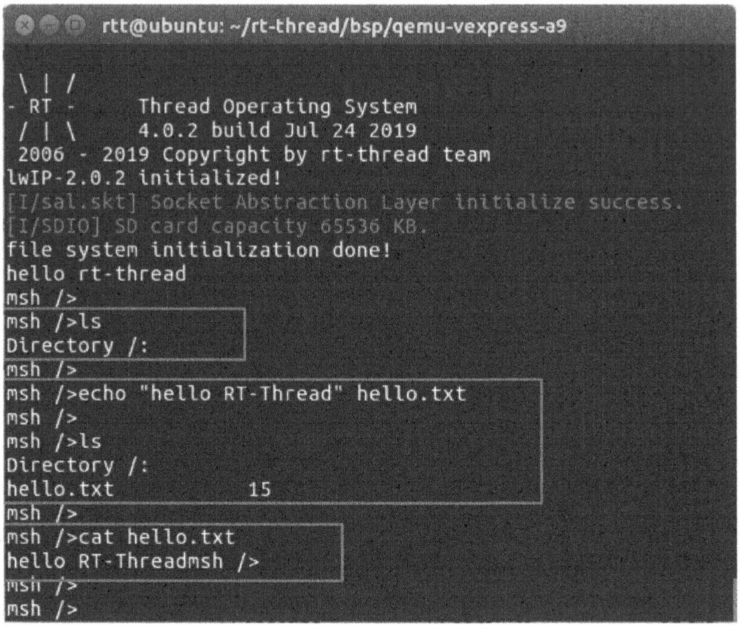

FIGURE C.21   File system commands.

the/directory has been added, the file system has been loaded, and then you can experience the file system with other commands provided by RT-Thread, as shown in Figure C.21.

- ls: Display the file and directory information
- cd: Switch to the specified directory
- rm: Delete files or directories
- echo: Write the specified content to the target file
- cat: Display the details of a file
- mkdir: Create folders

Please enter "help" to see more commands.

Run Network Sample Code

With QEMU, the configuration cannot be used to implement the sample in Chapter 16; instead, it requires an additional configuration.

Preparations

A TAP Network Card

Add a virtual TAP network card tap0 and bridge tap0 network card to the PC network card. Next, append the parameters to the qemu.sh file: -net nic -net tap,ifname=tap0. As shown in Figure C.22, save the qemu.sh file after modification.

- net nic: This is a required parameter, indicating that this is a network card configuration.
- -net tap: Using tap mode.
- ifname=tap0: "tap0" is the name of the tap network card.

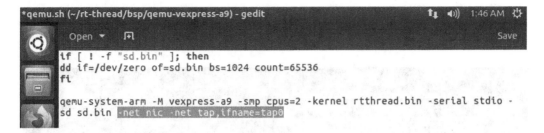

FIGURE C.22   Modify the qemu.sh file.

scons --menuconfig

Use the "scons --menuconfig" command to enter the configuration menu, select the network sample code, such as "tcp client" and "udp client", then save and exit.

```
-> RT-Thread online packages
  -> miscellaneous packages
    -> samples: kernel and components samples
      -> a network_samples package for rt-thread
          [*]    [network] tcp client
          [ ]    [network] tcp server (NEW)
          [*]    [network] udp client
          [ ]    [network] udp server (NEW)
          [ ]    [network] tcp client by select api (NEW)
          [ ]    [network] http client (NEW)
```

pkgs --update

Use the "pkgs --update" command to download the package to the packages folder in the BSP directory, as shown in Figure C.23.

```
rtt@ubuntu:~/rt-thread/bsp/qemu-vexpress-a9$ pkgs --update
Start to download package : network_samples-v0.3.0.zip
Downloaded 18 KB
Start to unpack. Please wait...
===============================> NETWORK_SAMPLES v0.3.0 is downloaded successful
ly.
```

FIGURE C.23   Download the package.

scons

Use the "scons" command to compile BSP, as shown in Figure C.24.

```
rtt@ubuntu:~/rt-thread/bsp/qemu-vexpress-a9$ scons
scons: Reading SConscript files ...
scons: done reading SConscript files.
scons: Building targets ...
scons: building associated VariantDir targets: build
CC build/applications/lcd_init.o
CC build/applications/mnt.o
CC build/drivers/board.o
CC build/drivers/drv_keyboard.o
```

FIGURE C.24   Compile BSP.

./qemu.sh

Use the "./qemu.sh" command to run QEMU, as shown in Figure C.25.

```
\ | /
- RT -     Thread Operating System
 / | \       4.0.3 build Jan 16 2020                              QEMU
 2006 - 2019 Copyright by rt-thread team
lwIP-2.0.2 initialized!
[I/sal.skt] Socket Abstraction Layer initialize success.
[I/SDIO] SD card capacity 65536 KB.
file system initialization done!
hello rt-thread
msh />
```

FIGURE C.25   Run QEMU.

Run Network Sample Code

Use the "help" command to view all system commands. It can be seen that tcpclient and udpclient examples have been successfully added to the system, as shown in Figure C.26.

```
msh />help
RT-Thread shell commands:
tcpclient         - a tcp client sample
udpclient         - a udp client sample
memcheck          - check memory data
memtrace          - dump memory trace information
list_fd           - list file descriptor
ifconfig          - list the information of all network interfaces
ping              - ping network host
dns               - list and set the information of dns
netstat           - list the information of TCP / IP
version           - show RT-Thread version information
list_thread       - list thread
```

FIGURE C.26   help.

Get the IP Address

In the console, the "ifconfig" command can be used to view the network, as shown in Figure C.27.

```
msh />ifconfig
network interface device: e0 (Default)
MTU: 1500
MAC: 52 54 00 97 32 78
FLAGS: UP LINK_UP INTERNET_UP DHCP_ENABLE ETHARP BROADCAST
ip address: 172.16.85.129
gw address: 172.16.85.2
net mask  : 255.255.255.0
dns server #0: 172.16.85.2
dns server #1: 0.0.0.0
msh />
```

FIGURE C.27   ifconfig.

Notes: If the obtained IP is 10.0.x.x, it is because the startup parameter "-net nic -net tap,ifname=XXX" was not added for QEMU.

At this time, the IP address **172.16.85.129** has been successfully obtained and the flag status is linked up, indicating that the network has been configured.

tcpclient Sample

Enable a TCP server on the PC before running the sample code, as shown in Figure C.28.

FIGURE C.28　Settings: enable a TCP server.

Then, in the FinSH console, enter the following command to enable the TCP client to connect to the TCP server:

```
msh />tcpclient 172.16.85.128 5000
Connect successful
```

When the console outputs the log information of "Connect successful," this means the TCP connection is successfully established.

Next you can communicate with the data in the network debugging tool window and send "Hello RT-Thread!," which means sending a piece of data from the TCP server to TCP client, as shown in Figure C.29. If sending "q" or "Q," the socket will be closed:

FIGURE C.29　Data.

And in the QEMU FinSH console,"Received data = Hello RT-Thread!" can be seen, as shown in Figure C.30.

FIGURE C.30　Received data.

udpclient Sample

Enable a UDP server on the PC before running the sample code. Select the protocol type UDP and fill in the local IP address and port 5000, as shown in Figure C.31.

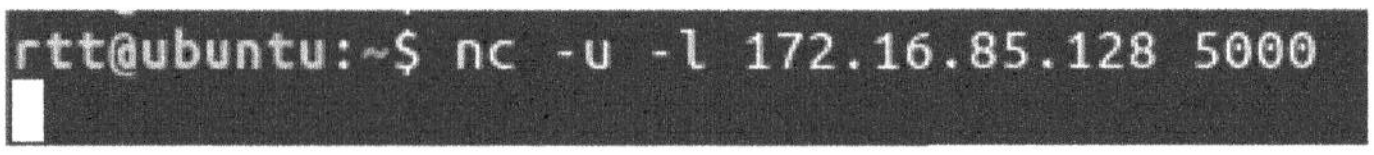

FIGURE C.31 Settings: Enable a UDP server.

Once the system is running, enter the following command in the QEMU FinSH console to get the sample code run.

```
msh />udpclient 172.16.85.128 5000
```

The server should receive 10 messages of "This is UDP Client from RT-Thread," as shown in Figure C.32.

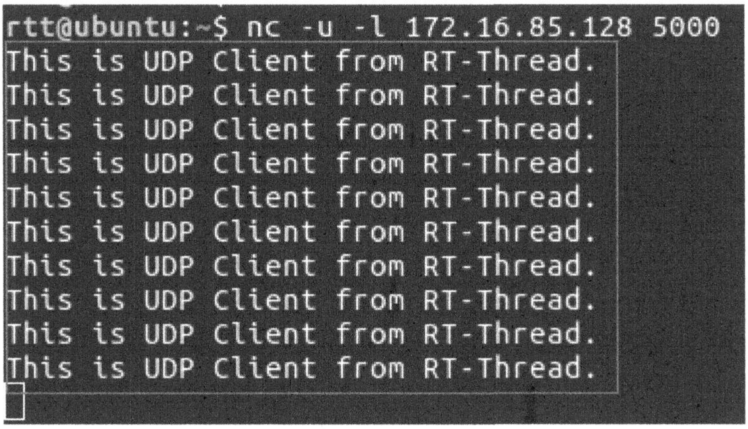

FIGURE C.32 The server received messages.

Appendix D: Getting Started with RT-Thread Studio (Windows)

RT-Thread Studio is a development and debugging software with a variety of features, as shown in Figure D.1. You can download the latest RT-Thread Studio software pack by visiting the official website at https://www.rt-thread.io.

This appendix will introduce how to build the RT-Thread full version project based on RT-Thread Studio, which helps developers get started with RT-Thread Studio quickly.

The main steps are as follows:

- Build an RT-Thread project.
- Modify the PIN information in "main.c".
- Compilation, downloading, and validation.

BUILD THE RT-THREAD FULL VERSION PROJECT

Use RT-Thread Studio to build the project based on v.4.0.2. The studio interface is as shown in Figure D.2.

The configuration process can be summarized as follows:

- Define your project name and the storage path of the project generation files.
- Select *Based on MCU* to create a project with the option of RT-Thread version v4.0.2.
- Select the vendor and chip series.
- Configure serial port information.
- Configure adapter information.

The full version of the RT-Thread project can be created by clicking the Finish button when the project is configured. Creating a nano version of the project is the same as the earlier discussed steps—just select the nano to get started.

MODIFY MAIN.C

PIN devices are automatically enabled when a project is created based on the full version of RT-Thread, and the following code is automatically generated in the "main.c" function:

```c
/* PLEASE DEFINE the LED0 pin for your board, such as PA5 */
#define LED0_PIN    GET_PIN(A, 5)

int main(void)
{
    int count = 1;
    /* set LED0 pin mode to output */
    rt_pin_mode(LED0_PIN, PIN_MODE_OUTPUT);

    while (count++)
    {
        /* set LED0 pin level to high or low */
        rt_pin_write(LED0_PIN, count % 2);
        LOG_D("Hello RT-Thread!");
```

```
        rt_thread_mdelay(1000);
    }

    return RT_EOK;
}
```

When using a PIN device, you only need to use GET _ PIN to get the corresponding PIN number and use functions such as "rt _ pin _ write" to operate the PIN number.

For example, the LED on the stm321475-atk-pandora development board is connected to PE7, so it is modified to:

```
#define LED0_PIN GET_PIN(E, 7)
```

COMPILE

Left-click to select the current project, and in the menu bar, click the button
 ✎ ˅ to compile the project. "quick-start" is the project name. Or you can compile the project by right-clicking on the project and selecting "build projects."

The results of the project compilation are as shown in Figure D.3.

DOWNLOAD

After the project compiles without errors, you can download the program to your board.

The program can be downloaded via the "Flash Download" button in the menu bar. You can also download it by right-clicking on the project and selecting "Flash Download." The program defaults to the download method selected when the new project is created. If you need to change this, click the reverse triangle icon button, which is next to the Flash Download button to choose the download method according to your debugger. Studio currently supports both ST-Link and J-Link download methods (as shown in Figure D.4). Here's the sample of selecting an emulator. Select the emulator and click the download button to download.

The program download result is as shown in Figure D.5.

FIGURE D.1 RT-Thread studio.

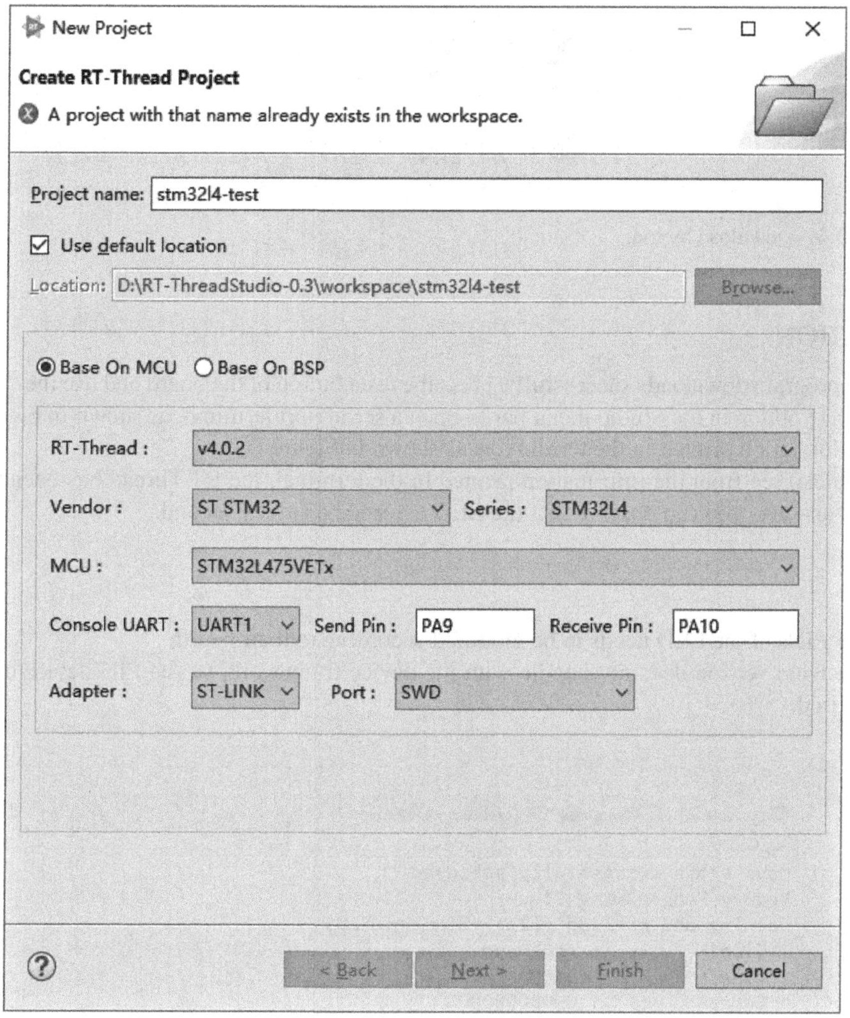

FIGURE D.2 new-project.

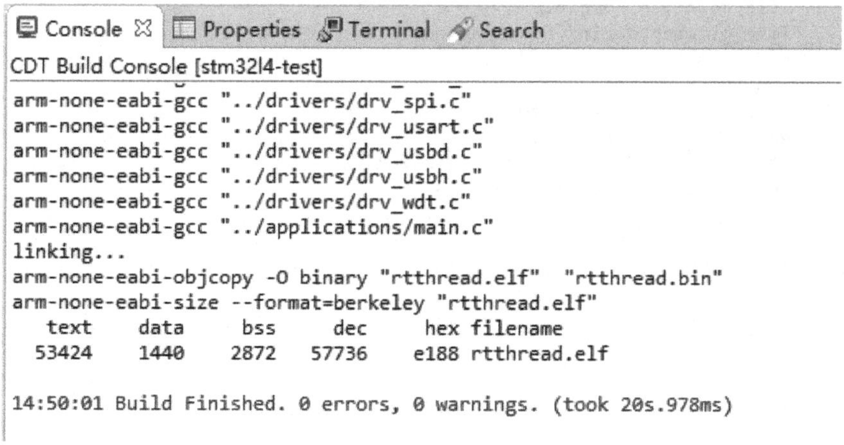

FIGURE D.3 Build.

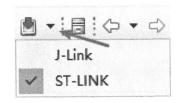

FIGURE D.4   Download button.

VALIDATION

After the program downloads successfully, press the reset button of the board and use the "Open a Terminal" option in the Studio menu bar to open a serial port terminal, as shown in Figure D.6.

The information printed in the terminal is as shown in Figure D.7.

As you can see from the information printed in the terminal, the RT-Thread has been successfully run, and also you can observe that the LED is turned on every second.

TIPS

- The PINs of the LED needs to be modified according to their board.
- The nano version does not bundle with the device framework, so the PIN device cannot be used.

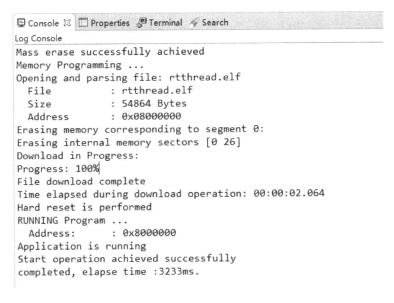

FIGURE D.5   Download result.

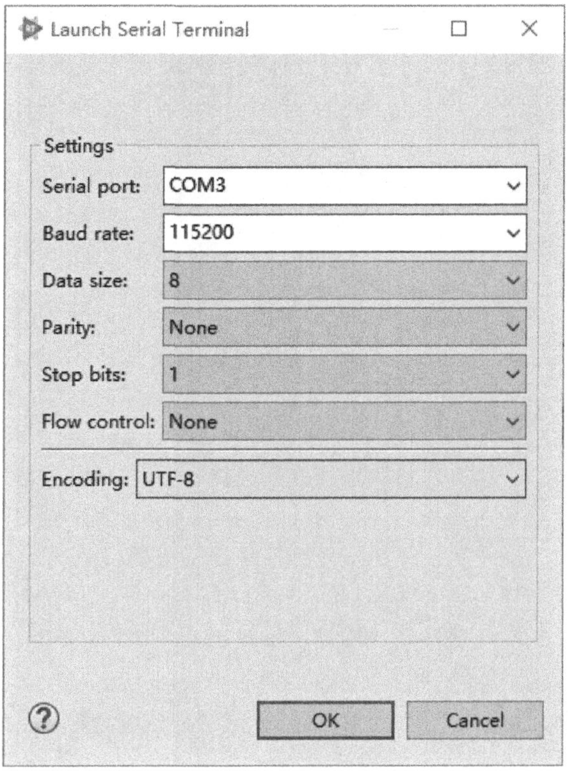

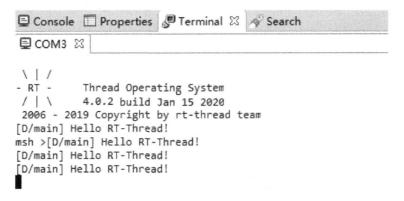

FIGURE D.6 Terminal.

FIGURE D.7 Print.

Index

Note: Page numbers in **bold** and *italics* refer to tables and figures, respectively

Printed in the United States
By Bookmasters